M R

CSR, Sustainability, Ethics & Governance

Series editors
Samuel O. Idowu, London Metropolitan University, London, United Kingdom
René Schmidpeter, Cologne Business School, Germany

More information about this series at http://www.springer.com/series/11565

Aneta Długopolska-Mikonowicz •
Sylwia Przytuła • Christopher Stehr
Editors

Corporate Social Responsibility in Poland

Strategies, Opportunities and Challenges

Springer

Editors
Aneta Długopolska-Mikonowicz
WropenUp Foundation
Wrocław, Poland

Sylwia Przytuła
Human Resource Management
Wroclaw University of Economics
Wrocław, Poland

Christopher Stehr
German Graduate School of Management
and Law and POLYMUNDO
Heilbronn, Baden-Württemberg, Germany

ISSN 2196-7075 ISSN 2196-7083 (electronic)
CSR, Sustainability, Ethics & Governance
ISBN 978-3-030-00439-2 ISBN 978-3-030-00440-8 (eBook)
https://doi.org/10.1007/978-3-030-00440-8

Library of Congress Control Number: 2018967306

This Springer imprint is published by the registered company Springer Nature Switzerland AG
The registered company address is: Gewerbestrasse 11, 6330 Cham, Switzerland

Preface

Corporate social responsibility (CSR) in general involves multiple areas, levels and actors such as stakeholders in each organisation. Responsibility itself is a broad concept framing the ethical code of conduct of a person, a state, (non-) governmental institutions or organisations. The most important questions still remain—even after a long tradition of public and scientific discussion—how to deal with this responsibility on an institutional and entrepreneurial as well as on an individual level? How is it possible to incorporate the ecological, the social and the economic responsibility within societies and enterprises in daily business to create win-win-win situations? These wider questions lead to wider answers.

Therefore, the following publication presents a really broad and fascinating spectrum of various aspects of CSR in Poland. It includes social and political perspectives, social media and corporate communication, financial aspects (financial sustainability, corporate financial performance, socially responsible investing), legal perceptions, ecological dimensions (green economy) as well as intercultural and individual issues (personal social responsibility).

The subject of CSR in Poland is shown not only in various dimensions of social responsibility, but also through the prism of the specific nature of the activities of various entities. Presented here are public administration institutions, NGOs (e.g. sports organisations), companies from the SME sector, international corporations and multinational enterprises (MNEs) representing various industries—with a specific focus of CSR as a strategic factor of competitive advantage in business operations.

The publication opens the introductory part of a general, historical nature concerning research on CSR, the legal status, the perspective of public administration and business in the field of socially responsible activities. Introduction closes the essay, which provokes critical analysis and reflection of each of the proposed chapters on the state of CSR in Poland.

Part I. Introduction

Introduction is launched by **Bolesław Rok**, who presents the integrated approach from CSR to corporate sustainability. Against the background of the historical overview of developing CSR topic in Poland, the author presents the achievements of business practice, which comes from social and ethical activities to treating CSR as a strategic factor of competitive advantage in business operations. Bolesław Rok lists barriers and limitations regarding the sustainable strategic management approach for three groups of companies. He outlines the key challenges and opportunities for business practice, which are mostly used by international corporations operating on the Polish market. He points out the need to actively engage and involve all actors and stakeholders in developing sustainable activities. B. Rok presented a model of the materiality matrix, as a useful tool for introducing possible changes in the business strategy of companies operating in Poland.

In the study of **Jacqueline Kacprzak,** four stages of CSR development in Poland were identified: the initial stage (1998–2009), the stage of proactive measures in the area of CSR undertaken by a wide range of stakeholders (2009–2013), the stage of research and development (2014–2015) and the stage of mature decisions and activities (2016—up till now). The author describes very preciously various activities carried out by public administration in the area of CSR and underlines the key role of CSR advisory boards in implementing CSR standards in Poland. She concluded that the role of public administration in this area is not only in the regulatory field, but also to inform and educate, disseminate standards and monitor the implementation of the responsible business conduct. Several examples of effective cooperation between government administration and business in the area of CSR in Poland were presented.

Sławonmir Magala, in an essayistic tone, conducts a discourse on socially responsible activities in the context of political changes in many countries of the world [Stalinism, the fall of communism in the countries of Central and Eastern Europe, the significance of inclusion and unity in the structures of the European Union, the growing importance of populist and right-wing movements in the political scene, separatist tendencies in many countries (Brexit, Catalonia)]. The author presents examples of various "socially irresponsible" actions, but politically motivated, e.g. concerning the Arab Spring refugees, civil wars and genocide in Rwanda and Cambodia, which are a manifestation of the moral collapse of political elites, but also failure in implementing the principles of CSR in the political, economic, cultural, religious and ethical dimensions. Slawomir Magala also places Poland against the background of these changes, indicating that one should draw on the historical experience of his own country and others. CSR requires transparency, trust and the involvement of many local and global market actors in building a holistic awareness of socially and politically responsible actions.

Part II. Financial Aspects of CSR

Part II is devoted to the analysis of socially responsible activities in relation to financial aspects.

Anna Witek-Crabb discussed the relationship between corporate social responsibility and corporate financial performance. She proposed a model based on correlation between CSR maturity levels (process, strategic, formal maturity) and financial sustainability construct in companies representing 9 sectors. She found that in the examined sample these two categories remain independent. The author concludes that Polish companies have implemented CSR at different levels; however, it does not impact their financial performance. In spite of the fact that Polish capitalism is relatively young, the dynamics of developing CSR concept is rather slow and Polish businesses and society are still on the CSR learning curve.

Karolina Daszyńska-Żygadło continues the topic of influence of CSR activities on financial performance. The purpose of this study was to recognise improvement or worsening of operating financial results of companies undertaking CSR activities. The research objective was to have all companies included in sustainability index RESPECT on Warsaw Stock Exchange. Results show that financial performance of companies improves significantly after inclusion in sustainability index. Performed research does not answer the dilemma whether financial performance of a company is being improved thanks to introducing CSR strategy, running CSR activities and getting verified by RESPECT index procedure or maybe that CSR strategy can be introduced due to financial performance improvement.

Łukasz Sułkowski and Justyna Fijałkowska present the issue of the global financial crisis, which significantly damaged the reputation and credibility of most financial institutions, including banks operating in Poland. CSR activities have become a remedy for improving the image and rebuilding trust in clients. The authors analyse various CSR activities carried out by several banks operating in Poland in recent years. These included saving resources, financing pro-ecological organisations, offering special financial products, actively participating in local socially appreciated initiatives and programmes and showing that they do care about their clients, employees, environment and local communities. It would seem that the effects of the banking crisis will be a warning for these institutions for many years. However, the authors cite another example of socially irresponsible activities of banks in Poland, which concerned credit agreements in CHD that regards 500 thousand agreements. It turned out that 95% of the cases contain illegal clauses. It may be concluded that in the area of ethical banking and responsible investment there is still much to do in Poland.

Part III. Legal, Financial and Environmental Aspects of CSR

The chapter of **Tomasz Słoński and Magdalena Ligus** is a "theme bridge" between the legal, financial and environmental aspects of CSR. The authors present the legal regulations developed by the European Commission (*Impact*

Assessment Guidelines), which are the reference point for the Polish legal framework regulating the activities of companies in the renewable energy sector (RES). Ecological activities in this sector are not free from price risk (prices are determined administratively) and also risk of volume component. These legal and administrative constraints cause perturbations in other sectors closely related to RES—companies integrated in the supply chain, customers, banks and other financial intermediaries.

Bożena Ryszawska in her chapter continues the subject of sustainable development and, in particular, raises the role and place of green economy in CSR 2.0. The author presents the transition of green economy through the prism of developing regulations and procedures regulating business activities in this area in the European Union. However, in real activities, the green economy in Poland has not gained visible recognition so far and compared to the EU countries is quite unfavourable in this respect. Nevertheless, since 2015 in governmental and industry planning studies such activities have been undertaken to fill the legal gap and provide financial instruments supporting the implementation of elements of the green economy in Poland.

Part IV. Practical Aspects of CSR

Part IV presents practical aspects of CSR activity in Poland in the dynamics of systemic transformation and from the perspective of different types of organisations like small enterprises, multinational corporations and sports organisations which can act as limited liability companies but also as foundations, associations or even student sports clubs. This part also presents the CSR reporting practice in Poland.

Rojek-Nowosielska presents the results of her empirical research of the CSR level of enterprises in Poland—before and after the transition of 1989. She presents the socially responsible activities of organisations created before the transition as obligations of the state which were shifted on enterprises to be performed under the then applicable provisions of law. The economic and social reforms of the new system resulted in a systemic change. The author claims that the character of the community-oriented activities of an enterprise was reactive, whereas CSR is proactive. The author proposes an analysis of the enterprises in the corporate social responsibility continuum model which corresponds with the models of potential performance perspective. The five general areas have been identified: employees, customers, suppliers, natural environment and local community.

Agnieszka Sokołowska-Durkalec attempts to clarify the instrumental dimension of social responsibility concept in small enterprises and presents the results of her research—selected problems, best practices and recommendations. She underlines the role of owners in building trust by sharing their values with the stakeholders and by institutionalising the values and the role of the owner/owners' expertise in overall CSR implementation and results. She discusses further that most entrepreneurs in the researched group recognise the existence of organisational culture in their enterprises, but they do not declare that it is based on universal ethical values, but rather more often on the pragmatism.

Bożena Dancewicz and Franziska Struve discuss in their chapter the influence of MNEs on CSR in Poland and the evolution of CSR in Poland based on the countries' specific historical background. They examine the role of MNEs, NGOs and the social society within the development process of CSR and ethical business behaviour in Poland. As an example, they take the development stages of the food retail branch and show based on the consumption industry the impact of ethical behaviour of MNEs in their local branches. The practical perspective is covered by the examples of Mars and Avon, which get evaluated based on qualitative interviews as well as based on their public communication.

Paweł Kuźbik presents in his chapter the results of his study on the current relationship between sport and social responsibility in its real, strategic dimension and analyses how the CSR concept is understood and implemented in sports organisations in Poland. The author analyses 85 sports clubs of professional nature and shows a broad perspective of their environment, function and changing conditions during the transformation in Poland. The cases presented reveal two ways of CSR in sports clubs: CSR through sport, where sport itself is supported, and CSR within sport, where sports organisations are socially responsible. The author claims that although the use of CSR in polish sports organisations has no strategic dimension, and its understanding is mainly focused on charitable activities, it is still possible to point out recurring procedures that allow to create an up-to-date model of activities in the area: sports clubs—CSR.

Part V. CSR Reporting

Justyna Fijałkowska and Małgorzata Macuda present the results of their empirical studies on voluntary CSR reporting practices in Poland among companies (including listed in the RESPECT Index), NGOs, associations, foundations and public institutions in the period of 2005–2017. The analysis embraces 357 CSR reports prepared by 109 organisations from 29 sectors in the last 13 years. The results of the research indicate that CSR disclosures are not very widespread in Poland and, even though evolving, are still in their early stage of development. This chapter concludes recommendations on how to improve the credibility and utility of CSR reports published by Polish companies.

Maria Aluchna and Maria Roszkowska-Menkes introduce the non-financial reporting in conceptual framework of CSR and stakeholder theory. Authors of this chapter underline that non-financial reporting has developed from CSR and environmental communication to sustainability disclosure towards integrated reporting. Based on the analysis of companies listed on the Warsaw Stock Exchange, they concluded that non-financial reporting is slowly becoming a mainstream practice in Poland. Yet, only 6% of the researched Polish companies published a CSR or environmental report, which placed Poland far behind the rest of Europe.

Part VI. Social Aspects of CSR

Part VI is a perspective on the social aspects of CSR in Poland including the personal social responsibility reflected in customer choices, in volunteering engagement and in the influence of the social media.

Marta Karwacka The author argues that especially in Poland the national history and its impact have to be considered while talking about CSR and the personal social responsibility (PSR). PSR is here described as the concept of minimising the negative impacts and maximising the positive impacts of each human being towards social, environmental and economical dimensions—which means much more than being just a sustainable or critical consumer. This PSR understanding of daily behaviour leads to claims for the companies' CSR measures and engagements. In Poland, the consumer awakening for becoming critical consumers was influenced by the Rana Plaza collapse 2013 in Bangladesh which strengthened in general the NGOs in Poland. The so-called millennials and the Generation Y with their new core values of ethical behaviour want consistency in their both worlds—the private life and the business world. These Polish consumers have to be differentiated and are here described within six different types. The author includes in the article her long experience as blogger on her page: www.howtowearfair.com.

Florian Urbański researched the different types of motivation among the Polish and the international employees for participating in intercultural CSR project. The study is based upon the psychological theory of intrinsic and extrinsic motivation of human behaviour and examines the motivation among the volunteers of different nationalities, gender and religions in an educational project about diversity. The results do not indicate any significant difference in motivation between nationalities or at least between Polish and international volunteers. The author underlines that volunteering is a relatively new concept in Poland due to the fact that volunteering did not exist as an organised strategy in Poland before the political and economic changes of the late 1980s. The author presents both the scale and the development of volunteering in a historical context. An important aspect of the study is also diversity management.

Justyna Szumniak-Samolej discusses in her chapter the potential ways of using social media in the field of creating and communicating the corporate social responsibility strategy and analyses the current status of using these tools in the area of CSR in Poland. She draws a context of technological and social changes in the corporate reality, pointing to Web 2.0. She indicates such social behaviour as transparency, dialogue, trust and the background for building socially responsible communication system in the company. Corporate communication based on social media can bring in benefits and threats for both stakeholders and society. Based on three case studies of multinational corporations operating in Poland, the author concludes that the level of practice, of taking the advantage of the potential of this technology to create and to communicate CSR activities, is very low. In the face of challenges posed by Web 4.0, the business practice will require both better tools and aware users of various media.

Part VII. Outlook on CSR in Poland and Visegrad

Part IV discusses possible development of CSR in Poland and presents solid recommendations.

Przemysław Wołczek presents the interdependence between strategy and CSR and discusses the impact of the character of the voluntary commitments on competitiveness and relations with stakeholders. The author describes commitments such as one-off, accidental or not related to the corporate strategy as well as permanent, long-lasting commitments included in the strategy of the company and its business model. The author explores if the competitive position of a given company and its relations with stakeholders depend on the character of CSR commitments. The conclusions include recommendations for all sectors.

The chapter by **Aldona Wiktorska-Święcka** is an outline of the possible development of legal aspects of the CSR activities in Poland based on experiences before and after the EU accession and current institutional arrangements. The author analyses the existing institutionalisation of the CSR concept in Polish public policies and proposes recommendations for the creation of a base for institutionalising CSR which should be based not only on business but also on strong institutions, state organisations and partnerships—which also include other stakeholders, including civil society and the media—between them.

The last chapter of this book about CSR in Visegrad countries by **Sylwia Przytuła, Sylvie Formánková, Iveta Ubrežiová and Anna Dunay** is a bridge between CSR knowledge and practice in Poland and the regional context within the Visegrad Group. Practices applied in the field of CSR in Poland should have their point of reference to be able to assess the level of development of these solutions relatively objectively. It cannot be related to the USA or all EU countries because of the differences in many areas of the economy, society and culture. In this case, the comparative perspective is made up of a group of Visegrad countries that are geographically close, originating from the same cultural cluster. Historically, Poland, the Czech Republic, Slovakia and Hungary joined the EU on 1st May 2004, which created many new opportunities for economic development, knowledge transfer and exchange of business practices. However, despite this common ground, the level of CSR knowledge and practices in these countries is at a very different level. The authors provide examples of CSR implementation in various industries explaining the specific legal and social context in these countries.

Wrocław, Poland — Aneta Długopolska-Mikonowicz
Wrocław, Poland — Sylwia Przytuła
Heilbronn, Germany — Christopher Stehr

Contributing Organization

SLOVAK UNIVERSITY OF AGRICULTURE
NITRA

FACULTY OF ECONOMICS AND MANAGEMENT
SUA NITRA

UNIVERSITY OF SOCIAL SCIENCES

WR OPEN UP

UNIVERSITY OF LODZ

Uniwersytet Wrocławski

FACULTY OF MANAGEMENT
University of Lodz

SZKOŁA GŁÓWNA HANDLOWA W WARSZAWIE
SGH

Wrocław University of Economics

KOZMINSKI UNIVERSITY

Collegium Civitas

RSM Erasmus ERASMUS UNIVERSITY

Contents

Editors and Contributors

About the Editors

Aneta Długopolska-Mikonowicz, She is a graduate of the University of Wrocław (German Studies MA and postgraduate studies in Public Relations), postgraduate programme of the Wrocław University of Science and Technology and *École* Centrale de Lyon (*Mastère Spécialisé* en Management des Grand Projects et Programmes). She is co-creator and president of the board of the WrOpenUp foundation developing since 2012 educational CSR programmes in cross-sector partnership (global business organisations operating in Poland, public administration and academia). She works as a manager in multinational business organisation where she also coordinates diversity programmes in Poland. She is member of SIETAR Poland (Society for Intercultural Education, Training and Research).

Email: aneta.dlugopolska@gmail.com
Wrocław, Poland

Sylwia Przytuła, PhD, is a professor in the HRM Dept. at the Wroclaw University of Economics. Her research and scientific interests include expatriation management, international HRM and cross-cultural management. She is an author and co-author of ca. 80 scientific publications (co-author: *Global Talent Management*, A. Al Ariss (ed.), Springer, 2014; *Intercultural Interactions in the Multicultural Workplace*, Springer, 2016). She is a member of SIETAR, HEF (Higher Education Forum) and IMISCOE. She participated in 30 international study visits and internship for academics in Europe, the USA and China. Due to her very extensive international network, she was engaged in several cooperative projects, e.g. Erasmus +, X-Culture.

Email: s_przytula@wp.pl

Katedra Zarządzania Kadrami, Uniwersytet Ekonomiczny we Wrocławiu, Wrocław, Poland

Christopher Stehr is professor for International Management at the German Graduate School of Management and Law (GGS). He studied political science and business administration at Ludwig-Maximilians-University of Munich. After receiving a scholarship of the German-Japanese-Academic Burse for his doctoral thesis on globalisation he worked as project manager at the Research Institute for Applied Knowledge Processing in Ulm with the focus on applied research within the topics of sustainability and entrepreneurial globalisation of enterprises. As an assistant professor at the Institute for Corporate Management of the University of Ulm he founded successfully Polymundo, a globalisation consulting company especially for small and medium-sized enterprises. The focus of his engagement is to accompany enterprises during different change processes in the areas of human resources, internationalisation, intercultural management and digitisation. He is an author/co-author/editor/co-editor of more than 100 scientific publications including several books on CSR (*CSR in China*, Springer/ 2015; *CSR und Marketing*, Springer/2017; *CSR in Brazil*, Springer 2018). He initiated the Heilbronn Declaration a voluntary self-commitment between different SMEs and MNEs with a focus on CSR and its successful implementation.

Email: Christopher.Stehr@ggs.de
German Graduate School of Management & Law (GGS), Heilbronn, Germany

About the Contributors

Maria Aluchna, PhD, is an associate professor, Department of Management Theory Warsaw School of Economics, and Director of Responsible Business Center and Post-graduate Studies on Products and Services Management at the Department of Management Theory Warsaw School of Economics. She specialises in corporate governance, strategic management and corporate social responsibility. She is fellow of Deutscher Akademischer Austauschdienst (research stay at Universität Passau), US-Polish-Fulbright Commission (research stay at Columbia University), Soros Foundation, Volkswagen Foundation and Foundation for Polish Science Development and visiting scholar at London Metropolitan Business School and Sydney University School of Business. She teaches corporate governance, responsible management (within the cooperation of the University of Illinois, Springfield, USA) and strategic management at MBA, postgraduate, PhD, MA and BA studies. She is member of European Corporate Governance Institute, International Corporate Governance Society, Finance Watch, European Academy of Management, Council of Management and Finance Collegium and editorial committees of *International Journal of Corporate Social Responsibility, European Journal of Economics and Management, Journal of Knowledge Globalization, Przegląd Organizacji* and *e-Mentor*. She has published in Poland and abroad five monographs, including two in English; five edited books, including four in English); and over 70 articles and conference papers (EURAM, AIB, ISBEE) and takes active part in international conferences.

E-mail: maria.aluchna@sgh.waw.pl
Szkoła Główna Handlowa w Warszawie, Warszawa, Polska, Poland

Bożena Dancewicz, PhD, is an academic teacher, business trainer and a cross-cultural specialist. She specialises mainly in negotiations and mediations, interpersonal communication, CSR, business ethics and diversity management. She has several years of experience of working with individual and corporate clients. She is co-author of books: *Intercultural Management Issues,* Prof. Małgorzata Rozkwitalska (ed.); *Przewodnik Praktycznego Zarządzania, Ekspatrianci w Polsce. Silne czy słabe ogniwo w zarządzaniu polskimi filiami*, Sylwia Przytuła (ed.); *Młodzież w społeczeństwie wielokulturowym*, Daniel Wiśniewski (ed.); *Migracje międzynarodowe i ekspatriacja. Perspektywa indywidualna, organizacyjna, społeczno-kulturowa*, Sylwia Przytuła (ed.), and also scientific publications regarding business ethics and CSR in cross-cultural context.

Email: bdancewicz@gmail.com
Nowy Dwór Mazowiecki, Poland

Karolina Daszyńska-Żygadło, PhD, holds position of assistant professor in the Corporate Finance Department at the Wroclaw University of Economics. Her fields of expertise are corporate valuation, financial planning and corporate sustainability. She authored and co-authored more than 40 scientific papers. In 2015, her book on scenarios approach in corporate valuation was published in Poland. Along with research and teaching, she deals with business consulting and mentoring, especially for early-stage business ideas and start-ups. She cooperates with Climate-KIC in the field of sustainability and renewable energy; she participated in Pioneers into Practice Programme. Recently, she took part as a data scientist in a project on assessing economic and climate impact of renewable energy investment with Porta Capena, an international smart IT solutions company.

Email: karolina.daszynska@ue.wroc.pl
Katedra Finansów Przedsiębiorstw i Finansów Publicznych (KFP), Uniwersytet Ekonomiczny we Wrocławiu, Wrocław, Poland

Anna Dunay, PhD, is an associate professor at the Department of Business Economics and Management of the Faculty of Economics and Social Sciences at Szent István University, Gödöllő, Hungary. Her main research fields are management of SMEs, entrepreneurship education, corporate social responsibility in different sectors and company competitiveness. She is author, co-author or editor of more than 150 scientific publications with nearly 250 independent citations (e.g. *Management and Organization: Concepts, Tools and Applications*, Pearson, 2017, with co-editors: Illés B.Cs., Nowicka-Skowron, M., Horská, E.; *The entrepreneurship in Poland and in Hungary. Future entrepreneurs education perspective* (with co-authors Illés B.Cs. and Jelonek, D.), Polish Journal of Management Studies 12. She is a visiting professor at the University of Silesia in Katowice and works in scientific and organising committees of international scientific conferences in CEE countries.

E-mail: dunay.anna@gtk.szie.hu
Szent István University, Gödöllő, Hungary

Justyna Fijałkowska, PhD, is a researcher and assistant professor at the University of Social Sciences in Lodz, Poland, and at Clark University Worcester, USA. She is also a researcher and assistant professor at the Faculty of Economy and Sociology, University of Lodz. Her scientific interests include CSR, integrated reporting, intellectual capital, intangibles and accountability of universities. She is an author and co-author of several scientific publications in the area of CSR, IR, intangibles, accountability and harmonisation and standardisation of accounting, published in international refereed journals, national journals and chapters in collective works published in Polish, English and Portuguese. She is a member of the European Accounting Association and Le Réseau universitaire des Pays du Groupe de Vysegrad (PGV Research Network). Justyna Fijałkowska teaches managerial accounting and foundations of business analysis at MBA Programme at Clark University; accounting, internal audit and control and managerial accounting at the University of Lodz; and cost accounting, managerial accounting and budgeting at the University of Social Sciences.

Email: jfijalkowska@san.edu.pl
Katedra Rachunkowości, Społeczna Akademia Nauk, Łódź, Poland

Sylvie Formánková, PhD, is an assistant professor at Mendel University, Czech Republic, where she teaches courses in both Czech and English on management, corporate social responsibility (CSR), managerial skills and European cultures and civilisations. Her research focuses on CSR and analysis of the wine industry. She is currently comparing different higher education institutions in terms of social responsibility, cooperating with several institutions in the Czech Republic and abroad.

Email: sylvie.formankova@mendelu.cz
Provozně ekonomická fakulta, Ústav managementu, Mendelova univerzita v Brně, Brno, Czech Republic

Jacqueline Kacprzak is counsellor to the Minister in the Ministry of Investment and Economic Development of the Republic of Poland. She is lecturer in the field of corporate social responsibility and responsible business conduct as well as business and human rights at the Collegium Civitas (Warsaw) and Kozminski University (Warsaw). She is working on her PhD thesis in sociology at the Collegium Civitas. Her research and scientific interests include responsible business conduct, business and human rights, changes in modernity and capitalism. She is also an international expert in the field of non-financial and sustainable reporting. She is member of jury at the CSR Reports Competition (Poland) and member of international Green Frog Award jury. She is also member of the Peer Review Committee in the process of translation into Polish the GRI G4 reporting standard. In 2017, she was a member of the jury of the Verba Veritatis Competition for the best BA, MA and PhD thesis in the field of business ethics.

Email: jacqueline.kacprzak@tlen.pl
Warszawa, Poland

Marta Karwacka, PhD, is a business sociologist. For over a dozen years, she has been dealing with the issues of sustainable development and corporate social responsibility (CSR). She focuses her interest on value management, corporate social responsibility, cross-sectoral cooperation, responsible marketing, sustainable consumption and closed-loop economy. She cooperates with companies, third sector and universities as a researcher and sustainability consultant. She is a member of the team of external experts of the Responsible Business Forum, the editorial team of the *CSR Compendium* magazine, the Verba Veritatis competition jury and the Sustainable Fashion Award 2018 jury. She is a co-founder of the RECONOMY Coalition for a Circular Economy. She is the author of the first book in Poland on the cooperation between business and non-governmental organisations in the CSR context. *Siła współpracy. Relacje przedsiębiorstw z organizacjami pozarządowymi w kontekście społecznej odpowiedzialności biznesu.* On her *How to Wear Fair blog?* she focuses on sustainable consumption in the clothing industry.

Email: markarwacka@gmail.com
Toruń, Poland

Paweł Kuźbik, PhD, is an assistant professor at the Faculty of Management of the University of Lodz. His scientific interests include topics related to sports management, social responsibility of organisations and organisational behaviour. In the years 2007–2015, he was a member of the Bi-NGO research team (communication of social activities on the Internet of the 500 largest companies in Poland). He is the author and co-author of several dozen scientific publications (e.g. Social responsibility in the organisational space of the football club, SIZ 2016; Social media and the development of corporate social responsibility in sports organisations, J. Reichel (ed.), CSR Trends. Beyond Business as Usual, CSiR Impact, 2014). From 2017, he is a member of an inter-university research project aimed at creating a CSR ranking of sports clubs in Poland.

Email: pawel.kuzbik@uni.lodz.pl
Faculty of Management, University of Lodz, Łódź, Poland

Magdalena Ligus, PhD, is an assistant professor at the Department of Corporate and Public Finance at the Wroclaw University of Economics. Her research interests focus on economic appraisal of investments in low-carbon energy sources and in particular on the evaluation of environmental externalities in such projects. Author and co-author of dozens of scientific publications (Evaluation of economic, social and environmental effects of low-emission energy technologies development in Poland: a multi-criteria analysis with application of a fuzzy analytic hierarchy process (FAHP), Energies 2017, 10(10), 1550; The analysis of competitiveness and potential of renewable energy technologies in Poland, Energy & Environment, Vol. 26, No. 8, 2015, 1247-1269). Since 2009, she has been the head of postgraduate studies "Investments in renewable energy sources" at the Wrocław University of Economics.

E-mail: magdalena.ligus@ue.wroc.pl

Katedra Finansów Przedsiębiorstw i Finansów Publicznych, Uniwersytet Ekonomiczny we Wrocławiu, Wrocław, Poland

Małgorzata Macuda, PhD, is an assistant professor in the Department of Accounting at the Poznań University of Economics and Business. Her PhD thesis received the second degree award in the competition of the Scientific Council of the Accountants Association in Poland for the best habilitation, doctoral and master theses in accounting. Her research interests include financial accounting, cost accounting, operational and strategic management accounting, corporate social responsibility and hospital accounting. She is an author and a co-author of ca. 40 scientific publications (monographs, articles, conference papers). She is a member of the European Accounting Association (EAA), International Association for Accounting Education & Research (IAAER) and Association Internationale des Economistes de Langue Française (AIELF). In 2016–2017, she was a secretary of the international journal *La Revue Internationale des Economistes de Langue Française* (RIELF).

Email: Malgorzata.Macuda@ue.poznan.pl

Katedra Rachunkowości, Uniwersytet Ekonomiczny w Poznaniu, Poznań, Poland

Sławomir Magala is a professor of cross-cultural management at the Rotterdam School of Management, Erasmus University Rotterdam, the Netherlands (1985–2015). Currently, he teaches at the Faculty of Management and Social Communication of the Jagiellonian University in Cracow, Poland. He has written books—*Cross Cultural Management* (2005), *The Management of Meaning in Organizations* (2009), papers and essays, and taught undergraduate and graduate courses ("Machiavelli and von Clausewitz for Managers", "Culture Vultures", "Identity in Work and Society"). He has supervised master and PhD theses, and tried to gauge changing cultural climate (also online: www.magala.nl). In 2006, he has succeeded David Boje as editor-in-chief of *Journal of Organizational Change Management*. Described as a generalist with a cause by his friends and colleagues.

Email: smagala@rsm.nl

Katedra Zarządzania Instytucjami Szkolnictwa Wyższego, Instytut Spraw Publicznych, Uniwersytetu Jagiellońskiego, Kraków, Poland

Magdalena Rojek-Nowosielska, PhD, is a professor in the Sociology and Social Policy Department at the Wroclaw University of Economics. Since 2001, she has been scientifically connected with the subject of corporate social responsibility. Other fields of her interest are non-financial reporting and cultural diversity management. She is a member of Technical Committee for Social Responsibility in Polish Committee for Standardization, Editorial Committee of Research Papers of the Wroclaw University of Economics, EBEN Poland (European Business Ethics Network) and PTPS (Polish Society for Social Policy) and member and also secretary of the audit commission. She is the originator, initiator and co-organiser of the annual scientific conference on social responsibility issues, as well as the author and editor of numerous publications on the issue of corporate social responsibility. Her latest book is *Corporate Social Responsibility. Model, Diagnosis, Evaluation*, Wroclaw, 2017.

Email: Magdalena.Rojek-Nowosielska@ue.wroc.pl

Wrocław, Poland

Boleslaw Rok, PhD, is a professor of Corporate Sustainability at Kozminski University and Director of Positive Entrepreneurship Research Lab, Department of Entrepreneurship and Ethics in Business. He is Academic Director of Corporate Responsibility Executive Programme; consultant and lecturer on ethics and compliance, corporate responsibility and sustainability; faculty at SUSTBUSY Research Center— Business & Society—Towards a Sustainable World in ESCP Europe; and founder of the Responsible Business Forum, affiliated to CSR Europe. Recent publications include *Sustainable business models - The case of collaborative economy, in: Sustainable Business Models: Principles, Promise, and Practice*, edited by Moratis, L. et al. (Springer, 2018); *Transition Redesigned. A Practical Philosophy Perspective*, co-edited with Gasparski W. W. (Routledge, 2017); and *Understanding Ethics and Responsibilities in a Globalizing World*, co-edited with Arruda de M.C. (Springer, 2016).

Email: brok@kozminski.edu.pl
Akademia Leona Koźmińskiego, Warszawa, Poland

Maria Roszkowska-Menkes, PhD, is an assistant professor in the Department of Management Theory in the Warsaw School of Economics (SGH). In research and teaching she specialises in issues of innovation management, corporate social responsibility and public relations, in particular with regard to new media and networked economy. She is secretary and lecturer at the SGH Postgraduate Course in Public Relations and Strategic Communicating in Companies; business trainer in the area of corporate social responsibility, public relations and innovation management; and the author of numerous research and business press articles concerning CSR, crowdsourcing and social media.

Email: maryrosz@gmail.com
Szkoła Główna Handlowa w Warszawie, Warszawa, Polska, Poland

Bożena Ryszawska, PhD, is a professor in the Department of Public and Corporate Finance at the Wroclaw University of Economics. Her research and scientific interests include green economy, sustainability transition, corporate sustainability and green finance. She is a member of Polish Association of Environmental Economists and Natural Resources and Sustainable Finance & Accounting Association in Poland. She is author and co-author of dozens of scientific publications (author of Green economy - theoretical aspects of the concept and measurement of its implementation in the European Union, Wroclaw 2013; Corporate Sustainability in Transition to Smart, Efficient, Low Carbon Economy in European Union in: Smart and Efficient Economy: Preparation for the Future Innovative Economy, Brno 2016, Sustainability transition needs sustainable finance. Copernican Journal of Finance & Accounting, 5(1) 2016.

Email: bozena.ryszawska@ue.wroc.pl

Katedra Finansów Przedsiębiorstw i Finansów Publicznych, Uniwersytet Ekonomiczny we Wrocławiu, Wrocław, Poland

Tomasz Słoński, PhD, is a professor in the Institute of Financial Management, Wrocław University of Economics (Poland), where he holds the chair of Corporate Finance and Public Finance. He is a co-founder of the Society of Firm's Appraisals in Poland. He delivers courses on corporate finance, mergers and acquisition and enterprise risk management. His current scientific research is focused on the evaluation and risk management in renewable energy projects and the cost of equity for small and medium enterprises.

Email: Tomasz.slonski@ue.wroc.pl

Katedra Finansów Przedsiębiorstw i Finansów Publicznych (KFP), Uniwersytet Ekonomiczny we Wrocławiu, Wrocław, Poland

Agnieszka Sokołowska-Durkalec, PhD, works as a professor and the chair of the Department of Enterprise Studies at the Faculty of Economics, Management and Tourism at the Wrocław University of Economics (Poland). Her major scientific and research interests focus on broadly understood problems of intangible assets (social responsibility of enterprises, intellectual capital management, knowledge management, modern organisation forms, organisational culture, etc.). Her scientific output covers over 100 scientific publications (articles, papers, book chapters, 2 monographs) in reviewed magazines, scientific studies, scientific bulletins, monographs, etc., as well as mostly active participation in about 90 national and international conferences. The author is a laureate of the competition for best Polish doctoral dissertations entitled "New trends in economic science and management" organised by the Foundation for the Promotion and Accreditation of Economic Courses and Polish Economic Society. Her doctoral thesis "Intellectual capital management in a small enterprise" was published and forwarded to scientific libraries throughout the country. In 2015, the Chapter of Scientific Society for Organization and Management TNOiK in the name of Prof. Karol Adamiecki rewarded the author's habilitation monograph titled *Social responsibility of a small enterprise. Identification-evaluation-directions of improvement*, published by the University of Economics in Wroclaw, Wroclaw 2013. The aim of the competition was to reward studies for being outstanding achievements in the field of organisational and management sciences, making a significant contribution to theoretical and practical solutions to improve the functioning of the economy.

Email: Agnieszka.Sokolowska@ue.wroc.pl

Wydział Ekonomii, Zarządzania i Turystyki, Katedra Nauk o Przedsiębiorstwie, Uniwersytet Ekonomiczny we Wrocławiu, Jelenia Góra, Poland

Franziska Struve is Project Assistant for International Study Tours and CSR at the German Graduate School of Management and Law GmbH. In addition, she is doing her part-time MBA in corporate management at the Hamburg School of Business Administration (HSBA). Moreover, she is the co-editor of the book *CSR und Marketing* published in 2017 by Springer and co-editor of the book *Corporate Social Responsibility* in Brazil to be published by Springer 2018.

Email: franziska.struve@ggs.de

German Graduate School of Management & Law (GGS), Heilbronn, Germany

Łukasz Sułkowski is a professor of economic sciences, specialising in management sciences. Currently, he is employed in the Institute of Public Affairs of the Jagiellonian University. He is vice-president for international affairs of the Polish Accreditation Committee and Director of Clark University Branch Campus at University of Social Sciences in Poland. He is also the President of PCG Polska. Since 2008, he is the editor-in-chief of the *Journal of Intercultural Management* (DeGruyter). In 2004–2011, he was the editor-in-chief of "Organizacja i Kierowanie" quarterly of the Committee of Organization and Management of the Polish Academy of Sciences "Organization and Management". Author or co-author of more than 300 publications, including 14 monographs published by such publishing houses as Springer, Lexington, Palgrave, Emerald, Lang, PWN and PWE. He is project manager or coordinator of 16 EU research grants (e.g. 6th Framework Programme, EQUAL), National Science Centre, Ministry of Science and Higher Education and State Committee for Scientific Research. He currently manages the "UNIFUT" project of the university of the future and organisational improvement of the university of the future. He is member of 4 international research associations (AAM, EURAM, IFERA, PGV) and since 2014 member of the Committee for Culture and Media Management of the Polish Academy of Arts and Sciences (PAU). He is chairman of cyclical conferences "Family Business. Managing Family Companies" (7 editions) and "Intercultural Management Congress" (11 editions) and reviewer in several dozen professorial, habilitation and doctoral proceedings.

Email: lsulkowski@san.edu.pl
Katedra Zarządzania Instytucjami Szkolnictwa Wyższego, Instytut Spraw Publicznych, Uniwersytetu Jagiellońskiego, Kraków, Poland

Justyna Szumniak-Samolej, PhD, is an assistant professor at the Department of Management Theory, Warsaw School of Economics, Poland. She works as a researcher, trainer, lecturer and conference speaker. Her key research areas are management in the new economy, corporate social responsibility, social media and innovative business models. Her academic publications include books (e.g. *Responsible business in the networked economy*, Poltext 2013; *Business potential of social media*, Poltext 2015; *Interviews with social business leaders in Poland*, Warsaw School of Economics Publishing 2016), refereed journal articles, book chapters and refereed conference papers. She is a member of the Working Group on Education and Popularization of CSR in the Ministry of Economic Development of the Republic of Poland. She is also a member of experts committee of "Philanthropist of The Year" competition organised by the Academy for the Development of Philanthropy in Poland.

E-mail: justyna.szumniak@sgh.waw.pl
Warszawa, Poland

Iveta Ubrežiová, Ing., PhD, is a professor at the Department of Management at the Faculty of Economics and Management, SUA, in Nitra, Slovakia. She is interested in international management and entrepreneurship, corporate social responsibility and small and medium-sized enterprises (supervisor and lecturer). She has a rich activity of publication. She is author or co-author of more than 200 publications. Her research activity is expressed in several national and international books, monographs, the scientific articles cited in Web of Sciences and

SCOPUS and more than 200 articles from domestic and international conferences. She was a main coordinator of the scientific VEGA project "Competitiveness of the Slovak agri-food enterprises on the European market" and "Corporate Social Responsibility (CSR) of the Slovak Enterprises in the Context of Internationalization in Business" as well as a co-coordinator of VEGA projects and a member of research team of national or international both scientific and educational projects. She attended universities like Erasmus+ teacher in Finland, Spain, Latvia, Lithuania, Hungary, Ukraine, Poland and Russian Federation. She participated in CEEPUS project as a teacher in Albania, Poland and Serbia. She is the main coordinator of the study programmes "Management of Enterprise" at the bachelor and master level "Management of Enterprise" at the Department of Management. She is a member of the Scientific Board at the Faculty of Economics and Management in Nitra (Slovakia) and a member of editorial boards of the scientific journals *Business Trends* (West Bohemian University in Pilsen, Czech Republic) and *Journal of Agribusiness and Rural Development* (Agricultural University of Poznan in Poznan, Poland).

Email: iveta.ubreziova@uniag.sk

DMan FEM, Slovak University of Agriculture, Nitra, Slovakia

Florian Urbański, MBA is project management officer in IBM Global Delivery Center Wroclaw (Poland), Intercultural Trainer and Workshop Coordinator of WrOpenUp Foundation. He studied business administration in Biberach University of Applied Sciences. His main scientific interests are intercultural communication, Polish-German business relations and projects of corporate social responsibility.

Email: F.Urbanski@gmx.net

Schelklingen-Gundershofen, Germany

Aldona Wiktorska-Święcka, PhD, is a professor of the University of Wrocław, MBA; head of unit of European Integration and Regional Development at the Department of European Studies (Faculty of Social Sciences); member of Regional Study Association, European Urban Research Association and International Public Management Network; and fellow of German Academic Scholars Exchange, European Union, Konrad Adenauer Foundation, US Department of State. Her field of expertise is public governance and public policy, territorial governance, sectoral governance, innovation in management and governance, normative assumptions of management and governance. She is author of monographs, papers and chapters in several domestic and international publications and leader of National Partners within international project consortia, financed upon Horizon 2020.

Email: aldona.wiktorska-swiecka@uwr.edu.pl
Wrocław, Poland

Anna Witek-Crabb, PhD, is an assistant professor in the Department of Strategic Management at the Wroclaw University of Economics, Poland. Her research interests include sustainable strategy, corporate sustainability, strategic management of businesses and non-profits and design thinking. She is a member of Strategic Management Society (SMS), EURAM (European Academy of Management) and BAM (British Academy of Management). She is author of over 50 publications, among others "Warsztat stratega" ("Strategist's toolkit"), Wolters Kluwer, 2013, and a National Science Centre grant holder in "Sustainable strategy of enterprises". She is participant of numerous international conferences and a visiting scholar at the University of Malaga.

Email: Anna.Witek@ue.wroc.pl
Katedra Zarządzania Strategicznego, Uniwersytet Ekonomiczny we Wrocławiu, Wrocław, Poland

Przemysław Wołczek, PhD, is an assistant professor in the Department of Strategic Management at the Wroclaw University of Economics. His research and scientific interests include strategic management, strategy implementation, corporate social responsibility and sustainability and ESG analyses. He authored and co-authored more than 50 scientific publications (e.g. co-author of Strategiczny CSR, Texter, 2016; author of Communicating CSR in the Practice of Polish Companies, GSTF Journal on Business Review, Global Science and Technology Forum, 2015, Vol. 4, No 2; The Concept of Corporate Social Responsibility and Sustainable Development, Problems of Sustainable Development, 2014, Vol. 9, No. 1). Since 2017, he is a member of the Working Group on Education and Popularization of CSR of the Team for Sustainable Development and Corporate Social Responsibility at the Polish Ministry of Development. In 2016, he participated in Pioneers into Practice Programme. He is a member of ISDS (International Society for Development and Sustainability). Along with research and teaching, he deals with business consulting and tutoring.

Email: przemyslaw.wolczek@ue.wroc.pl

Katedra Zarządzania Strategicznego, Uniwersytet Ekonomiczny we Wrocławiu, Wrocław, Poland

Part I
Introduction

Transition from Corporate Responsibility to Sustainable Strategic Management

Boleslaw Rok

1 Introduction

Corporate social responsibility (CSR) has a long history, but the remarkable dynamics of the various theoretical concepts and practical applications can be seen especially in recent years. There are many studies that present a detailed history of the development of different frameworks for CSR (Crane et al. 2008; Rok 2013). When analysing the development of the CSR at the company level it can be stated that general declarations of a moral nature and discussions about the scope of the principle of stewardship were replaced with a more holistic vision of corporate responsibility or corporate sustainability as the foundation of a new business paradigm, grounded to a certain degree in management theory. In the context of the evolution of CSR—as an umbrella concept—towards corporate sustainability or corporate strategic responsibility, businesses are confronted now with the imminent challenge of adapting deep strategic orientation toward Sustainable Development Goals (SDGs) (Avery 2017; Kolk et al. 2017).

Corporate responsibility can be one of the most important dimensions—if not the foundation—of an effective strategy, resulting from the objectives and priorities adopted by business leaders and managers. It is often said that incorporating CSR principles into a company strategy may lead not only to social benefits for stakeholders, but also—at least in some cases—to tangible business benefits. Some researchers in Poland are pointing out that the corporate social responsibility can be a source of competitive advantage of a company, provided that the activities are rightly undertaken. Main business benefits are: the boost of the firm's confidence, the improvement of its reputation on the market and the increase of employees' loyalty,

B. Rok (✉)
Akademia Leona Koźmińskiego, Warszawa, Poland
e-mail: brok@kozminski.edu.pl

A. Długopolska-Mikonowicz et al. (eds.), *Corporate Social Responsibility in Poland*, CSR, Sustainability, Ethics & Governance,
https://doi.org/10.1007/978-3-030-00440-8_1

what can lead to the improvement of companies' long-term performance and competitive advantage (Ceglinski and Wisniewska 2017). On the global scale, in a research study conducted already a few years ago it was noted that out of 100 randomly selected companies from the Fortune 500 only one third declared in its mission that its goal is to maximise value for shareholders, and the remaining declared that their goal is to maximise the well-being of all stakeholders (Agle et al. 2008).

The question remains, however, how to incorporate CSR into business strategy? Perhaps the CSR strategy should be developed independently and only refer to the business strategy to some extent, i.e. supplement it? CSR strategy could be a part of a business strategy or just the opposite (Bove and Empson 2013). In some cases, one can easily recognize that business strategy is contradictory to CSR strategy. Therefore, this conceptual paper is aimed at exploring the possibilities for integration of corporate responsibility or corporate sustainability into strategic business framework, providing a theoretical analysis based on the existing literature and the management practice in companies operating on the Polish market.

2 Corporate Responsibility Context in Poland

In the literature on the history of economic and business thought in Poland many statements can be found about corporate responsibility to society as well as the responsibility of business owners, even before the concept of corporate social responsibility really started being analysed scientifically. Typically, reference was made to the concept of philanthropy, although this was not always correct—it would have been better to describe this approach as the social engagement of entrepreneurs. In Poland, this "philanthropic" activity of entrepreneurs developed in earnest already in the second half of the nineteenth century, with industrial development, mainly inspired by positivist ideas (Leś 2001). The activities of people like Kronenberg, Bloch, Norblin or Wawelberg do not only determine the history of industrial development in Poland, but also demonstrate that the idea of corporate social responsibility—understood as activities carried out by entrepreneurs—is not exclusively an "imported product". Entrepreneurs, together with the representatives of the then Polish aristocracy, outstanding noble families and intellectuals, supported the construction of factory estates, the development of vocational education, healthcare and culture or summer holiday activities for children of poor parents.

But almost till the end of the 1980s issues of corporate social responsibility were discussed nearly exclusively in the US market. It was not until the 1990s that the process of CSR internationalisation and diversification began. From this perspective we can say that CSR came to Poland with only a slight delay compared to other European countries—primarily in the academic community of business ethicists. Already in the late 1990s the first public discussions were held on this topic in the business community—often inspired by non-governmental organisations, while 10 years later the term CSR was used practically on all the websites of the biggest companies in the Polish market. Starting with review articles in magazines for managers (Rok 2000), corporate social responsibility has also become an interesting topic for the general public (see more in the article of Kacprzak in this book). The European policy in the field of CSR has played a major role in the dissemination of this concept. Initially this was because the preparations for EU accession and subsequently—since 2004—on account of participating in the development and implementation of various European policies.

Existing scientific research suggests that CSR is becoming an over-simplified and peripheral part of corporate strategy. Rather than transforming the main corporate discourse, it is argued that corporate responsibility and related concepts are limited to rhetoric defined by narrow business interests and "serve to curtail interests of external stakeholders" (Banerjee 2008). When reading the strategy documents of some of the biggest companies on the Polish market CSR priorities already appear at the level of the mission, vision or corporate values statement. It was a part of the bigger research project conducted by author, so the names of these organizations were hidden, but the dominance of rhetoric was clear at this stage of development in Poland (Rok 2013). For example, the mission of the construction company X refers to the way of conducting its business activities, namely "in accordance with the highest global quality standards, considering the principles of sustainable construction, guided by the principles of professional ethics and attention to user comfort and satisfaction". In the company Y the mission was formulated as follows: "By setting new trends, offering unique and effective solutions in the interests of human health and life, we focus on those business opportunities that are socially responsible, environmentally friendly as well as economically valuable". Such a mission can be the basis for a company strategy, according to which the realised business projects are limited to only those that are in line with the principles of sustainable development, although no information confirming such a hypothesis could be found.

Elements of CSR can also be seen at the level of values in company X, as the basic value that is often adopted is responsibility, which can be understood in the following way: "As a socially responsible and conscious firm we work in accordance with our motto 'people first' and take care to ensure the safety of our workers. We are also an environmentally friendly company, caring for the natural environment at every stage of our development. We implement specific measures to improve the state of our planet". In the corporate values statement of company Y, we can read: "We respect our customers, shareholders, the natural environment and local communities". Such corporate values statements are increasingly being expanded, creating a formal code of ethics, for example.

These statements, however, are often followed by a description of the business strategy, where corporate social responsibility is no longer expressed in any way. Business strategies of companies operating on the Polish market are usually based on certain pillars or overarching objectives, such as building shareholder value, followed by certain priorities or areas within the framework of which the strategic objectives (e.g. maximising the use of existing infrastructure or strengthening the competitive advantage while reducing external risks), planned or already undertaken actions and measures are presented. Usually the mentioned strategic objectives do not include any objectives that directly refer to the principles of corporate social responsibility or corporate sustainability, and that means that CSR is not incorporated into the business strategy.

In the academic literature there is a growing evidence that CSR is beneficial to the bottom line, at least when using a long-term and value creation perspective (Husted et al. 2015), although empirical studies of the relationship between CSR and corporate financial performance have reported mixed findings (Brooks and Oikonomou 2018). For progressive managers on the global market CSR is no longer a purely moral or "social" responsibility for the common good, but increasingly a potential strategic resource to be leveraged to improve corporate financial performance and customer relationships. The concept of CSR has expanded to entail both social, environmental and economic interests at the macro level as well as the organizational level.

According to Porter and Kramer sources of strategic competitive advantage are broadened to include social aspects and social influence (2011). The attractiveness of the CSR and diffusion of related concepts is leading to the amplification of the increasing convergence between CSR and strategy. This strategic turn in the conception of CSR is likely to bring about a shift in corporate practices from reactive attitude and passive compliance with societal expectations to more proactive engagement with social issues, particularly among global market-oriented MNCs who are under monitoring in relation to various aspects of value creation (Yin and Jamali 2016).

Some companies in Poland, usually those that have a comprehensive mission that includes the principles of ethical conduct and social responsibility in greater detail, also have a CSR strategy or a sustainable development strategy. For example, in company Z the mission contains the following statement: "Setting the interest of our shareholders, customers and employees as our main priority, we wish to focus on our reliability and transparency as a partner and to develop the Group's values in accordance with the principles of sustainable development". This statement is further expanded as part of the "Sustainable Development and Corporate Social Responsibility Strategy". This means that the company Z has at least two strategies, i.e. a business one, which focuses on the main objectives, and another one, which concerns corporate social responsibility, which in a way is a secondary area of focus.

"Strategic CSR" is a term used quite often recently not only in the academic literature but in practice as well. For Chinese researchers "Strategic CSR is the very type of CSR most firms are taking in the practice because the act of a firm's social contribution is not only good for the society but also good for the investor's long-term profitability and development" (Wenzhong and Yanfang 2017). Some researchers (Bocquet et al. 2017) are examining two types of CSR behavior: strategic and responsive, using the "strategic CSR model" described by Porter and Kramer (2006). In this approach CSR can deliver more business opportunities and innovations, enabling the firm to build competitive advantages through social progress, redesigning products, markets, and productivity in the value chain.

CSR strategies are usually constructed in the same way as business strategies, which means that they describe several important areas of focus, priorities, strategic objectives, actions and measures. Unfortunately, however, they are often similar in different companies, even from different industries. Even though it is understandable that certain elements—like important stakeholder groups—may overlap, the main strategic determinants should be the basis for building a unique position in the market. It is possible to determine the approach to corporate social responsibility based on the strategic documents of various companies as well as their CSR reports, although then we must acknowledge the fact that these reports are usually a bit "tailored" to the needs of company's stakeholders. Moreover, such statements often suggest a holistic approach based on the integration of business strategy and CSR strategy, while practice and the examples given prove otherwise. For example, in the report of company XX (Rok 2013) three main objectives are given: portfolio development, expansion and the exploration of new business areas. These objectives are being pursued to ensure a stable income, long-term growth and a continuous improvement of the company's value. In terms of CSR the mentioned report of XX also adds that "in Poland we observe a growing awareness that profits cannot be achieved without a credible dialogue with the social, environmental and economic environment. We also operate according to this idea". Thus, we can see that the operational definition of CSR in this case includes only two elements: a dialogue with the stakeholders as a tool for strengthening the generated profits and an emphasis on three aspects of sustainable development, particularly within the context of risk management.

In corporate report of company YY (Rok 2013) we can read that "corporate social responsibility is a tool that should contribute to responsible company management, improve the image and credibility of the brand and, consequently, improve the financial results in the long term". Here we have both the operational element concerning management systems and the social element focused on image. In another report of YY, 2 years later, the given statements are less ambiguous. The company mission is formulated as "maximising the benefits to society by providing services of the highest quality", which clearly contains strategic objectives regarding the implementation of CSR priorities and ensures their operationalization through a clear CSR management structure covering all areas of functioning of the company YY.

The report of bank ZZ, as another example, states that "corporate social responsibility is inherent in the nature of our Bank, which with its activities makes a significant contribution to the economic, social and environmental development of the country". Thus, for bank ZZ corporate social responsibility is—on the one hand—"an approach to the management of the organisation, which allows it to contribute to the sustainable development of society", which is basically the core business activity resulting from the "nature" of the bank, and on the other hand, it assumes "going beyond the obligations required by law and caring for: ethical principles, building positive relationships with employees, the social environment and the natural environment" (compare article of Sułkowski and Fijałkowska in this book). Can we therefore assume, based on the information from the cited reports, that in these companies the operational definition of CSR is explicit enough to clearly formulate their priorities? It seems that it is not, after all. In recent reports from the global market the challenge with an operational definition of CSR is solved by replacing term CSR with sustainability, which is much better defined, integrating social, environmental and economic performance (Schaltegger and Wagner 2017).

3 Barriers to Sustainable Strategic Management Approach

From the point of view of management practitioners in Poland, corporate social responsibility is still predominantly formulated in a language that is inconsistent with modern strategic management, whether in the form of radical postulates of a social activism nature, or merely—moral obligations. Discussions about sustainability or CSR are too abstract or—on the contrary—too simplified, detached from the real challenges that arise in management here and now (Aluchna and Rok 2018). The main underlying factor is the low level of social trust, significantly hindering the development of the concept of CSR, which should predominantly consist in partnership and cooperation, allowing to determine the actual social responsibility, i.e. a participatory management model that considers the participation of various groups of stakeholders. Some positive outcomes can come with an ongoing expansion of The UN Sustainable Development Goals—the 17th Goal recognizes the importance of partnerships and collaborative governance but organizing large multi-stakeholder partnerships requires sophisticated implementation structures for ensuring collaborative action (MacDonald et al. 2018). Now we can observe an information chaos, where questionable knowledge combined with a fixed view of the economic reality become the basis for developing policies on corporate social responsibility by various companies, sometimes even acting in good faith. At best, the consequences are like after taking a placebo—neutral for the quality of management systems and social performance.

In the Polish market a systematic pursuit of implementing comprehensive systems of corporate social responsibility can be seen primarily in three groups of companies (Gasparski et al. 2010). The first group includes transnational companies considering the various aspects of corporate social responsibility, that in the global market are assessed both by investors and e.g. consumers or other stakeholders. Usually, after the first stage of development of CSR in the home market, such corporations are beginning to expect that their principal subsidiaries operating in other markets will implement the same guidelines for the management of various areas, as well as reporting that considers the globally adopted indicators. This group of companies on the Polish market is growing the fastest in terms of quality of CSR management, at first mainly influenced by the guidelines from their parent companies, and later often independently of these guidelines, once the business benefits of CSR become more obvious. Initially, however, the pressure clearly comes from the parent company (See article of Dancewicz in this book).

The second group of companies—smaller in terms of numbers—consists mainly of large and rapidly developing private companies with domestic financial capital. Typically, they begin to seek CSR solutions when they already have solid business partners who have implemented CSR standards in international markets and based on these standards have undertaken to verify their suppliers and other business partners in this regard. A similar situation occurs when such companies try to gain international business partners—sometimes also investment or financial partners—and within the framework of their future cooperation they must document not only the quality of their products, but also the quality of their management—that is, if those partners attach great importance to sustainability in the supply chain. In such cases, the pressure comes from business partners.

The third group primarily consists of the largest companies of the State Treasury, where the management is more sensitive to pressures of a political nature, especially coming from the European Commission within the framework of European economic policy, an increasingly important part of which are corporate sustainability, circular economy, social inclusion, etc. Often such companies are also motivated by the desire to improve their management standards, to be able to strongly emphasise that they match the international standards in their industry (See more in the article of Kacprzak in this book, Chap. 2).

Corporate sustainability is becoming a subject of discussion among managers in Poland in recent years, although this is a difficult discussion. Mainly because sustainability is a very vast subject and if it cannot be linked to actual examples from the life of the company, then the discussion becomes abstract. The emerging controversies in Poland around the term itself as well as the entire concept of CSR were caused by many reasons. The first important factor causing various controversies was the memory of our recent history, when all companies were, in a sense, "socially responsible" by definition. This term was and often still is associated with a return to the times of the discredited socialist, or as it is called today, communist economy (See also article of Rojek-Nowosielska in this book). At the same time, however, being aware of the evaluative nature of the term "corporate social responsibility", the representatives of most companies eagerly began using this term, although they claimed that the essence of the social responsibility undertaken by their business is primarily compliance with the law and concern for economic performance. Therefore, respondents more strongly identified themselves with the vision of a business where the social consequences of the core business activity are indeed acknowledged, but no additional, voluntary commitments of a social or environmental nature are made (Rok 2013).

The second factor that is specific to the Polish market, affecting the emerging controversies, is mainly associated with the difficulties in defining the various positive effects of corporate sustainability or strategic responsibility. The acknowledged measures of success still primarily include financial indicators, rather than long-term benefits for the natural environment, for employees or the local community. One can think that failing to see the relationship between the growth of social welfare, combating climate change and company performance is caused to a certain extend by the fact that even the managers of the largest companies are convinced that social results or social benefits resulting from company activity cannot be measured. That is why dealing with corporate responsibility is often entrusted to the staff of communications departments, who often use simple indicators and in addition to the incurred expenditures they report the number of positive press coverage. The result is a kind of "vicious circle" that strengthens managers in their belief that corporate responsibility only means costs that need to be borne to build an image and no strategic reorientation of management priorities is needed here. More research in this field is necessary, using methodologies specific for psychological or sociological inquiry.

The applied CSR rhetoric often has little to do with the real social or environmental challenges associated with the conducted business activity. But some practitioners from business are putting already important questions during MBA courses or different trainings: How can strategic priorities in terms of CSR be pursued, to maximise the benefits for both the company and society at the same time? How should managers and other employees be trained in order for them to respond adequately to the changing social and environmental expectations? How should CSR be managed to effectively build a competitive advantage on the global market? That is why it is important to set new frameworks for the strategic approach to corporate responsibility.

Managers have been intuitively recognizing that the integration of corporate responsibility and sustainability is an important topic, but they just rarely consider it in strategic management. In the academic literature corporate sustainability goes beyond corporate growth and profitability and includes a firm's contribution to societal goals, environmental protection, social justice and equity, and economic development (Schwartz and Carroll 2008). Considering corporate sustainability or strategic responsibility in business strategies and processes has become recently a promising way to cope with the global challenges on the market (Bhattacharyya 2010; Hahn et al. 2017) under the label of strategic sustainable management or "sustainable strategic management" (Rego et al. 2017). Since decisions related to corporate sustainability are taken at a strategic level, which is essential for generating a sufficient level of company-wide commitment, there has also been increased interest in the subject related to the possible integration of corporate sustainability in a company's strategy.

There is still a lack of empirical—quantitative and qualitative—studies on the integration of corporate sustainability into strategic management (Engert et al. 2016). Some conceptual papers are discussing different options and motivations behind implementing sustainability into the corporate strategy. Different options are considered: to adjust the corporate strategy to include objectives regarding economic, ecological and social performance; to define a specific sustainability strategy as part of the corporate strategy; or to redefine the corporate strategy based on the premise of creating a sustainable strategic management system (Oertwig et al. 2017).

4 Materiality Matrix: A Tool for Aligning Corporate Sustainability with Business Strategy

Corporate responsibility approach is implemented primarily by larger enterprises once certain choices have already been made, such as the selection of a market domain. Therefore, the important question is whether CSR can lead to the need to modify or even significantly change the business strategy. The CSR concept clearly defines this change: striving to optimise the value for all stakeholders. Companies that implement corporate responsibility have various ways of formulating their business goals, but often they refer to—at least in the global market—creating stakeholder value, shared value or sustainable value. Its basis is the principle formulated by C. Laszlo, among others: "sustainable value occurs only when a company delivers it to shareholders while not destroying value for other stakeholders" (Laszlo 2003). Corporate sustainability is conducting business with a long-term goal of maintaining the well-being of the economy, environment and society.

From a stakeholder perspective, a company's activities cannot be seen only in terms of value creation for that company, but also in terms of sharing that value between various stakeholder groups and consequently achieving an adequate level of consistency. Creating sustainable value means accepting a commitment describing a specific vision of the desired future that the company strives for by merging the company's superior shareholder value with stakeholder value. In the management practice of sustainable companies, value optimisation is often based on Freeman's stakeholder strategy matrix (Freeman et al. 2007), or on other similar tools, such as an assessment map of the company impact on stakeholders and shareholders (Laszlo 2003). Prioritisation is an important element influencing the formulated strategic objectives and it relies heavily on the various results achieved in the process of involving stakeholders.

The degree to which stakeholder expectations are considered is a very important business issue, although it often comes down to identifying the strength of individual stakeholders and the potential threats to company activities—thus, it is considered in terms of risk management. The materiality matrix balancing relevance for shareholders and other stakeholders can be defined in a variety of ways (Fig. 1), particularly when it comes to the actual practice of developing or modifying a strategy, rather than just some acceptable, usually simplified, form that can then be presented in a public CSR report.

For example, on the shareholder axis, instead of the traditional business case-based approach, i.e. the profitability for the company, measures are used that refer to the significance of the economic, environmental and social—including long-term—impacts, level of importance for company sustainability success, or relevance from external stakeholders' point of view. On the axis of other stakeholders also certain changes are introduced, going from a typically reputational approach, when we determine the impact on the assessment made by stakeholders, to an assessment of the importance attached to the objectives and the resulting future actions that can be undertaken by the stakeholders, and ending with the relevance of the actual challenges concerning various aspects of sustainable development. The materiality matrix (see Fig. 1) for strategic priorities allows to focus on the area that is essential both for shareholders and stakeholders, e.g. primarily undertaking actions that concern the most important forms of company impact on the natural and social environment, while also responding to the greatest challenges of sustainable development.

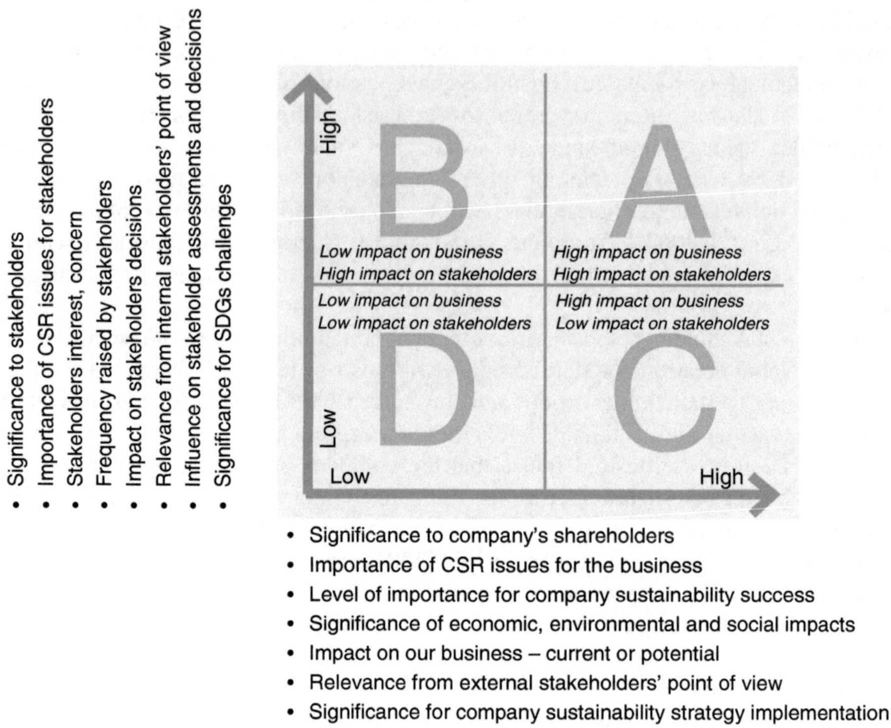

Fig. 1 Materiality matrix about here

Therefore, in companies where the conclusions gathered in the process of dialogue with stakeholders serve not only to assess the CSR strategy or the undertaken measures, but also to introduce possible changes in the business strategy, a properly used materiality matrix can be a useful tool.

Sustainability issues of particular significance are initially selected based on GRI reporting standard and its own sustainability themes. After comprehensive review made in the process of stakeholder dialogue, internal and external, with the active participation of top managers from all business units, the matrix can be build, with a clear and proper assignment of issues to the particular part of the matrix: A, B, C, D. The materiality of each of selected issues—concerning for example sustainability in the supply chain, environmental aspects in production, social aspects in operations, sustainable products delivering—should be scored and plotted in the materiality matrix, using two scales representing importance to business and importance to stakeholders. To implement real sustainability strategy company is starting from the A part—high impact on business and high impact on stakeholders.

This process also contributes to strengthening the organisational learning process, and even if at a given stage, for whatever reason, not all challenges can be addressed, it is important to be aware of the relevance and to know the long-term approach. This is also an important factor in shaping the motivations of company leaders, resulting in a broadened perspective and a subsequent transformation of the organisational culture recognising the importance of the company taking on sustainability challenges. When reconstructing this approach, several elements need to be focused on that are important in the process of creating business and social value simultaneously.

The starting point is usually the identification of a significant social challenge, which in the perspective of further company development—if not taken on—could be a significant barrier to e.g. increasing sales in the future. We are not talking about purely market challenges here, but about those that concern broadly understood sustainable development, i.e. a desired state of equilibrium that ensures a socio-economic and environmental order, justice between generations and within generations. Such challenges are often the focus of attention of both the external and internal stakeholders of a company. They may be due to demographic trends (e.g. aging population, which may indicate the need to adapt the manufactured products to the future expectations of the growing social group of seniors) or current events affecting possible changes in customer attitudes (e.g. criticism of aggressive marketing of financial products). While it is often claimed that CSR is a response to stakeholder expectations, what matters is to what extent the expectations "fit in" with trends, important topics and events. It is also important to what extent managers can predict probable future expectations, not yet expressed directly. That is why at this stage it is important to provide a clear answer to the following question: why is this challenge important to the company development?

The second element of the presented model is the decision-making process resulting from a certain type of perception and motivation. The "change agents" are often CSR managers, provided that their position within the organisational structure allows them to influence important business decisions. This is also dependent on the market domain and the challenge itself that has been identified—some challenges are easier to recognise as very important both for society and for the company (e.g. e-education), others seem unrealistic (e.g. reducing global hunger). Decision-making can be influenced by how others perceive the alignment with the company strategy or code of values, the benefits of compliance with "soft regulations", the possibility for further developing an existing product or programme, profitability, moral and social norms, trends and political correctness, enhancing employee loyalty or achieving individual goals. At this stage, the main question is: what do we want to achieve with this challenge?

The third element is the answer to the question: what exactly do we do and how do we do it? Here it is important to develop a coherent business strategy, to define how to achieve the planned goals. This strategy can be linked to the strategy of corporate sustainability—if the company has one. The bottom line is to plan the necessary resources, implementation tools, and indicators for measuring the business results in relation to the expectations and the set goals. Thus, we are evaluating the effectiveness, i.e. the ratio of effort to results. It is from the point of view of effectiveness that innovative solutions such as partnerships with NGOs or public organisations are considered at this stage to increase the chances of reaching the target group while minimising costs.

The last element of this model is measuring the impact, i.e. finding the answer to the question: how do we know that this benefited the target social groups? Impact indicators should be specific and measurable, where possible, and not in the form of e.g. "overall increase in the quality of life of the inhabitants of the region". The impact can be indirect or direct, achieved in the short or long term, it may concern economic factors (e.g. the creation of high quality jobs in the region), environmental factors (e.g. a reduction of the total emissions of pollutants at city level), or social factors (e.g. increased availability of selected products for marginalised groups).

Impact measurement is a rapidly developing new reporting area, one that is very important for increasing sustainable value (Crane et al. 2017). With these types of projects, it is not just about business results—or at least no loss—but it is also about an attempt to assess the actual contribution to reducing a given social problem, e.g. whether the introduced innovative service for previously marginalised people has reduced the scale of exclusion in the target group. Achieving measurable success motivates to broaden the scope of activity, to modify important tools and to maintain such projects in the long run. Thus, it allows for effective operationalization while gaining new allies among the management staff. Because nothing persuades managers better than effectiveness—also in terms of solving social problems using business methods.

5 Strategic Challenges for Sustainability

The well-known strategic management scholar K. Obloj points out that every change in strategy must be linked to changes at the level of four essential processes: understanding and interpreting the business environment, setting goals, defining company boundaries, and building a business model (Obloj 2010). In all these processes it is necessary to broaden the perspective, primarily to understand the market challenges related to the changing environment in the context of sustainable development, and to formulate or change the strategic objectives and the business model. If we were to transfer these considerations to a more general level, we would be dealing with a certain broadening of the strategic perspective in the direction of corporate sustainability strategy. The management executives in companies that implement corporate responsibility should assume a broader perspective at the stage of interpretation and giving meaning to changes, sense-making. Based on, among other things, the conclusions drawn from an analysis of the materiality matrix as well as a PESTLEE strategic analysis, which considers political, economic, social, technological, legal, environmental and ethical factors, they focus primarily on the most important sustainability challenges that may be relevant to the company's further development.

The starting point in this process is to identify and propose business solutions to—or ways to at least contribute to some extent to minimising—significant challenges in the environment, rather than just trying to understand trends that we have no influence on anyway. This makes it possible to create real value for shareholders and at the same time for other stakeholders, creating a set of complementary goals that allow for the sustainability or optimisation of value for all stakeholders. This, in turn, should be ensured by an adequate business model, enabling sustainability innovation, e.g. for the efficient, economic and ethical use of material and non-material resources. That is why not only the assessment indicators of the key financial results (KPIs—key performance indicators) are important here, but also indicators of economic, social and environmental impact (key impact indicators). And finally, in terms of defining the boundaries of a company, broadening the perspective involves answering the question about the limits of corporate responsibility, i.e. what we take responsibility for and what we do not. It is pointed out more and more often that the boundaries of a company do not coincide with the boundaries of responsibility.

In practice it can hardly be expected that consciously implementing corporate responsibility in a radical way, virtually overnight, will lead to a significant change in priorities at company level, or will cause a significant change in the company strategy, leading to a full merger of business and social goals. The experience in implementing the principles of corporate responsibility and sustainability as an important management dimension in major companies shows that the changes are permanent and based on the continuous improvements, not incremental ones. Sustainable strategic management requires an accurate diagnosis conducted by senior managers regarding the organisation itself and its environment. The uncertainty and unpredictability of changes in the technological, economic or political environment cause an increased emphasis on flexibility and the ability to respond quickly to social expectations.

However, broadening the perspective in the process of developing an integrated business strategy contributes to several important changes, above all at the business model level, especially in the case of new ventures. A business model that incorporates CSR principles is a form of organisational innovation, truly provides value for shareholders and other stakeholders at the same time (Schaltegger et al. 2016). From a strategic perspective this means that this type of innovation of the business model should lead to creating a competitive advantage through a higher value for the customer, contributing to sustainable development at both the level of the company and of society.

6 Conclusion: Managing as if CSR Mattered

CSR is not just a form of doing good deeds and not just a better way to maximise profits, but rather a growing awareness that the role of business in society is not just about benefiting business owners or shareholders. If there is the belief that the sole purpose of companies is to maximise profit for shareholders, CSR will only be implemented once it is proven that it is a tool for cost reduction (e.g. access to capital) or better risk management. It is still the case in Poland, where companies lack a strategic approach with respect to corporate sustainability integration.

Adopting a broader perspective, i.e. considering not only financial performance but also social and environmental results—with the belief that business also creates environmental and social value—leads to the search for justifications based on elements of competitive positioning, such as building new brand attributes and gaining new markets.

It is worth noting that corporate responsibility is formed in a process of dialogue. By collaborating with stakeholders and establishing, through a process of dialogue, the significant areas of responsibility and main sustainability challenges, an organisation could expand its ability to create its own future in the best possible way. Active stakeholder engagement would also make it possible to address the changing context of the functioning of a company, to define the scope that is consistent with materiality matrix. Companies cannot co-create sustainable value without an active cooperation, because competitiveness requirements make cooperation necessary. In other words, the development and the growth of value of a responsible company should translate into increased satisfaction and quality of life for all stakeholders.

It can be assumed that discussions about corporate responsibility are in principle a form of response to one of the most important dilemmas in management theory as well as in business ethics. Does the only role of business come down to effectively engaging resources to maximise profits, without violating the law, or is the role of business much broader nowadays, primarily consisting in looking for sustainable solutions that are beneficial to all stakeholders? If we assume that the role of business is to create sustainable value, then we can say that corporate responsibility is becoming a theory of corporate ethics, while the practice of sustainability consists in finding the best ways to build a dynamic equilibrium in the time and space of stakeholders. This approach is about finding a balance in the economic, social and environmental dimensions, in relationships with current stakeholders, but also in the process of protecting and developing the resources needed in the future. The main pillar of this approach is sustainable strategic management system and the whole process of integrating business strategy with sustainability strategy.

For companies operating on the Polish market it is the biggest challenge for next few years. There are some pioneers in this field already, mostly among multinational companies (Compare article by Dancewicz in this book). But it is also the biggest challenge for the strategic management scholars—the new research area devoted to real integration of sustainability into business decisions is not well defined yet. Further steps in this area of sustainable strategic management framework are nevertheless needed to extend the scope towards integrated management systems to evaluate and control the sustainability performance of companies. Overall, the new approach to sustainability management clarifies the interplay between classical business strategy and corporate responsibility and highlights the opportunity for future development not only on the Polish market.

References

Agle, R. B., Donaldson, T., Freeman, R. E., Jensen, M. C., Mitchell, R. K., & Wood, D. J. (2008). Dialogue: Toward superior stakeholder theory. *Business Ethics Quarterly, 18*(2), 153–190.

Aluchna, M., & Rok, B. (2018). Sustainable business models: The case of collaborative economy. In L. Moratis, F. Melissen, & S. O. Idowu (Eds.), *Sustainable business models: Principles, promise, and practice* (pp 41–62). Cham: Springer.

Avery, G. (2017). Top leaders shift their thinking on corporate social responsibility. *Strategy & Leadership, 45*(3), 45–46.

Banerjee, S. B. (2008). Corporate social responsibility: The good, the bad and the ugly. *Critical Sociology, 34*(1), 51–79.

Bhattacharyya, S. S. (2010). Exploring the concept of strategic corporate social responsibility for an integrated perspective. *European Business Review, 22*(1), 82–101.

Bocquet, R., Le Bas, C., Mothe, C., & Poussing, N. (2017). CSR, innovation, and firm performance in sluggish growth contexts: A firm-level empirical analysis. *Journal of Business Ethics, 146*(1), 241–254.

Bove, A., & Empson, E. M. (2013). An irreconcilable crisis? The paradoxes of strategic operational optimization and the antinomies of counter-crisis ethics. *Business Ethics: A European Review, 22*(1), 68–85.

Brooks, C., & Oikonomou, I. (2018, January). The effects of environmental, social and governance disclosures and performance on firm value: A review of the literature in accounting and finance. *The British Accounting Review, 50*(1), 1–15.

Ceglinski, P., & Wisniewska, A. (2017). CSR as a source of competitive advantage: The case study of Polpharma group. *Journal of Corporate Responsibility and Leadership, 3*(4), 9–25.

Crane, A., McWilliams, A., Matten, D., Moon, J., & Siegel, D. S. (2008). *The oxford handbook of corporate social responsibility*. Oxford: Oxford University Press.

Crane, A., Henriques, I., Husted, B. W., & Matten, D. (2017). Measuring corporate social responsibility and impact: Enhancing quantitative research design and methods in business and society research. *Business & Society, 56*(6), 787–795.

Engert, S., Rauter, R., & Baumgartner, R. J. (2016). Exploring the integration of corporate sustainability into strategic management: A literature review. *Journal of Cleaner Production, 112*, 2833–2850.

Freeman, R. E., Harrison, J. S., & Wicks, A. C. (2007). *Managing for stakeholders: Survival, reputation, and success*. New Haven: Yale University Press.

Gasparski, W., Lewicka-Strzałecka, A., Rok, B., & Bąk, D. (2010). In search for a new balance: The ethical dimension of the crisis. In B. Fryzel & P. H. Dembinski (Eds.), *The role of large enterprises in democracy and society* (pp. 189–206). Houndmills, Basingstoke: Palgrave Macmillan Ltd.

Hahn, T., Figge, F., Aragón-Correa, J. A., & Sharma, S. (2017). Advancing research on corporate sustainability: Off to pastures new or back to the roots? *Business & Society, 56*(2), 155–185.

Husted, B. W., Allen, D. B., & Kock, N. (2015). Value creation through social strategy. *Business & Society, 54*(2), 147–186.

Kolk, A., Kourula, A., & Pisani, N. (2017). Multinational enterprises and the sustainable development goals: What do we know and how to proceed? *Transnational Corporations, 24*(3), 9–32.

Laszlo, C. (2003). *The sustainable company: How to create lasting value through social and environmental performance*. Washington, DC: Island Press.

Les, E. (2001). *Zarys historii dobroczynnosci i filantropii w Polsce*. Warszawa: Wydawnictwo Proszyński i S-ka.

MacDonald, A., Clarke, A., Huang, L., Roseland, M., & Seitanidi, M. M. (2018). Multi-stakeholder Partnerships (SDG# 17) as a means of achieving sustainable communities and cities (SDG# 11). In *Handbook of sustainability science and research* (pp. 193–209). Cham: Springer.

Obloj, K. (2010). *Pasja i dyscyplina strategii. Jak z marzen i decyzji zbudowac sukces firmy*. Warszawa: Wydawnictwo Poltext.
Oertwig, N., et al. (2017). Integration of sustainability into the corporate strategy. In R. Stark, G. Seliger, & J. Bonvoisin (Eds.), *Sustainable manufacturing: Sustainable production, life cycle engineering and management* (pp. 175-200). Cham: Springer.
Porter, M. E., & Kramer, M. R. (2006). The link between competitive advantage and corporate social responsibility. *Harvard Business Review, 84*(12), 78–92.
Porter, M. E., & Kramer, M. R. (2011). Creating shared value: How to reinvent capitalism – and unleash a wave of innovation and growth. *Harvard Business Review, 89*(1–2), 62–77.
Rego, A., Cunha, M. P., & Polónia, D. (2017). Corporate sustainability: A view from the top. *Journal of Business Ethics, 143*(1), 133–157.
Rok, B. (2000). Odpowiedzialnosc spoleczna w zarzadzaniu. *Personel i Zarzadzanie, 22*, 7–11.
Rok, B. (2013). *Podstawy odpowiedzialnosci spolecznej w zarzadzaniu*. Warszawa: Wydawnictwo Poltext.
Schaltegger, S., & Wagner, M. (Eds.). (2017). *Managing the business case for sustainability: The integration of social, environmental and economic performance*. London: Routledge.
Schaltegger, S., Hansen, E. G., & Lüdeke-Freund, F. (2016). Business models for sustainability: Origins, present research, and future avenues. *Organization and the Environment, 29*, 3–10.
Schwartz, M. S., & Carroll, A. B. (2008). Integrating and unifying competing and complementary frameworks: The search for a common core in the business and society field. *Business & Society, 47*(2), 148–186.
Wenzhong, Z., & Yanfang, Z. (2017). Research on ethical problems of Chinese food firms and implications for ethical education based on strategic CSR. *Archives of Business Research, 5*(7), 76–84.
Yin, J., & Jamali, D. (2016). Strategic corporate social responsibility of multinational companies subsidiaries in emerging markets: Evidence from China. *Long Range Planning, 49*(5), 541–558. https://doi.org/10.1016/j.lrp.2015.12.024.

CSR Impact on Polish Economy: A Public Administration Perspective

Jacqueline Kacprzak

1 Introduction

The issue of corporate social responsibility is still a new phenomenon in Poland. There is not much scientific literature in this field, and most of the papers are reportable. The aim of this chapter is to present the process of CSR development in Poland from the public administration's point of view. Information that can be found in different sources as studies, reports and the Internet referring to the governmental CSR advisory boards require organizing and prioritizing. However, at the beginning, it is important to indicate the scope of the conceptual term of CSR. In Poland the English term corporate social responsibility (CSR) is commonly used as social responsibility of business (społeczna odpowiedzialność biznesu). However, in official translations into Polish of European Commission documents, CSR is translated as social responsibility of enterprises (społeczna odpowiedzialność przedsiębiorstw) and this Polish term is also used by public administration. A definition which describes CSR as the responsibility of enterprises for their impacts on society was adopted by the European Commission in 2011.

In its strategy for CSR for 2011–2014 the Commission defines also the role of public administration in the area of CSR: Public authorities should play a supporting role through a smart mix of voluntary policy measures and, where necessary, complementary regulation, for example to promote transparency, create market incentives for responsible business conduct, and ensure corporate accountability. Description of the process of CSR development in Poland from the perspective of public administration is presented in this chapter in two parts. The first is devoted to

J. Kacprzak (✉)
Collegium Civitas, Warsaw, Poland
e-mail: jacqueline.kacprzak@tlen.pl

A. Długopolska-Mikonowicz et al. (eds.), *Corporate Social Responsibility in Poland*, CSR, Sustainability, Ethics & Governance,
https://doi.org/10.1007/978-3-030-00440-8_2

discussing public administration activities through CSR advisory boards, while the second is an overview of the different instruments for implementing CSR in practice.

2 Development of CSR in Poland

The activity of Polish public administration in disseminating standards of social responsibility of enterprises should be dated from year 1998, when the OECD National Contact Point (NCP) was founded. It was a result of honouring commitments made within the Organization for Economic Co-operation and Development (OECD), which Poland joined in 1996 committing to implementing, among others, provisions of the Declaration on International Investment and Multinational Enterprises which contains Guidelines for Multinational Enterprises, the oldest international standard of responsible business behaviour. In Poland the end of the 1990s was a period of constructing a new system. It proceeded the period of real socialism and would be sometimes called political capitalism (Lewicka-Strzałecka 2006). The accumulation of capital, which took place in the initial period of system transformation, resulted from transferring the assets of state-owned enterprises to new economical entities known as public-private enterprises. The directors' posts of such enterprises were usually filled by the owners of the private enterprises, and the personnel worked both in the public and the new private enterprises (Lewicka-Strzałecka 2006). Simultaneously, the end of the 1990s was a period of intense foreign capital investments in Poland and of the development of western corporations. Through opening subsidiaries and initiating business activities in Central and Eastern Europe, these corporations brought along their organizational/corporate culture, including CSR. Corporate social responsibility became noticeable in behaviours which were a bit unusual for Polish enterprises, e.g. in considering environmental and social issues. Apart from enterprises implementing corporate social responsibility standards, new entities, which were increasingly engaging in CSR issues, appeared on the Polish market.

Authors researching CSR issues have various approaches to classifying stages of CSR development in Poland. An interesting one was presented by Janusz Garbara, who along with colleagues from Romania carried out a comparative analysis of CSR in Poland and Romania (Grabara et al. 2016). He has indicated five stages of CSR development in Poland:

> The stage of silence (1997–2000), when still no one spoke of CSR in Poland and Polish enterprises showed an absolute lack of interest regarding CSR;
>
> The awakening phase (2000–2002), when the concept of CSR was rather met with criticism and scepticism, both from the side of authorities and business;
>
> The stage of interest (2002–2004), when first signs of the need to implement the ethical behaviour and CSR approach in Polish enterprises appeared;
>
> The stage of proactive measures in matters of CSR and of implementing first projects with substantive positive results (2004–2006) and
>
> The period which authors call the stage of advanced implementation of CSR in practice, (2006–2016) when managers of large and medium enterprises tried to adapt their activities to standards observed in western entrepreneurial practices.

I propose a different classification, from the perspective of public administration's involvement in the process of developing CSR in Poland:

The Initial Stage (1998–2009) This time frame is defined by establishing the OECD National Contact Point (NCP) for disseminating OECD Guidelines for Multinational Enterprises in 1998. Since 2001, the NCP has begun operating within the structure of the Polish Information and Foreign Investment Agency. In 2004, in the former Ministry of Labour and Social Policy, first steps were made towards creating an entity for CSR issues in public administration. Some activities in this area were also visible in other departments: the Ministry of Treasury, the Ministry of Economy and the Environment, the Institute of Labour and Social Studies and the Office of Competition and Consumer Protection.

The Stage of Proactive Measures in the Area of CSR Undertaken by a Wide Range of Stakeholders (2009–2013) This was the period when most projects funded by the European Social Fund under the 2007–2013 financial perspective (including those focused on implementing and promoting CSR) were carried out. In 2009, the first CSR advisory board was established. Also in 2009, the first index of responsible companies in Central-Eastern Europe, named Respect Index, was launched at the Warsaw Stock Exchange. The Polish version of the ISO 26000 norm referring to social responsibility in organizations was published in 2012.

The Stage of Research and Development (2014–2015) The new period of promoting standards of CSR by public administration was initiated by establishing the second CSR advisory board as a subsidiary body to the Minister of Economy. The period included many meetings, exchange of experiences and project planning. Due to European Union's regulation as regards disclosure of non-financial and diversity information (directive 2014/95/UE) coming into force at the end of 2014, work on transposing the directive into national law begun.

The Stage of Mature Decisions and Activities (2016–Up Till Now–2017) In 2016 the advisory board for sustainable development and social responsibility of enterprises, a subsidiary body to the Minister of Economic Development and Finance was established. Public administration is increasingly active in promoting responsible business conduct (RBC), a concept of running businesses in a responsible manner, implemented by OECD. In June 2016, the OECD National Contact Point has been moved from the Polish Information and Foreign Investment Agency to the Ministry of Economic Development. Also in July 2016, a Polish translation of GRI G4 Sustainability Reporting Guidelines was published, then, in October 2016, a Polish translation of the OECD Due Diligence Guidance for Meaningful Stakeholder Engagement in the extractive sector was issued. In May 2017, the preparation of the Polish National Action Plan for the Implementation of the United Nations Guiding Principles on Business and Human Rights 2017–2020 (NAP on BHR) concluded as the NAP on BHR was accepted by the Council of Ministers.

As we see in the above differentiation a significant impact on the development of Polish CSR had the governmental activity through its CSR advisory boards. The first CSR advisory board (2009–2013) was a result of the meetings of the informal inter-ministerial Group on CSR, taking place in the Ministry of Labour and Social Policy since 2004. It was the first platform for exchanging experiences in the area of CSR among representatives of public administration, social organizations and businesses. The aim of this Board, being a subsidiary body to the Prime Minister, was primarily to coordinate activities and disseminate good practices in the area of CSR, and facilitate dialogue between the administration, businesses, social partners and NGO's.

Over the period of 2009–2013, 14 Board meetings took place and between the board meetings four working groups were active:

1. the working group for the system of promoting CSR in Poland,
2. the working group for responsible investments,
3. the working group for sustainable consumption and
4. the working group on CSR and education.

The main effects of the board's work include organizing conferences, seminars and trainings promoting CSR in the period of 2010–2013 and common projects and publications carried out by partners engaged in the board's activities, e.g. ESG reporting analysis in state-owned and partly state-owned companies, and guidelines for investors in the area of broader consideration of non-financial factors in investment strategies.

Through dialogue and cooperation, representatives of various communities, working together in the working group for the system of promoting CSR in Poland, in 2013 created recommendations for determining directions of supporting CSR under the 2014–2020 New Financial Perspective, stating, among others: We believe that social responsibility of enterprises builds up the competitive advantage of Polish enterprises and strengthens their market position. It also enables the implementation of innovative solutions and mechanisms which ensure that human rights are being respected and high standards of work and life quality maintained, which is substantially dependent on the professional stage of life. We are calling for means to implement the concept of social responsibility in small, medium and large enterprises. Despite expectations of experts from the working group the new EU fund financial perspective for 2014–2020 did not indicate CSR in operational programmes, such as the Operational Programme Knowledge Education Development (POWER), the Operational Programme Smart Growth, the Operational Programme Infrastructure and the Environment, the Operational Programme Digital Poland, Rural Development Programme and the Regional Operational Programmes of particular voivodships. However, this does not mean that there is no possibility to implement projects concerning CSR. The components of CSR education or implementation of CSR rules should be included in general projects descriptions, mainly at the projects with the aim to support entrepreneurship.

The proceedings of the first CSR advisory board influenced the inclusion of matters connected with CSR in the National Reform Program, which was the main tool used in implementing the EU "Europe 2020" Strategy at that time. The National Reform Program included an entry stating that principles of social responsibility contribute to a sustainable economic growth conducive to social inclusion. As a result of Board members' activities, the norm ISO 26000 was adjusted for implementation into the system of Polish norms and translated into Polish in 2012. The norm includes guidelines for organizations which define tools for implementing the concept of CSR and is applicable in all kinds of organizations, public, private and non-profit, regardless of their size and location.

The first CSR advisory board operated in 2009–2013 and was dissolved in November 2013. To continue the promotion of CSR, in 2014 the second CSR Board, bearing the same name: the advisory board for Social Responsibility of Enterprises, was established as the subsidiary body to the Minister of Economy. The thematic scope of the working groups has changed slightly:

1. the working group for implementing CSR principles,
2. the working group for sustainable consumption and production,
3. the working group for monitoring CSR trends and
4. and the working group for education and promotion of CSR.

Still this governmental initiative provided a space for multi-stakeholder forum, where the representatives of public administration, employers organizations, trade unions, social organizations, academia and representatives of business and trade organizations could exchange their experience in the field of CSR. Among the results of this Board's activities are many various meetings, trainings and conferences as well as common initiatives and projects. An important effect of the Board members' activities was also establishing a partnership for translating the GRI G4 Sustainability Reporting Standards into Polish. These standards, created by an international organization Global Reporting Initiative, is the most popular sustainability reporting standard. In connection with the 2014/95/UE directive entering into force in 2014, as regards disclosure of non-financial information, the working group for monitoring CSR trends created recommendations for the Minister of Finance, which were included in the preparatory process of the draft law on changing accountancy law implementing the aforementioned directive into Polish law. Moreover, the CSR advisory board began working on the National Action Plan on CSR in 2015, which is a requirement for all EU member states. In 2015, Polish public administration also began working on the National Action Plan on Business and Human Rights (NAP on BHR).

In 2016, the third CSR advisory board was established: the advisory board for Sustainable Development and Social Responsibility of Enterprises as the auxiliary body to the Minister of Economic Development and Finance. The main Board's task is to maintain a social dialogue, exchanging experiences between public administration, businesses, social partners, NGO's and non-governmental academic and research institutions working in the area of sustainable development and CSR. An important aspect of this CSR Boards work is also to take into account CSR issues in the perspective of possible actions in support of the Strategy for Responsible Development, the main strategic document indicating the directions for Poland's development up to 2020. Similarly to the activities undertaken by the first two boards, the advisory board for Sustainable Development and Corporate Social Responsibility, between its proceedings, operates through working groups: (1) the working group for CSR strategy, (2) the working group for business ethics and responsible business conduct standards and (3) the working group for education and popularization of CSR. The working group for CSR strategy continues to work on the National Action Plan on CSR. The aim of the working group for business ethics is to prepare a report on the current state of business ethics in Poland as well as a guidebook on ethical standards for entrepreneurs. The topics of education and popularization of CSR were met with most public interest and more than 100 representatives of businesses, academia, social organizations, trade organizations and local authorities actively take part in the proceedings. This group deals with a few important topics from the point of view of education and CSR popularization, including stating objectives for a CSR social campaign; preparing educational tools for teachers in the area of CSR; determining how to address CSR issues in small and medium enterprises; compiling educational initiatives in the area of CSR from 2015–2017 and implementing CSR on the local administrative level. Members of working groups include representatives of various communities, among others—business, trade organizations, employer organizations, trade unions, social organizations, representatives of local authorities and the academic community.

The mentioned activities of CSR advisory boards are an example of the engagement of public administration in disseminating knowledge about CSR. The operation of the two first boards was handled by the former Ministry of Economy. The activity of the advisory board for sustainable development and social responsibility of Enterprises is organized by the Ministry of Economic Development. The activities undertaken by these administrative bodies manifest the role of public administration defined by the European Commission as creating favourable conditions for shaping norms and forms of cooperation between partners in the public arena which would facilitate voluntary commitment to sustainable development and CSR. The activities of each board and its working groups required, and still do, substantial engagement both from the administrative and social partners. The active participation of representatives of various organizations, institutions and enterprises therefore indicates the importance of the topics dealt with, which are worthy of devoting time and resources to. It can be considered interesting that each of the Boards had slightly different working groups. Clearly visible that each of the Board has been tackling the problem of education in CSR. It is still a very important issue in the field of CSR. On the other hand, topics such as responsible investment or sustainable consumption and production, which are also important areas of CSR after preparing some publication by the working group or after the series of meetings, the topics have been exhausted in some sense. As a new direction in the CSR board's work in 2017, we can see ethical aspects of responsible business conduct.

Despite the engagement of Boards and working groups members, they have also met with some criticism. Although the critical opinions, it seems that the continuous engagement and interest of various communities in the proceedings of the advisory board for sustainable development and social responsibility of enterprises indicates a responsible approach of many Polish entrepreneurs and their social partners to implementing the concept of CSR in practice. The public administration's openness and initiative have a significant impact on shaping the public's awareness of the responsible approach to conducting business. That is also the uniqueness of the Polish public administration approach to implement CSR by actively involving multi-stakeholder representatives into joint initiatives.

3 Instruments Supporting CSR Development in Poland

An important instrument in supporting CSR development in Poland were EU funds of the 2007–2013 financial perspective. These were mainly resources coming from the European Social Fund (such as the Human Capital Operational Programme) and the required domestic resources, which served to finance the support for broadly-defined entrepreneurship. The impact of EU funds on Polish economy, including their influence on the shape of main macroeconomic indicators, undergoes recurring assessments carried out by the Ministry of Economic Development (former Ministry of Infrastructure and Development and Ministry of Regional Development) based on chosen indicators of goal completion defined in main strategic documents: the 2004–2006 National Development Plan, 2007–2013 National Cohesion Strategy, 2014–2020 Partnership Agreement. The impact of EU's cohesion policy on the socio-economic development in Poland was assessed in 2014 as unambiguously positive (2014 National Strategic Reference Framework Implementation Report). European funds significantly influenced economic growth, investments and the Polish labour market, as well as the internal and external economic balances. Thanks to EU funds, the GDP increase accelerated and as a result Poland experienced the consequences of the global financial crisis to a lesser extent. Taking advantage of EU resources caused a noticeable increase in employment and a drop in unemployment. The data concerning results of projects carried out in the area of CSR can be found in evaluation reports of EU projects, mainly addressed to entrepreneurs and those supporting entrepreneurial development. Despite the widely-shared opinion that many projects devoted to CSR have been carried out within the 2007–2013 financial perspective, an evaluation of the EU fund impact only in the context of CSR has not been carried out. Below, as an example, I have chosen a few projects realized under the 2007–2013 Human Capital Operational Programme:

1. Regional Support Programme for CSR for small and medium-sized enterprises and NGO's, including: educational activities of the CSR Manager Academy, the Small Grant Programme for enterprises wanting to carry out projects created by business representatives with NGO's and activities of the Lower-Silesian Federation of Socially Responsible Enterprises which brings together enterprises engaged in CSR in Lower Silesia.

2. Warmia-Mazurian Forum for CSR, primarily aimed at raising awareness of management in the micro, small and medium enterprise sector about the impact CSR has on boosting competitiveness and ensuring a stable development of any organization, as well as intensifying cross-sectoral cooperation between representatives of local administration, NGO's, business environment institutions and enterprises for the purpose of disseminating the concept of CSR.
3. Promotion of CSR Standards and Promotion of greater entrepreneurial involvement in building social capital in Poland with the use of CSR tools, two projects carried out by the Polish Confederation of Private Employers „Lewiatan". The projects were complementary to each other, and so the first one structured and organized knowledge about CSR, including goal setting in the process of creating CSR business strategies and a methodical guidebook has been created titled The CSR development model. The second project resulted in creating an IT tool used for mapping stakeholders.
4. The project Together for the development of craftsmanship was an interesting way of using EU funds for determining needs and expectations connected with the concept of CSR, carried out by the Wielkopolska Chamber of Crafts in Poznań in 2012–2013. The research carried out among craftsmen attempted to answer the question, if CSR in craftsmanship is possible. Polish craftsmanship, understood as small and medium enterprises, was definitely one of areas needing support in acquiring knowledge about corporate social responsibility and about benefits resulting from implementing this concept of conducting business in practice. This example of a project addressed to craftsmen clearly shows that CSR should not only be associated with large corporations. It is the small and medium enterprises that have a significant potential, which may be used for the common good.

The most comprehensive assessment of the state of implementing CSR standards in Poland was carried out in 2011 within the research conducted under the Operational Programme Innovative Economy, financed by the European Regional Development Fund, ordered by the Polish Agency for Enterprise Development. The most quoted result of this research was the statement that the term CSR is known only to 31% of business representatives operating in Poland—information encountered in many studies devoted to corporate social responsibility in Poland. This research is interesting because it encompassed both micro, small, medium enterprises and big ones, divided into two subgroups: nation-wide and those co-financed under the Operational Programme Innovative Economy. The mentioned level of CSR awareness referred to the nation-wide research sample, as in the group of entrepreneurs realizing activities under the Operational Programme Innovative Economy knowledge about CSR was much higher—in this group 63% of enterprises confirmed having encountered the term. Business representatives from the nation-wide sample who declared knowing the term "corporate social responsibility" identified it primarily with such aspects as maintaining good relations with employees, caring for the natural environment, integrity in dealing with business partners, caring for the employee welfare and activities benefiting local communities. Among the beneficiaries of the Operational Programme Innovative Economy, the term was understood differently: it was associated more with activities for local communities, then with caring for the natural environment and employees. It also turned out that CSR is known mainly to representatives of large (70% of indications) and medium enterprises (50% of indications), and to a smaller extent to representatives of small and micro enterprises (36 and 26% of indications). Among business representatives who declared knowing the term corporate social responsibility from the nation-wide sample, only one in five knew any recommendations or standards in the area of implementation.

At this point of consideration it should be also add that one of the results of the activities of the first CSR advisory board (2009–2013) was the preparation of objectives and implementing two systemic projects realized by The Polish Agency for Enterprise Development (PARP): (1) Corporate social responsibility and (2) Enhancement of regional competitiveness through Corporate Social Responsibility measures. The first project was co-financed by the Operational Programme Innovative Economy (carried out in 2010–2015). 257 enterprises (26 micro, 180 small and 51 medium) have received help and support in preparing and implementing responsible business strategies. A research on a sample of managers from micro, small and medium enterprises (SMEs) was also carried out in this project. Interviews were conducted with a group of 1000 representatives of SMEs, mainly decision-making managers. Main questions referred to the familiarity with the term “corporate social responsibility”. The answers demonstrated that two thirds of respondents didn’t know the term CSR. The second PARP project, Enhancement of regional competitiveness through Corporate Social Responsibility measures, was carried out under the Swiss-Polish Cooperation Programme. The aim of the project was to raise awareness and knowledge about CSR among representatives of small and medium enterprises, Marshall Offices and Regional Investor and Trade Centres. Three grant competitions for pilot projects in the SME sector were announced and later an evaluation carried out which stated that this project, not only raised awareness and knowledge and helped beneficiaries understand what CSR is, but also helped define, organize and formalize the management of this area, both through determining strategic goals and appointing people in enterprises responsible for CSR. The fact that 90% of firms continue activities initiated in the programme or carry out further projects in the CSR area also confirms the efficacy and usefulness of this project (PARP 2015). The above examples of projects carried out by the Polish Agency for Enterprise Development indicate that devoting resources to supporting entrepreneurs in raising knowledge about responsible business conduct bears fruit.

At the end of this chapter I would like also to draw attention to an interesting example of an effective cooperation between government administration and business in the area of CSR. The project Vision 2050: The New Agenda for Business in Poland was carried out by the Ministry of Economy, Responsible Business Forum, and Deloitte in 2011–2015. In June 2012, during the Earth Summit in Rio de Janeiro, Poland presented the report Vision 2050: The New Agenda for Business in Poland, which described challenges facing Polish businesses in the near future. The challenges were presented in six areas: (1) social capital, (2) human capital,

(3) infrastructure, (4) natural resources, (5) energy and (6) state and institutions' quality. The main programming document defining ten crucial challenges, in which businesses should engage to support socio-economic change were signed by 99 Polish companies. Vision 2050 defines the role that business should play in the coming decades. Some of the enterprises which took part in preparing Vision 2050 signed also a partnership declaration for implementing the 2030 Agenda in Poland, initiated by the Ministry of Economic Development in June 2017.

4 Conclusions

From a number of studies, projects and activities of various institutions, organizations and companies that have been carried out in Poland since 2009, we may infer that the concept of CSR is developing. The direction of this development and its effectiveness may be disputed from the point of view of validity, efficacy, importance etc.—however, it's impossible to say that this topic was absent in Poland. In the same time, it is difficult to infer, based on the available data, what influence it has on Polish economy. It seems that it would be useful to carry out in-depth research in the near future, which would not only verify the level of CSR concept awareness, but also its impact on the economy. Such research should focus not only on the question of the level of CSR knowledge among entrepreneurs, but on the attempt to assess the impact of particular aspects. At present, a very important source of information may be the non-financial disclosure by big companies. After the entry into force of the provisions of the EU directive in this filed in Poland about 300 companies will be obliged to disclose information in five key areas: social, environmental, human rights and anti-corruption aspects of their activity. However to implement this reporting obligation requires from companies some efforts and sometimes to implement new strategies and policies. Again, role of the public administration in this area is not only in the regulatory field but also to inform and educate, disseminate standards and monitor the implementation of the responsible business conduct.

References

Grabara, J., Dura, C., & Drigă, I. (2016). Corporate social responsibility awareness in Romania and Poland: A comparative analysis. *Economics and Sociology, 9*(4), 344–359.

Lewicka-Strzałecka, A. (2006). *Odpowiedzialność moralna w życiu gospodarczym.* Warszawa: Wydawnictwo IFiS PAN.

Documents

Communication from the Commission to the European Parliament, the Council and the European Economic and Social Committee and the Committee of the Regions — A renewed EU strategy 2011-14 for corporate social responsibility (COM(2011) 681 final, 25.10.2011).

Sprawozdanie z realizacji w 2013 roku Narodowych Strategicznych Ram Odniesienia na lata 2007–2013, Ministerstwo Infrastruktury i Rozwoju, Instytucja Koordynująca Narodowe Strategiczne Ramy Odniesienia, Warszawa 2014.

Sprawozdanie z wdrażania Programu Operacyjnego Kapitał Ludzki 2007–2013, w 2013 roku, Ministerstwo Infrastruktury i Rozwoju, Warszawa 2014.

The OECD Guidelines for Multinational Enterprises, OECD (1976) 2011.

Red Roots of Corporate Irresponsibilities (Corporate Social Responsibility with a Historical Twist)

Sławomir Magala

Slogans were empty suits draped on the corpse of an idea
(Viet Thanh Nguyen 2015, 267)

1 Introduction to the Academic Landscape of the Post-communist Research

The fall of communism breeds analogies with the fall of feudalism. English gentry fenced off the commons at the beginning of the capitalist revolution. "Tragedy of the commons" is a contemporary ideological echo of an attempt to justify this post-feudal fencing. In the late twentieth century fencing Russian, Chinese or Polish communist gentry fenced off state property at the beginning of the communist devolution. Tracking and removing the post-communist networks from public life should have been both a political and an academic, cognitive, scientific challenge. This challenge has not been met. The tracking of the communist injuries and their reconstruction (or compensation) is hampered by a historical neglect. Organized academic programs and schools need time and funding and are vulnerable to sponsoring pressures. Post-communist establishments have a formidable network of ideological constraints on social research. Chinese communist elites are not exposed to a reliable research, nor are the Russian ones. To a certain extent, the Polish, Czech, Romanian, Bulgarian or Hungarian ones are—but the process is hampered, not in the least by the "useful idiots" among the western, primarily

S. Magala (✉)
Katedra Zarządzania Instytucjami Szkolnictwa Wyższego, Instytut Spraw Publicznych, Uniwersytetu Jagiellońskiego, Kraków, Poland

Rotterdam School of Management, Erasmus University Rotterdam (Emeritus), Rotterdam, The Netherlands
e-mail: smagala@rsm.nl

A. Długopolska-Mikonowicz et al. (eds.), *Corporate Social Responsibility in Poland*, CSR, Sustainability, Ethics & Governance,
https://doi.org/10.1007/978-3-030-00440-8_3

German, media professionals. German media conglomerates owe 80% of the Polish local press, TV and social media and they clearly prefer a deal with the post-communist networks to their exposure to judgement. This unwillingness to face the post-communist problems is defended with many alibis.

Let us consider a simple case in point. Why would the mainstream media in the European Union (say, The Guardian, Die Welt, Le Monde) care for Muslim immigrants, low percentage of women in executive suites and Lesbian single mothers and not about the entire populations of their fellow-citizens recovering from the colonial (Africa, Asia, South America) and neocolonial (all of the above plus central and eastern Europe, which had been colonized first by Soviet Russians and then by core EU economies) oppression, discrimination and terror (the communist one in central and eastern Europe, Asia, Cuba)? Well, because the former communists managed to network for maintenance of privilege in central Europe and China and offer a ready-made virtual monopoly of the media to the relative newcomers to the elite recruited from the anti-communist opposition. Well, because a choice of political allies in post-Gaddafi Libya or Syria torn by a civil war is difficult and it is even more difficult to explain to large audiences. In the Polish case the former communist elites managed to strike deals with parts of emergent the "Solidarity" elites and with the European Commission and the European Parliament. When the Polish "Law and Justice" party seriously advanced the de-communization of the Polish state institutions, when Poland and Hungary started questioning the Europe of two velocities, liberal and left parliamentarians in the European Parliament started opposing it as an attack at democracy (as if democracy was a divine property of ex-communists promising continuity of power in colonized central and eastern Europe with some usefully co-opted liberals and neo-liberals).

This offensive of ideological correctness enabled the EU ideologues working through the mainstream media in the EU member states to not to notice 1.5 million Ukrainian migrants working in Poland (who make Poland one of the most hospitable countries in the EU with the most diverse and successfully integrated workforce) and focus on principled resistance of the democratically chosen government to a compulsory allocation of new migrants. The new migrants, invited single-handedly by the German government, rapidly emerged as a burden on the rich countries of the EU. The decision to dump this burden on the new members states of the European union from central and eastern Europe demanded a systematic and media-managed process of organized forgetting (a process supported by the evolution of preferences of citizens under democracy, cf. Achen and Bartels 2016).

2 CSR in New Public Management of the EU

Tacit and systematically overlooked causal link of the colonial and neocolonial inequalities surfaced in spite of the mainstream media conspiracy of silence. The core capitalist states of western Europe became rich because they had profited (and still profit) from colonial and neocolonial exploitation and from the recent historical fact, namely that they had also happened to be on the western side of the iron curtain during the Cold War. Older stories also re-surface (in case of the Dutch colonial history the “Robber State” stands out, cf. Vanvugt 2017). Likewise, the acknowledgment of Belgian genocide in Kongo should result in removal of the statues of king Leopold directly responsible for this crime from public spaces (as it did in removal of the monuments of heroes of the Confederates in civil war in the USA and in removal of the monuments of Red Army soldiers from public spaces in Central and Eastern Europe). Acknowledgment of the moral right of the Poles to demand German compensation for genocide and destruction of property in World War II should result in a reversal of the policy of calling German concentration camps—the Polish ones, and avoiding reparations after 1945. Calling upon central European states to become the buffer zones for the immigrants invited by Germany and heading primarily to Germany and the United Kingdom exposed the double asymmetry. First, neither Poles, nor Hungarians ever had colonies, so they should not be included in the post-colonial compensation package. Second, it was definitely Germans, and not mythical Nazis who evaporated into historical vacuum, who started WWII, killed six million of the Polish citizens, and destroyed property on a mass scale, so they should stop avoiding the Cold War power games and alibis—and face their responsibilities as far as compensating victims goes. These asymmetries have not been studied and they have been downplayed in the mainstream media. Hence the revenge of historical neglect—inability to comprehend contemporary problems and a refusal of the political elite of the European Union to see the crisis.

For the sake of the present paper let us focus not on African colonies or German guilt, but on the post-communist contribution to the crisis. The crisis is closely related to the post-communist takeover of the Russian state (Putin is not formally a communist but he emerged from the communist secret services and undermined both state administration and the democratic representation) and the post-communist alliances with the emergent political elites and media professionals in central and eastern Europe. Former communists and their neoliberal allies tacitly agreed on colonizing the post-communist countries of central and eastern Europe and offering this new colony to the power brokers in Brussels. When the Hungarians and the Poles started questioning this neocolonial policy—their leaders became the scapegoats of the EU media. Media may obscure the real sources of the crisis by scapegoating politicians who pioneered its recognition, but academic researchers should not be as blind as targeted media audiences.

Can this inability of the academic and political elites in the EU and of the media professionals to reject the left illusions—the inability to analyze contemporary problems of the EU—be overcome? If we apply the principles of CSR to the working of the European Commission and the European Parliament—yes, we do stand a chance. Grasping this chance should begin with the radical democratization of the European Union. Such democratization of the EU is as necessary as the systematic de-communization (including the neutralization of the post-communist power networks) in Central and Eastern Europe. Both are the keys to further successful development of the European community experiment. This time with CSR as part and parcel of managerial package.

3 Corporate Social Responsibility and Historical Memories

One of the necessary steps in dealing with historical memory of the Europeans is the creation of the international tribunal of justice for the public review and legal processing of the communist genocide. Communism has never been tried by a court of law established by a global civil society interested in allocation of responsibility and the punishment of the perpetrators of the communist genocide (repetition of genocidal projects in Rwanda and Miramar demonstrates that the deterring effect is worth creating). We should learn from our failures to come to terms with genocide perpetrated in and by Russia, which never de-stalinized and is currently re-stalinizing. Similar cases in Cambodia and Rwanda have also demonstrated failure of moral imagination on the part of the civilized world.

Communist crimes—including Russian genocide—had not yet been morally and legally processed by International Tribunal of Justice nor by the international public opinion and this is the prime evidence for a red corporate irresponsibility even in face of clearly criminal record. What I mean by red corporate irresponsibility is the organized campaign of cover-up of Stalinist and later crimes by the state bureaucracies commanded by the communist parties. German war crimes and genocide under Hitler were at least exposed to a legal scrutiny in Nuremberg—but only due to a military defeat in World War II, not due to a routine working of international law, nor due to the effectiveness of the League of Nations, not due to the ability of the world public opinion to bring those guilty of genocide to justice. Nor was the de-nazification deep enough (or mostly avoided—like in Austria, whose citizen, Kurt Waldheim, who committed war crimes as a German officer, even served as the highest functionary of the United Nations). Nor were Germans required to compensate the country they damaged most—namely Poland.

Time to do the historical homework in the European Union and time to see what influence could CSR policies have in preventing the corporate giants from opportunistic and immoral exploitation of such windows of opportunity as tax haven in Luxembourg or elsewhere, as secret nuclear armaments program in North Korea or elsewhere, as construction of energy infrastructure in the Crimean Peninsula occupied by the Russian military under Putin. Siemens, a German-based multinational corporation should not have been allowed to sell power generating turbines for the Russian occupied Crimea, and German government should not have allowed it. It could be prevented, if Siemens did have working CSR policies and if German government had also acted according to the ethical principles. But is CSR policy a sufficient instrument for safeguarding moral conduct and refusal to cooperate with moral suspects?

OECD Guidelines for Multinational Enterprises and United Nations Guiding Principles on Business and Human Rights should be complemented with ethical guidelines for dealing with the past injustices, of which the colonial exploitation before WWII and the neocolonial exploitation after WWII stand out as problems we should not avoid tackling any more. Neglected moral homework of mankind should be acknowledged and completed. Genocide conducted by former communist elites (especially in Russia, China, North Korea and Cambodia) should be processed and remembered in all history handbooks, as is the German genocide committed mostly in German concentration camps. History handbooks of all member states of the EU should devote more attention to the historical tragedy of Yalta agreements. From the point of the EU project, the Central and Eastern European states had been betrayed in Yalta by the western allies, missing both the post-war reconstruction within the Marshall plan and the emergent common market initiatives which resulted in the creation of the European Union. The Yalta tragedy should be significantly more exposed in handbooks of history of all world's schools. Hitler and Stalin started WWII together by jointly invading Poland (on September 1st and September the 17th, 1939, respectively). They both planned attacking one another, but Germans attacked first. Stalin joined the western allies demanding colonies for his communist experiment—and got them. After the fall of communism his political children, former political successors ("round table" negotiators in case of Poland) managed to secure a dominant position in post-communist political orders. These post-communist networkers made skillful use of the ebb and flow of political activism and survived early attempts at de-communization (cf. Kaminski 2014; Nauka.gov. 2017). Is CRS enough to purge them?

4 Is CRS Enough?

Corporate Social Responsibility, according to the most democratic definitions provided by Wikipedia, is a recent form of an organizational self-regulation exercised inside business organizations and frequently transferred to organizations managed by public authorities. Even Google tries not to be evil, even Enron did strategic charities, so why shouldn't a national government ministry or a public university follow the suit? As of the present writing—they have just done so. New Public management is not only about streamlining efficiencies but also about transparent displaying of responsibilities. Success is mixed. If CSR can be called a success—it means that corporate social responsibility can be exercised not only in business companies but also in professional bureaucracies, especially if they had been subjected to the reforms often labelled as "new public management". Synonyms of CSR include corporate conscience, corporate citizenship and responsible business—since we associate a democratic manifestation of civic responsibilities with an appeal to higher values and a contribution to the goals of broader stakeholders of business corporations and public bureaucracies—citizens, employees, consumers, investors, special interest groups, religious and national communities, protected minorities, etc. Theoretically, corporate social responsibility should provide a missing link between a routine daily scrutiny of managerial decision making and a more thorough but less frequent moral audit of the broader and more lasting consequences of collective activities influenced by the very same managerial decision-making—considered both from the point of legality and moral worth.

CSR is often discussed in the context of globalization (which enables networks of states and markets to influence their members through, say, the European Union or a North Atlantic Treaty Organization) and neoliberalism (which weakens local regulations and professional controls, encouraging migrations and the rise of the "precariat"). Both globalization and neoliberalism frequently result in a clash of concerned corporate citizens with the neoliberal professionals and mainstream media. Concerned corporate citizens would like to prevent collective agencies from committing crimes of neocolonial defense of status quo and neglecting to respond to dubious practices of unfair exchange. The neoliberals usually claim that the business of business is business, and business requires putting profits above any other possible goals of a business corporation or a state. Does CSR offer a chance of taming the corporate greed and of turning this conflict between ethical values and quasi-military priorities of market competitors into creative channels of desirable social transformations? And does it stand a chance if the usual neoliberal interpretation of the logic of capitalist markets is coupled with the co-optation of the post-communist neo-capitalists and quasi-democrats? We are asking this question in multi-mediated Plato's cave, in which the ideological sugar-coating is made of the leftist hopes of the 1968 generation, but infrastructure is a solid market capitalism embraced by increasing affluence of evolving societies.

5 CRS in World's Former "Red" Zones

CSR is very rarely discussed in the context of the long red shadows thrown on the post-communist societies by the former communist power elites. Viet Thanh Nguyen wrote about both sides of the Vietnamese war while embedded in the US society and professionally successful in creative cultural institutions (University of Southern California, US literary prizes establishment). No comparable quality literature emerged in the communist-ruled Vietnam (and if it did, it was not directly available in the multi-mediated cultural communications in the West, a fact which Nguyen himself regrets, cf. Viet Thanh Nguyen 2017). Former top functionaries of the communist mono-party, of the party-controlled government, of the police and of the military secret services managed to transfer their elitist privileges (organized impunity and irresponsibility guaranteed by the communist society's organizational matrix and secrecy) into the post-communist ownership and command of democratic institutions and market companies. This phenomenon led to the creation of the post-communist oligarchs in Russia and to the similar, although less extreme, developments in other post-communist countries. Parts of the former communist elites managed to transform former state property into their private assets, the process still relatively unknown and gradually detected, publicly described and legally uncovered. A sequel to Viet Thanh Nguyen's "The Sympathizer" could very well be written among the new business tycoons of the post-communist Vietnam fighting Kenyan and Brazilian coffee producers, and competing successfully against the Chinese and Indian tea exporters.

Here is the pro-communist irony of fate: the Polish success in undoing this corruption after the elections of 2015, which broke the post-communist oligarchy (of which Donald Tusk was a prominent member) is criticized by the EU power elite prompted by Tusk and networked with the post-communist politicians. Red roots of this systematic undermining of the institutions of the reconstructed civil society by former communist power elites are relatively easy to detect, but much more difficult to remove. The removal of the post-communist corruption is hampered by the fact that communist top henchmen have never been brought to a court of international law. Neither Stalin, nor Pol Pot, nor Mao Dze Dong, nor their collaborators, had their Nuremberg trial.

We are quite aware that the communist take-overs and their rule of terror were as bloody as fascism/Nazism and could compete on the dead body count in the past century. And yet, they had not been brought to justice. Stalin and Beria were killed by their close collaborators, but Mao died in bed, as did Honecker or Jaruzelski. Gulag archipelago, the long history of Russian, Chinese, Cambodian or North Korean genocide and crimes against mankind had not yet been morally and legally processed by International Tribunal of Justice and by the international public opinion. In a sense, only victims of a German regime of Hitler have been granted a status of victims of genocide, of a crime against mankind (but not a status of a victim that a German state should compensate). A victim of a Russian Soviet (or Cambodian, or Chinese) genocide is still unmentioned in mainstream left-liberal media and tacitly counted as an indispensable social cost of political experiments, unfortunate by-product of large-scale historical processes, not an unhappy human victim, wronged, tortured and murdered (for a reconstruction of the communist ride on the waves of popular unrest, cf. Kolakowski 2008).

Corporations and states learn from the totalitarian experiences of the past. Corporations know today that IBM should not have sold its office equipment to Germans, who used it to manage genocide more efficiently and effectively. Socially responsible corporations know that they should not have avoided paying taxes to the EU member states by misusing a Luxembourgian tax heaven run by a certain Jean Claude Juncker. Corporate Social Responsibility is an applied ethical initiative and when it emerges in a given context, it should be analyzed from the point of shadows of this context's past. Colonial and neocolonial past in the case of the core EU countries. In the Central and Eastern European member countries of the EU one should also acknowledge and process the hidden injuries of unpunished neo-colonization of the post-communist countries (in which the former communists were eager complices). We cannot—CSR says we should not—ignore the moral context of CSR's implementation, the shadow of crimes without punishments. Only if we face them, can we go on discovering, inventing, designing the evolving democratic institutions and patterns of collective action (cf. Smith 2016).

OECD Guidelines for Multinational Enterprises and United Nations Guiding Principles on Business and Human Rights should be complemented with ethical guidelines for dealing with the former communist elites and their successors (very frequently their offspring). Determining their responsibility for the past genocide and present corruption will require a screening of the perpetrators of the daily banality of the communist evil. This initiative is necessary if global threats posed by those who deny the communist holocausts are to be successfully faced and removed in order to further sustainability.

6 CSR and Oligarchic Risks

The fall of ENRON is a case in point: irresponsible practices had been exposed thanks to the morally brave whistle blower and those guilty of crimes and misdemeanors, even if from the very top of the corporate executive hierarchy, have been duly punished. In spite of some conspiracy theories about the same being presently done by the Googles, the Facebooks, the Amazons and the Microsofts or Apples (cf. Peretti 2017), we are not completely "rebooted" yet. Can the same defense against secret "rebooting" be raised in the case of the institutions of market democracies subtly undermined by the elites of the communist system trying to prolong their influence beyond the breakdown of the communist administrations?

Corporate Social Responsibility of a public institution assumes and requires transparency. Transparency requires uncovering the continuity of privilege of the former communist professional power brokers (who had been distributed through an array of different networks and institutions prior to the formal dissolving of the communist state) in Central and Eastern Europe. Interestingly enough, not many political scientists ventured a difficult reconstruction of the policies of the European Union impacted by the Polish "Solidarity"—as it broke the communist hold on the post-Yalta Europe in 1980 and led to the "round table" peaceful dismantling of the communist state. This happened around 1988. A year later East Germans removed the Berlin Wall (this happened in 1989), consigning the German Democratic Republic (1989) and Soviet Union (1991) to the dustbin of contemporary European history. Multicultural policies of Brussels and Berlin, of the German and EU political elites, culminating in the Maastricht treaty, were not particularly transparent at least as far as the consulting activities with the large multinational corporations (Dutch-British Royal Dutch Shell is the case in point) went. The same could be said about the EU policies generated in the years 2015–2017; they lacked transparency, were partisan in supporting leftist-liberal ideological "cocktail" and sought no democratic foundation in popular, genuinely negotiated support. This is why they ultimately failed, not because the progressive elites in Brussels or Berlin were sabotaged by Syrian human traffickers, Islamic radicals or conservative European constituencies. They failed, because in September 2017, when the allocation principles expired, no member state met its obligations (except for Malta, which accepted 137 refugees against 131 planned). In this sense, exposing institutions and corporations working for multicultural allocation of illegal migrants smuggled through the Mediterranean Sea might be aided by CSR. The legitimacy of the EU decisions would be well served by the Corporate Social Responsibility policies applied to the EU bodies, which are perceived as a decaying project of bureaucratic elites cutting all ties with their constituencies and with their historical roots (cf. Nguyen 2015 for a complex network of historical compromises between citizens, states and secret services). CSR policies should also be applied to the NGO's, which frequently collaborated with smugglers of immigrants, providing—for instance—rubber boats. In a sense, fear of patriotism made the elites fear national pride and go for emotionally flavorless cosmopolitan blandness of civic participation.

Reading the European "constitution" (rejected by the French or the Dutch but applied by Timmermans—illegitimately—against Poland) one cannot fail to notice that the ruling bureaucracies of the EU are afraid of mentioning the role of either Greek philosophy, or Christian religion or the Roman law in shaping the European project for over 2000 years. However, the very negotiation of the future of the European Union requires facing the real challenges (Brexit, Catalonia, French subversion of financial rules of the EU) and not salvaging of the neocolonial inequalities within the Europe of two velocities. This neocolonial stabilization of the new, enlarged EU, has already been undermined by the Russian military invasion of Ukraine and Russian corruption of German industrial establishment (Nordstream 2 is the case in point. German politicians are rewarded, as former chancellor Schroeder, with lucrative positions in Russian oil and gas exporting corporations, where no attempts to introduce CSR are ever made). The European project has to be re-negotiated among all member states, old and new, not re-allocated by the old among the new. This is the revenge of the democratic deficit on the EU project. This is the deficit, which introduction of CSR policies inside the institutions of the European Union can help Europe, Inc. survive in contemporary world.

References

Achen, C. H., & Bartels, L. (2016). *Democracy for realists. Why elections do not produce responsive governments*. Princeton, NJ: Princeton University Press.

Kaminski, A. Z. (2014). *Dezercja elit. Konsolidacja ustroju politycznego w pokomunistycznej Polsce* [Desertion of the Elites. Consolidating Political Regime in the Post-communist Poland]. Warsaw: Institute of Political Studies.

Kolakowski, L. (2008). *The main currents of Marxism. The founders-the golden age-the breakdown*. New York: Norton.

Nauka.gov. (2017). http://www.nauka.gov.pl/aktualnosci-ministerstwo/23-szkoly-wyzsze-podpisaly-deklaracje-spolecznej-odpowiedzialnosci-uczelni.html

Nguyen, V. T. (2015). *The sympathizer*. New York: Grove Press.

Nguyen, V. T. (2017). *Nothing ever dies. Vietnam and the memory of war*. Cambridge, MA: Harvard University Press.

Peretti, J. (2017). *Done. The secret deals that are changing our world*. London: Hodder & Stoughton.

Smith, G. (2016). *Democratic innovations. Designing institutions for citizen participation*. Cambridge: Cambridge University Press.

Vanvugt, E. (2017). *Roofstaat (compact). Wat iedere Nederlander moet weten*. Amsterdam: Nijgh & van Ditmar.

Part II
Financial Aspects of CSR

CSR Versus Business Financial Sustainability of Polish Enterprises

Anna Witek-Crabb

1 Introduction

"Business sustainability" can be interpreted in two ways. Usually this term is identified with an entrepreneurial activity related to sustainable world development through involvement in the practice of corporate social responsibility (CSR). CSR is then a tool for mitigating the negative impact of the company on its stakeholders and a way to creating shared value (Porter and Kramer 2006). In this case the question about sustainability is a question about well-being, balance and long-term benefits for the entire world.

The other interpretation of "business sustainability" refers to a micro scale. In this approach it is the company which is subject to development and it is its well-being, perseverance and growth that are important in the long term (Visser 2007). The question about "business sustainability" is in this case a question about activities ensuring longevity and success of the company.

The main research area of the chapter is the intersection of the two approaches: do companies committed to pursuing values of sustainable development become sustainable as businesses? To address this issue the relationship between CSR maturity and financial sustainability of a company is analysed.

In this chapter two main research questions and four auxiliary questions were posed:

1. How is corporate social responsibility implementation related to the financial sustanability of the firm in Polish listed companies?

A. Witek-Crabb (✉)
Department of Strategic Management, Wroclaw University of Economics, Wrocław, Poland
e-mail: Anna.Witek@ue.wroc.pl

A. Długopolska-Mikonowicz et al. (eds.), *Corporate Social Responsibility in Poland*, CSR, Sustainability, Ethics & Governance,
https://doi.org/10.1007/978-3-030-00440-8_4

- *What kind of CSR implementation (incidental, tactical or strategic) could be associated with financial sustainability of businesses and their sustainability?*
- *Is strategic CSR more likely to increase business financial sustainability than incidental or tactical CSR?*

2. What are the other factors influencing CSR maturity in Polish listed companies?
   - *How does company size influence CSR maturity?*
   - *How does industry the company's in influence CSR maturity?*

The research sample consisted of 93 Polish companies listed on the Warsaw Stock Exchange, representing 9 industries and present in WIG indices. During the analysis two aggregated measures were applied: CSR maturity index and financial sustainability index. The structure of these measures has been discussed in the later part of the elaboration and constitutes, to a large extent, own contribution of the author. In order to verify the results Tobin's Q ratio was also used. The collected data related to the period of 2007–2017.

2 Corporate Social Responsibility and Corporate Financial Performance Debate

The idea of corporate social responsibility (CSR) has existed in management science for over 50 years. At the end of the 1970s less than a half of companies from the Fortune 500 list mentioned CSR in their annual reports (Min-Dong 2008), in 2016 82% of S&P 500 companies where publishing Corporate Sustainability Reports (G&A Institute Flash Report… 2017). In late 1990s the CSR concept was already accepted and sanctioned by most international institutions (such as the UN or the World Bank), governments, corporations, non-governmental organizations and consumers. In Poland the development of the CSR concept fell on verge of the 20th and 21st centuries. The last 15 years have brought popularisation of the concept and increased awareness both among institutions, consumers, academics and entrepreneurs. During the 15 years the Responsible Business Forum has extended the network of affiliated companies from a few to over 400, popular contests promoting the CSR attitude in business have been held in Poland (e.g. "Fair Play Company" or "Benefactor of the Year"), and the Stock Market since 2009 has run an index of corporate responsible businesses (i.e. Respect Index) and promotes the principles of corporate governance. It is worth to remember, however, that although the CSR concept is becoming increasingly well-known, it is too soon to say it is widely used.

CSR in Polish enterprises has been the subject of academic research for a dozen years. Nevertheless, studies that verify the relationship of CSR with corporate financial performance of Polish enterprises are largely absent. As exceptions we can indicate the works of Bek-Gaik and Rymkiewicz (2014, 2015) and Otola and Tylec (2016). The results of all these research studies indicated to inconsiderable positive relationship between CSR and corporate financial performance, however, the CSR measurement was very simplified (boolean in the case of Bek-Gaik and Rymkiewicz or based on declared CSR expenses in the case of Otola and Tylec). In this paper, the assessment of CSR will be more complex and will enable the classification of businesses that use incidental, tactical or strategic CSR.

Although in world literature the discussion on the relationship between corporate social responsibility and corporate financial performance has been present for nearly 40 years, it has not brought clear answers. Hundreds of research studies with variable results have been carried out—some of them indicate to a positive relationship, some to negative relationship, and some to the lack of link between the studied categories. The reviews of the literature concerning the relationship between CSR and CFP have been carried out by, among others: Griffin and Mahon (1997), Orlitzky and Benjamin (2001), Orlitzky et al. (2003), Margolis and Walsh (2003), Margolis et al. (2009), Kitzmueller and Shimshack (2012), Lu et al. (2014) or Wang et al. (2016). The reasons for the equivocalness of results are most often seen in methodological shortcomings, the lack of a consistent way of CSR measuring, ambiguity regarding the direction of CSR and CFP impact (Hong et al. 2012) or the omission of relevant variables (Margolis et al. 2009).

The positive relationship between CSR and CFP has been indicted in the results of, for example, Cheng et al. (2014), who have shown that companies with better CSR experience less capital restrictions. Verschoor (2003) and Godfrey et al. (2009) believe that using CSR leads to value creation for the owners, Brown (1998) and Schnietz and Epstein (2005) claim that the reputation arising from using CSR affects financial performance, and Webley and More (2003) declare that the presence of ethical codes correlates with financial performance indicators. Orlitzky et al. (2003), in their review research, confirm the weakness of the positive correlation between CSR and CFP, which is two-directional.

Some authors have also found a negative correlation between CSR and CFP. Brammer et al. (2006) claim that CSR can cause the decrease in the rate of return on shares, and Izzo and Donato (2015) have shown a negative impact of CSR on stock prices. Also such researchers as Moore (2001) or Nelling and Webb (2009) have discovered a negative relationship between CSR and CFP.

The lack of link between CSR and CFP has been described by, among others, Surroca et al. (2010), who claimed that there is no direct link between these variables, only an indirect link through company's intangible resources. Mittal et al. (2008) indicate to the lack of link between CSR and company accounts, while Fiori et al. (2007) between CSR and market outcomes. The lack of dependence is also demonstrated by Brine et al. (2006).

In this study an attempt to analyse the relationship between CSR and corporate financial performance of Polish companies was made. Particular attention has been put on creating a comprehensive CSR measure, which would enable the assessment of CSR maturity. During the construction of the CSR maturity measure many CSR stage models have been referred to (Witek-Crabb 2017). These theoretical models demonstrate a step-wise character of CSR maturity in a company. The models differ from one another, however, they usually indicate progression from the "do nothing" attitude towards the "do a lot" attitude (compare (Carroll 1979), from reactive to proactive CSR (Maon et al. 2010), sometimes referring to CSR as elementary, basic, satisfactory, and advanced (Baumgartner and Ebner 2010) or reactive/ defensive, compliant, and proactive/ strategic/ transformational (compare Wilson 1975; Zadek 2004; Maon et al. 2010; Visser 2014). In this paper, however, the most important role has been played by the model proposed by van Marrewijk and Werre (2003). The authors have adapted the theory of spiral dynamics (Beck and Cowan 1996; Graves 1974) to the topic of corporate social responsibility by creating a 6-step evolutionary model. The model shows different possible levels of CSR implementation, dependent on ambitions, awareness of contexts of managerial activity and their current systems of values (van Marrewijk and Werre 2003).

3 Methodology and Research Design

The main goal of the research was to analyse the relationships between two variables concerning "business sustainability". The first variable was business financial sustainability and the second variable was maturity of corporate social responsibility. Both these constructs will be described below.

The research sample of 93 listed companies consisted of 8 small companies (up to 50 employees), 21 medium (51–250 employees) and 64 big companies (more than 250 employees), including 14 medium big (251–500 employees), 34 big (501–5000 employees) and 16 very big (more than 5000 employees).

The companies represented 9 sectors: banking (10 companies), building (16 companies), chemical (5 companies), developers (11 companies), energy (5 companies), IT (20 companies), media (8 companies), food (13 companies) and telecommunications (5 companies). The sample contained all WIG-indexed companies from the chosen sectors, which remained in that index in years 2007–2013.

The conducted study relates to the period of 2007–2017, where financial data relate to the period of 2007–2013 (database verification in 2017), and data relating to CSR were collected in the period of 2016–2017.

In order to find the answers to the research questions, the quantitative analysis (financial sustainability assessment) and the qualitative analysis (CSR maturity assessment) were matched together, which was aimed at finding relationships between these two groups of variables.

3.1 Financial Sustainability Construct

Among numerous financial indicators it is difficult to indicate a single one, which would best illustrate the development potential of a company. In the present study, the Author wanted to find such an indicator of financial health that could act as a good predictor of corporate sustainability.

In order to determine the financial developmental potential of the 93 examined companies, a step-wise financial analysis has been carried out:

- An aggregate financial sustainability measure was created, aimed at performing a synthetic assessment of corporate financial potential. For the main part of the research the accounting indicators were used as described below.
- A linear ordering method was applied (Słoński and Zawadzki 2015) with the use of "the financial sustainability benchmark", created on the basis of best practices of companies representing a given sector. This helped in determining a detailed ranking of enterprises within individual sectors for 2007–2013.
- Tobin's Q ratio was also used as an auxiliary measure. The first ranking method was created on the basis of the linear ordering method, with the use of accounting indicators of banks and companies from outside the financial sector. The basis of the second ranking method was the classification of companies according to Tobin's ratio.

An aggregate financial sustainability measure, illustrating the multi-dimensional situation of a company was created. Among the 22 analysed financial indicators from Reuters DataStream, 6 were selected—the least correlated ones and thus characterised by the largest information capacity. These indicators are:

1. Operational profitability,
2. Current liquidity,
3. Receivables turnover in days,
4. Turnover of short-term liabilities,
5. General debt,
6. Fixed asset financing.

These indicators represent all fields of financial analysis: profitability (indicator 1), liquidity (2), operating efficiency (3 and 4) and debt (5 and 6).

For the banking sector, due to the differing financial reporting rules, the aggregate financial measure was created on the basis of other indicators. Following the analysis of the coefficients of correlation between indicators and the analysis of cause-effect relationships, the complete list of indicators, available in Reuters DataStream, has been limited to five:

1. Net return on sales,
2. Cost level indicator, efficiency indicator,
3. Debt ratio.
4. Share of liquid assets to total assets,
5. Ratio of revenue bearing assets to total assets,

Tobin's Q ratio (Tobin and Brainard 1968) was used as an auxiliary measure for determining the financial viability of companies. Supporters of market indicators believe that investors are capable of assigning value to all assets, both tangible and intangible. Tangible assets do not include only those assets which are recorded and reported by the company (such as trademarks, licenses, patents, etc.), as well as those which have not or could not have been disclosed by the company (relational capital, brand capital, future effects of R&D), but creating company's goodwill. Tobin's Q as a market measure was used to verify the initial results based on accountancy measures.

3.2 Maturity of Corporate Social Responsibility Construct

The assessment of CSR maturity in the examined companies was based on qualitative content analysis contained on relevant companies' websites.

Content analysis constitutes one of qualitative research methods, placed next to grounded theory, phenomenology and ethnography (Burnard 1995). It is an approach that allows to interpret meanings and sense on the basis of the content present in the communication environment (textual, image-based, interview-based, observation-based or derived from printed or electronic media (Kondracki and Wellman 2002). These data, encoded and categorised, allow the exploration of patterns and regularities, which leads to better understanding of a phenomenon. Qualitative content analysis is defined as research method for subjective interpretation of the content text data through the systematic classification process of coding and identifying themes or patterns (Hsieh and Shannon 2005).

In the case of CSR study, the analysis was based on the information contained on the websites of relevant companies. The purpose of the study was to assess the level of CSR practice maturity, understood as the degree of excellence of these practices in relation to a given pattern.

During the process of encoding and interpreting data contained on the websites of relevant companies, recourse was made to the existing theories of process maturity (CMM—Capability Maturity Model) and various progression models of CSR. It is therefore reasonable to assume that a deductive approach was made, based on the theory (Potter and Levine-Donnerstein 1999; Berg 2001), complying with the requirements of directed approach to content analysis (Hsieh and Shannon 2005).

In the paper is has been assumed that the websites of public companies may serve as a suitable source of information concerning the maturity of applied CSR practices. Transparency and communication constitute important values contained within the concept of corporate social responsibility. Communication and reporting associated with this type of activity is one of the first signs of the growing maturity of these practices.

In order to assess the degree of maturity of CSR practices, aggregate CSR maturity measure was created. It consisted of the qualitative assessment of a company in three areas:

- CSR process maturity—content of the websites were analysed on the basis of the Capability Maturity Model (Persse 2001) theory, adapted to CSR specificity. According to this model, the maturity of CSR is accompanied by increasing predictability, repeatability and process stability. The more mature CSR, the more processes are planned, measured and dynamically optimized.

- CSR developmental maturity—content on the websites was analysed on the basis of the integral model contained in van Marrewijk and Werre (2003) Multiple Levels of CSR theory. In this area the content analysis was aimed at encoding the values, objectives and motivation underlying the CSR activity of companies. In this model, the CSR maturing process goes together with a certain phenomenon, where company values change from egoism to altruism and holism, objectives become more long-term and relate to a wider group of stakeholders, while motivation changes from being instrumental to being targeted on the quality of life and shared value.
- CSR formal maturity—company documents contained on the websites of relevant companies were analysed, while searching for symptoms of CSR presence in the formal layer of company communication and the coherence of CSR activities with its strategy. The assessment was made in respect to the adopted benchmark. The analysis was based particularly on the following documents: the mission of the company, ethical code, CSR reports, CSR strategy, company strategy, documentation concerning quality systems and good corporate governance practices, and the adhesion of the company to different CSR-supporting initiatives (e.g. Global Compact, Global Reporting Initiative, Responsible Business Forum, awards for CSR activity, presence in the Respect Index).

The 3 above-mentioned perspectives of CSR assessment were not fully separable. However, each of the applied models brought either additional knowledge about the company or verified the already-acquired knowledge through the analysis of another area.

The measurement carried out with the aggregate CSR measure made it possible to assess the level of CSR maturity in companies and categorise them according to the following CSR groups: incidental, tactical or strategic (Table 1).

Companies applying **incidental CSR** are characterised by low activity in this area. Minor CSR projects are implemented in a disorderly manner and ad hoc. The efficiency of these actions is usually not monitored and they are not planned over a long-term. These activities result from charitable motives or willingness—and within the scope—prescribed by law and social expectations. Companies using CSR at the incidental level are only in a minor extent engaged in communicating this area in company documents, such as the mission, values, strategy or reports.

Companies applying **tactical CSR** manage this area in a rational way. The CSR projects are largely planned and repeatable. The company is usually driven by instrumental motivation. Engaging in CSR is considered an investment which will pay for itself. Such a company integrates economic, environmental and social objectives, however, CSR does not constitute an important element in the whole company strategy. In case of tactical CSR, the company reports on its activities and develops a CSR policy.

The highest level of maturity is represented by companies applying **strategic CSR**. In these companies CSR is incorporated in its entire operational activity and strategy. The company engages in CSR for normative reasons, it acts in the basis of a clearly defined system of values and shares solidarity with a wide group of stakeholders.

Table 1 CSR maturity levels—theory based model

	Process maturity (based on CMM model)	Developmental maturity (based on van Marrewijk-Werre model)	Formal maturity (based on documents analysis)
Incidental CSR	• Incidental or no CSR activity • Little or no CSR awareness • Ad hoc implementation • Chaotic, unpredictable procedures and processes • No measurement or control of the processes • Zero to few projects	• Motivation: CSR undertaken out of duty and in order to comply with legal and social requirements • Activities: some charity responsiveness • Values: order, correctness, justice	• Mission statement focusing on customers • No code of ethics or explicit value declaration • No CSR reporting • No CSR strategy • CSR not part of a business strategy • Corporate governance practice in a minimal degree • Declared quality orientation
Tactical CSR	• Defined and repeatable CSR activities • Basic management procedures, some documentation and standardisation • Basic level of CSR monitoring • Lots of CSR projects but no long-term strategy or planning • Ad hoc resourcing, driven by enthusiasts, no top management commitment	• Motivation: profit through CSR implementation. Business case • Activities: integrating social, economic and ecological factors in order to gain effectiveness • Values: care for people, planet and profit	• Mission statement focusing on customers • Explicit values declaration • Basic CSR reporting • CSR policy • CSR not part of a business strategy • Mostly following corporate governance good practice • Declared quality and sometimes CSR orientation
Strategic CSR	• Managed • Methodical implementation of CSR, long-term goals and planning • Measurement, monitoring and analysis of CSR effectiveness and impact • Double-loop learning, optimisation of the CSR processes for stability and effectiveness • Best CSR practices, CSR leadership • Full integration with company strategy, clear commitment	• Motivation: sustainability important in itself • Activities: CSR permeating all the aspects of company's functioning and strategy. Triple bottom line. • Values: wellbeing of all the stakeholders, shared value, solidarity, harmony, interconnectedness, community	• Mission statement focusing on different stakeholders • Code of ethics and values declaration • Advanced CSR reporting (i.e. GRI) • CSR strategy in place • CSR as part of a business strategy • Following corporate governance good practice Declared quality and CSR orientation

Source: Own work

4 Results and Discussion

4.1 CSR Maturity Versus Financial Sustainability

The analysis of the maturity level of CSR practices in the 93 listed companies, has shown that the sample was composed of 44 companies applying incidental CSR (47%), 29 companies using tactical CSR (31%) and 20 companies using strategic CSR (22%).

The first research question, which the study was about to answer, concerns the relationship between the level of CSR maturity and the financial sustainability of enterprises. It was formulated as follows:

How is corporate social responsibility implementation related to financial sustainability of the firm?

- *What kind of CSR implementation (incidental, tactical or strategic) is associated with business financial sustainability?*
- *Is strategic CSR more likely to increase business financial sustainability than incidental or tactical CSR?*

In order to answer such posed questions it was checked whether there is a monotonic relation between the two groups of variables: CSR maturity and financial sustainability. In order to do this, Spearman correlation coefficients have been determined between:

(a) CSR maturity measure and financial sustainability measure,
(b) CSR maturity measure and company's Tobin's Q, over the years (Table 2)—Tobin's Q applied as complementary information on the relationship between CSR and financial standing of the company.

Subsequently, financial sustainability scores were compared for companies applying strategic, tactical and incidental CSR. For this purpose, mean financial sustainability results and mean Tobin's Q results were presented for each of the 3 groups (Table 3). Mean coefficients of variation for each group were also compared, as financial sustainability relies not only on the mean financial result but also on its stability in time.

Table 2 Spearman correlation coefficients between CSR maturity and financial measures (financial sustainability and Tobin's Q)

		N	CSR maturity (aggregate index)
Financial sustainability (aggregate index)	Mean (2007–2013)	93	−0.071
	Coefficient of variation	93	−0.109
Tobin's Q	Mean (2007–2013)	86	0.166
	Coefficient of variation	86	−0.259*

*Dual correlation significance = 0.016
Source: Own work

Table 3 Financial measures (financial sustainability and Tobins Q) vs. incidental, tactical and strategic CSR

		Incidental CSR	Tactical CSR	Strategic CSR	Population average
Financial sustainability (aggregate index)	Mean (2007–2013)	0.411	0.416	0.428	0.416
	Coefficient of variation	0.276	0.224	0.24	0.252
Tobin's Q	Mean (2007–2013)	1.255	1.377	1.197	1.279
	Coefficient of variation	0.346	0.328	0.196	0.309

Source: Own work

The analysis of Spearman correlation coefficient indicates that there is no relationship between the financial sustainability of a business and maturity of its corporate social responsibility ($p = -0.071$). No matter what the CSR maturity level of a company is, it does not affect its financial sustainability. Therefore, we can say that in the studied companies CSR practices does not translate into better business sustainability. Financial sustainability measure presents the degree of organisation's financial well-being and its developmental potential. Thus we cannot say, by the case of the examined sample, that using more advanced CSR practices affects the sustainability of company's development, nor can we say that companies of a greater financial potential are more likely to implement mature CSR practice.

The analysis of mean financial sustainability in groups of companies that apply strategic, tactical and incidental CSR shows that the differences between values are scarce and statistically irrelevant (Table 3). Firms implementing incidental, tactical and strategic CSR have similar results in financial sustainability.

The use of Tobin's Q ratio also shows that there is no connection between company's market value and CSR maturity ($p = 0.166$). Companies applying more advanced CSR practice are not rated higher by investors. This may indicate that CSR awareness in Poland is yet relatively low and the market does not reward social involvement of businesses.

The examination of the correlation between the coefficients of variation of financial sustainability/ Tobin's Q and CSR maturity indicates the lack of relationship in the first case ($CV = -0{,}109$) and a minor negative correlation in the second case ($CV = -0{,}259$). This means that, in the examined period, companies of a more mature CSR were characterised by a more stable Tobin's Q ratio in time (less variability). This may mean that greater CSR maturity is reflected in the stability of investor confidence towards the company and consistency of its valuation. The average variability of Tobin's Q in the examined period is 19.6% for companies applying strategic CSR, 32.8% for companies using tactical CSR and 34.6% in case of companies using incidental CSR (Table 3). Companies using strategic CSR are then characterised by a significantly more stable Tobin's Q ratio in time than companies applying tactical or incidental CSR. This regularity does not occur in relation to financial sustainability measure. There is no connection between the level of CSR and the variability of financial sustainability results. No matter what the CSR level is, the variability of financial sustainability over the examined years varies between 22% and 27% (Table 3).

To answer the first research question it should be emphasized that in the examined sample of 93 Polish companies there is no correlation between financial sustainability and CSR maturity of enterprises. Companies using incidental, tactical and strategic CSR achieve similar average financial sustainability results. Their Tobin's Q market valuation is also similar. Thus, CSR activity in the examined group of listed companies contributes primarily to world sustainability. Nonetheless, the relationship between CSR activity and corporate (business) sustainability could not be found. It should be highlighted that companies using tactical or strategic CSR—i.e. focused on gaining competitive advantage and consistent with the strategy—did not achieve better financial results than companies which used incidental CSR.

4.2 *CSR Versus Company Size and Industry*

The second research question draws attention to other factors which may affect the level of CSR maturity in the examined Polish companies. The results will be discussed below in brief, as this topic does not constitute the main subject of this work. Research question:

What are the other factors influencing CSR maturity?

- *How does company's size influence CSR maturity?*
- *How does company's industry influence CSR maturity?*

The analysis of demographics data against CSR maturity results indicates that factors such as the size of the company and the industry it operates in are significant when measuring the level of CSR activity.

There is a strong correlation ($p = 0.651$) between the level of CSR maturity and the size of the company, measured by the number of employees (Table 4). The average employment in a company using strategic CSR amounts to 7874 employees, in a company using tactical CSR—3960 employees, and in a company with incidental CSR—715 employees. There is only one medium-sized company among the companies applying strategic CSR. Other companies in that group are big and very big (Table 5).

The size of the company is a feature that strongly determines the way of CSR practice implementation. At larger companies—with a greater number of stakeholders and larger potential impact on the environment—the need to implement a CSR strategy is identified sooner and more complexly. The company wants to grow

Table 4 Spearman correlation coefficients between CSR maturity and number of employees

	N	CSR maturity
Number of employees	90	0.651*

*Dual correlation significance = 0.005
Source: Own work

Table 5 CSR maturity in companies of different sizes

CSR Maturity:	Size of the company (number of employees) Small (1–50)	Medium (51–250)	Medium big (251–500)	Big (501–5000)	Very big (> 5000)	Total
Incidental CSR	7	17	6	13	1	44
Tactical CSR	1	3	8	12	5	29
Strategic CSR	0	1	0	9	10	20
Total	8	21	14	34	16	93

Source: Own work

Table 6 CSR maturity in different industries

	Industry	Mean CSR maturity for the industry
Incidental CSR (0–0.32)	Telecommunication	0.26
	IT	0.28
Tactical CSR (0.33–0.65)	Developers, Media	0.33
	Food	0.40
	Building	0.44
	Chemistry	0.48
	Banking	0.64
Strategic CSR (0.66–1)	Energy	0.7

Source: Own work

and reaches for more formal management systems. Standardisation and process stabilisation, which usually occurs at a medium company and accompanies its growth later on—in the case of this study also relates to CSR processes.

Another feature identified by the research—relevant for the level of CSR maturity—is the industry in which the company operates. Mean CSR maturity scores for industries indicate that different industries have different CSR standards. The highest CSR standards in the examined sample were represented by companies from the energy and banking sectors, while IT and telecommunications sectors display the lowest standards in this respect (Table 6).

Industries are also characterised by different variability of CSR maturity. Companies from the energy sector are the most similar to one another with regard to their CSR standards. The coefficient of variation for that group is only 12%. The largest difference, on the other hand, occur in the media and telecommunications sectors, where the coefficient of variation is respectively 70% and 120%. This means that in these sectors there are both companies with very advanced CSR practices and companies which apply CSR only at a basic level. The coefficient of variation in other industries is about 50%. This may lead to a conclusion that in Poland, apart from a few exceptions (here: the energy industry), strict standards relating to CSR practice do not yet exist. Such activities are rather discretionary.

5 Summary and Conclusions

Analyses concerning the relationship between CSR maturity and company's capability of building its own financial development potential (financial sustainability) lead to the conclusion that in the studied sample these two categories remain independent. Polish companies, represented in the research sample by 93 businesses from 9 sectors, have implemented CSR at different levels, however, it does not impact their financial performance. The study has shown that CSR maturity is not associated with financial sustainability or market valuation of companies, but it depends on the size of the company and the industrial sector it operates in. Financial sustainability is therefore neither caused by CSR, nor does it influence it. Enterprises implementing CSR—no matter whether they use it strategically or incidentally—in effect support world sustainability but not their own. Business sustainability, understood as the ability of a company to persist and grow in the long term (Visser 2007, p. 445), remains in the examined companies independent from CSR activity.

An important factor determining the level of CSR in a company is company's size and sectoral affiliation. Larger companies use CSR in a more strategic way. Incidental CSR predominates in small and medium-sized businesses. It is not a surprising result. It shows that CSR, similarly to other activities and processes occurring in a company, becomes more stable, predictable and formalised alongside with company's growth.

The applied methodology using two aggregated measures (for CSR maturity and financial sustainability)—on the one hand provides the possibility of comprehensive evaluation of CSR practices and financial sustainability, while on the other hand may cloud the interpretation due to complexity of the measure and impede the search for specific links between the different aspects of CSR and financial well-being. In the next research step it would be legitimate to disaggregate the indicators and search for the links between individual CSR aspects (e.g. dialogue with stakeholders, communication, environmental performance, investment in human capital) and different accounting and market measures.

Polish capitalism is relatively young. The CSR concept has been present in Poland for about 20 years and we are slowly learning it. On the one hand thanks to CSR we improve Polish business practice and the implementation of a new culture of organization management. On the other hand we change our attitudes and social expectations. The results of this study may indicate that the CSR culture in Poland even though developing, is still quite shallow. The examined companies do not see advantage of financial annuities derived from using CSR, while customers and investors seem not to be ready to reward such activity yet, even though in declaration the consumers expect CSR business involvement (Żychlewicz 2016). Even the most CSR engaged Polish companies are more likely to report input than efficiency measures when it comes to CSR involvement (Waniak-Michalak 2017), and their quality of reporting is varied (Rogowski and Łabędzki 2016), with the foreign capital companies showing more advancement in the matter than domestic capital ones (Chojnacka and Wiśniewska 2016). Polish businesses and society are on the CSR learning curve. The business practices and the society awareness are not yet mature and companies in Poland are yet to see the full integration of social, ecological and financial value.

Acknowledgement The project has been financed by the National Science Centre as a research project No. 2012/07/D/HS4/00531.

References

Baumgartner, R. J., & Ebner, D. (2010). Corporate sustainability strategies: Sustainability profiles and maturity levels. *Sustainable Development, 18*(2), 76–89.

Beck, D., & Cowan, C. (1996). *Spiral dynamics: Mastering values, leadership and change*. Malden, MA: Wiley-Blackwell.

Bek-Gaik, B., & Rymkiewicz, B. (2014). Społeczna odpowiedzialność przedsiębiorstw a finansowe miary dokonań. *Finanse, Rynki Finansowe, Ubezpieczenia, 67*, 137–151.

Bek-Gaik, B., & Rymkiewicz, B. (2015). Wpływ CSR na finansowe miary dokonań jednostek. *Finanse, Rynki finansowe, Ubezpieczenia, 73*, 151–165.

Berg, B. L. (2001). *Qualitative research methods for the social sciences*. Boston, MA: Allyn and Bacon.

Brammer, S., Brooks, C., & Pavelin, S. (2006). Corporate social performance and stock returns: UK evidence from disaggregate measures. *Financial Management, 35*(3), 97–116.

Brine, M., Brown, R., & Hackett, G. (2006). Corporate social responsibility and financial performance in the Australian context. *Economic Round-up, 3*(4), 47–58.

Brown, B. (1998). Do stock market investors reward companies with reputations for social performance? *Corporate Reputation Review, 1*(3), 271–280.

Burnard, P. (1995). Interpreting text: An alternative to some current forms of textual analysis in qualitative research. *Social Sciences in Health, 1*, 236–245.

Carroll, A. B. (1979). A three-dimensional conceptual model of corporate performance. *The Academy of Management Review, 4*(4), 497–505.

Cheng, B., Ioannou, I., & Serafeim, G. (2014). Corporate social responsibility and access to finance. *Strategic Management Journal, 35*, 1–23.

Chojnacka, E., & Wiśniewska, J. (2016). *Raportowanie danych CSR w Polsce* (Vol. 436, pp. 55–63). Wrocław: PN Uniwersytetu Ekonomicznego we Wrocławiu.

Fiori, G., Donato, F., & Izzo, M. F. (2007). *Corporate social responsibility and firms performance – An analysis on Italian listed companies*. Available at https://ssrn.com/abstract=1032851

G&A Institute Flash Report. Research Results: S&P 500 Companies Sustainability Reporting. (2017). Accessed August 01, 2017, from https://www.ga-institute.com/

Godfrey, P., Merrill, C., & Hansen, J. (2009). The relationship between corporate social responsibility and shareholder value: An empirical test of the risk management hypothesis. *Strategic Management Journal, 30*(4), 425–445.

Graves, C. W. (1974). Human nature prepares for a momentous leap. *The Futurist, 8*(2), 72–87.

Griffin, J. J., & Mahon, J. F. (1997). The corporate social performance and corporate financial performance debate twenty-five years of incomparable research. *Business and Society, 36*(1), 5–31.

Hong, H., Kubik, J., & Scheinkman, J. (2012). *Financial constraints on corporate goodness* (NBER Working Papers, no. 18476). National Bureau of Economic Research.

Hsieh, H., & Shannon, S. E. (2005). Three approaches to qualitative content analysis. *Qualitative Health Research, 15*(9), 1277–1288.

Izzo, M. F., & Donato, F. (2015). The relationship between corporate social responsibility and stock prices: A study of an Italian market. *Corporate Ownership and Control, 12*(2), 608–617.

Kitzmueller, M., & Shimshack, J. (2012). Economic perspectives on corporate social responsibility. *Journal of Economic Literature, 50*(1), 51–84.

Kondracki, N. L., & Wellman, N. S. (2002). Content analysis: Review of methods and their applications in nutrition education. *Journal of Nutrition Education and Behavior, 34*, 224–230.

Lu, W., Chau, K. W., Wang, H., & Pan, W. (2014). A decade's debate on the nexus between corporate social and corporate financial performance: A critical review of empirical studies 2002–2011. *Journal of Cleaner Production, 79*, 195–206.

Maon, F., Lindgreen, A., & Swaen, V. (2010). Organizational stages and cultural phases: A critical review and a consolidative model of corporate social responsibility development. *International Journal of Management Reviews, 12*(1), 20–38.

Margolis, J. D., & Walsh, J. P. (2003). Misery loves companies: Rethinking social initiatives by business. *Administrative Science Quarterly, 48*, 268–305.

Margolis, J. D., Elfenbein, H. A., & Walsh, J. P. (2009). *Does it pay to be good. . .and does it matter? A meta-analysis of the relationship between corporate social and financial performance*. Available at https://ssrn.com/abstract=1866371

Min-Dong, P. L. (2008). A review of the theories of corporate social responsibility: Its evolutionary path and the road ahead. *International Journal of Management Reviews, 10*(1), 53–73.

Mittal, R. K., Sinha, N., & Singh, A. (2008). An analysis of linkage between economic value added and corporate social responsibility. *Management Decision, 46*(9), 1437–1443. https://doi.org/10.1108/00251740810912037.

Moore, G. (2001). Corporate social and financial performance: An investigation in the UK supermarket industry. *Journal of Business Ethics, 34*(3–4), 299–315.

Nelling, E., & Webb, E. (2009). Corporate social responsibility and financial performance: The "virtuous circle" revisited. *Review of Quantitative Finance and Accounting, 32*(2), 197–209.

Orlitzky, M., & Benjamin, J. D. (2001). Corporate social performance and firm risk: A meta-analytic review. *Business and Society, 40*, 369–396.

Orlitzky, M., Schmidt, F. L., & Rynes, S. R. (2003). Corporate social and financial performance: A meta-analysis. *Organization Studies, 24*, 403–441.

Otola, I., & Tylec, A. (2016). Społeczna odpowiedzialność biznesu a wyniki finansowe przedsiębiorstw. *Zeszyty Naukowe Politechniki Częstochowskiej Zarządzanie, 24*(1), 81–89.

Persse, J. R. (2001). *Implementing the capability maturity model*. New York: Wiley Computer Publishing.

Porter, M. E., & Kramer, M. R. (2006). Strategy and society: The link between competitive advantage and corporate social responsibility. *Harvard Business Review, 84*(12), 78–92.

Potter, W. J., & Levine-Donnerstein, D. (1999). Rethinking validity and reliability in content analysis. *Journal of Applied Communication Research, 27*, 258–284.

Rogowski, W., & Łabędzki, M. (2016). Ocena jakości informacji zawartych w raportach społecznych spółek wchodzących w skład indeksu RESPECT. *Organizacja i Kierowanie, 1*, 115–130.

Schnietz, K., & Epstein, M. (2005). Exploring the financial value of a reputation for corporate social responsibility during a crisis. *Corporate Reputation Review, 7*(4), 327–345.

Słoński, T., & Zawadzki, B. (2015). Finansowa miara rozwoju spółek notowanych na GPW w Warszawie. Unpublished paper.

Surroca, J., Tribó, J. A., & Waddock, S. (2010). Corporate responsibility and financial performance: The role of intangible resources. *Strategic Management Journal, 31*, 463–490. https://doi.org/10.1002/smj.820.

Tobin, J., & Brainard, W. (1968). Pitfalls in financial model building. *American Economic Review, 58*(May), 99–122.

van Marrewijk, M., & Werre, M. (2003). Multiple levels of corporate sustainability. *Journal of Business Ethics, 44*(2), 107–119.

Verschoor, C. (2003). More evidence of better financial performance. *Strategic Finance, 85*(5), 18.

Visser, W. (2007). Sustainability. In W. Visser, D. Matten, M. Pohl, & N. Tolhurst (Eds.), *The A to Z of corporate social responsibility*. Chichester: Wiley.
Visser, W. (2014). *CSR 2.0. – Transforming corporate sustainability and responsibility*. Heidelberg: Springer.
Wang, Q., Dou, J., & Jia, S. (2016). A meta-analytic review of corporate social responsibility and corporate financial performance. *The Moderating Effect of Contextual Factors, Business & Society, 55*(8), 1083–1121.
Waniak-Michalak. (2017). Mierniki działalności społecznie odpowiedzialnej ujawniane przez firmy w Polsce. *Zeszyty Teoretyczne Rachunkowości, 95*(151), 97–115.
Webley, S., & More, E. (2003). *Does business ethics pay? Ethics and financial performance*. London: The Institute of Business Ethics.
Wilson, I. H. (1975). What one company is doing about today's demands on business. In G. A. Steiner (Ed.), *Changing business-society interrelationship*. Los Angeles: Graduate School of Management, UCLA.
Witek-Crabb, A. (2017). *Evolutionary models of CSR – Between the paradigms of moral and economic value creation*. Conference Paper, EURAM 2017. ISSN: 2466-7498.
Zadek, S. (2004). The path to corporate responsibility. *Harvard Business Review, 82*(12), 125–132.
Żychlewicz, M. (2016). Corporate social responsibility on large and medium-sized companies in Poland. *Journal of Corporate Responsibility and Leadership, 3*(1), 69–80.

Impact of CSR on Operating Financial Results: The Case of Companies from RESPECT Index

Karolina Daszyńska-Żygadło

1 Introduction

The issue of corporate social responsibility (CSR) has drawn the attention of business and academics for the last two decades. Significant increases in CSR investments called Socially Responsible Investing (SRI), sustainability equity indexes on various stock exchanges, issuance of growing number of CSR reports, and in-depth scientific studies have established CSR as an important topic in business literature and practice. The vast majority of academic research concentrates on verification of the relationship between CSR activities (disclosure) of companies and economic and financial performance. Substantial evidence can be found in existing literature that CSR activities can play a significant role in driving positive changes in a company's value. Although, it is important to notice that CSR-related research is still emerging and there are many research papers with contradictory and inconsistent results (Rodgers et al. 2013; Baird et al. 2012; Peng and Yang 2014; Galant and Cadez 2017) At the same time, empirical studies are mainly conducted on big samples from developed stock exchanges and developed markets and there is a lack of evidence from less developed markets, especially Central and Eastern European countries. In most studies corporate environmental and social activities, CS (Corporate Sustainability) or CSR are approximated by the sustainability stock indices and they are widely recognized as appropriate indicators to measure effectiveness of CSR activities. It is assumed that companies included in such an equity index are recognized for their exceptional CSR performance, better than other publicly traded companies from the same market. In this study it is also assumed that

K. Daszyńska-Żygadło (✉)
Department of Corporate Finance and Public Finance, Wrocław University of Economics, Wrocław, Poland
e-mail: karolina.daszynska-zygadlo@ue.wroc.pl

A. Długopolska-Mikonowicz et al. (eds.), *Corporate Social Responsibility in Poland*, CSR, Sustainability, Ethics & Governance,
https://doi.org/10.1007/978-3-030-00440-8_5

this is the best available option of measuring effectiveness of CSR activities, although there is an ongoing discussion questioning the quality and representativeness of available CSR ratings (including ratings conducted for the purposes of SI indices construction) (Scalet and Kelly 2010; Saadaoui and Soobaroyen 2017), keeping in mind that there also exists the risk of ratings being constructed on the basis of unreliable data. Most ratings rely on surveys whose response rates are statistically insignificant, as well as on self-reported company data that have not been verified externally (Porter Kramer 2006).

The research question of this chapter is related to fundamentals and not only to the perception of investors, more precisely it is put as: can we recognize improvement or worsening of operating financial results of companies undertaking CSR activities in a long-term perspective.

The contribution to the science of the chapter is twofold. First it is to expand existing literature by results originating from Polish equity market in order to show whether growing sustainability concerns develop in similar directions to Western Europe markets. Warsaw Stock Exchange provides a sustainability equity index called RESPECT. Therefore, it has been used as a CSR activities proxy for assessment of operating financial performance in correspondence with these activities. The second contribution is to extend existing literature by verifying long-term effects of CSR activities of companies that are included in the sustainability index by the methodology that has never been used in correspondence with CSR.

The methodology applied in this research is based on event studies that employ accounting-based performance measure, in details described by Barber and Lyon (1996), widely used in research analyzing operating performance after mergers and acquisitions (for example in the paper of: Healy et al. 1992; Ghosh 2001; Linn and Switzer 2001; Perepeczo 2009).

The structure of this chapter is as follows; after an introduction, the second part presents a discussion in the literature about defining CSR and different motives of companies introducing it. The third part is dedicated to analysis of extant literature researching the relationship between CSR performance and financial performance of companies. The RESPECT index is shortly introduced in part four. Methodology and data is presented in part five, while discussion and concluding remarks are in the last, sixth part.

2 Definition of Corporate Social Responsibility and Corporate Sustainability

Even though corporate social responsibility (CSR) has become indispensable in modern business discourse and many companies and academics are practicing and studying it, identifying and defining the concept (phenomenon) is still open to contest. One could say that there is a need for a singular and universal definition in order to enable evaluation of its implementation, but on the other hand it seems unachievable that the variety of factors and dimensions addressed under the scope of CSR would yield one definition used by everyone (Okoye 2009).

The debate around defining CSR is present in literature, Dahlsrud (2008) identified 37 definitions from 27 authors. Definitions primarily originated from Europe and US, but there were also some from India and Canada. Research results showed that CSR is being defined based on five dimensions, but not all of them are used in every definition, while three or more dimensions are mentioned in 31 definitions and constituting 97% of the total frequency count. Those five dimensions encompass: environmental dimension (natural environment), social dimension (relationship between business and society), economic dimension (socio-economic or financial aspects, including describing CSR in terms of business operation), stakeholder dimension, voluntariness dimension (actions not prescribed by law).

The definition that is used most frequently is of the Commission of the European Communities (2001), that includes all five dimensions and states that CSR is the concept whereby companies integrate social and environmental concerns in their business operations and in their interaction with their stakeholders on a voluntary basis. This definition fits most accurately to what is being researched in this study, therefore, whenever CSR or CSR activities are mentioned, they are understood according to the definition of the Commission of the European Communities (2001).

In a later study, Rahman (2011) analyzed the evolution of defining CSR and identified milestones of every decade starting from 1950, noticing that at the beginning CSR was focused only on the dimension of obligation to the society, while twenty-first century perspective broadened a lot and includes integration of social and environmental concern, voluntariness, ethical behavior, economic development, improving quality of life of citizens, human rights, labor rights, protection of environment, fight against corruption, transparency and accountability.

Additionally, van Marrewijk (2003) recommends to keep a distinction by associating Corporate Social Responsibility with the communion aspect of people and organizations and Corporate Sustainability with the agency principle even though nowadays many consider CSR and CS as synonyms. In detail, it means CSR is related to transparency, stakeholder dialogue and sustainability reporting i.e. communication, while CS is connected with strategic issues related to value creation, environmental management, production systems and human resources management.

The broad definition of both concepts—CSR and CS—is that they refer to company voluntary activities demonstrating the inclusion of social and environmental concerns in business operations and in interactions with shareholders.

The differentiation between particular levels of ambitions of companies meeting sustainability goals which have to be aligned with perceived challenges offered in the environment made by van Marrewijk (2003) is important from the perspective of this study. The emphasis is put not on the self-determination of a company in meeting ambitious sustainability goals but on the fact that it operates in a specific business environment and realizing those goals has to be adjusted to changes in the environment. Otherwise the company is risking business failure. Those five definitions of CS/CSR proposed by van Marrewijk (2003) according to motives and ambition levels of companies are as follows: (1) Compliance-driven CS/CSR consists of providing welfare to society, within the limits of regulations from the rightful authorities, possibly supplemented by charity and stewardship considerations. Motivation for these activities is perceived as obligation (duty) or correct behavior (2) Profit-driven CS/CSR consists of integrating social, ethical and ecological aspects into business operations and decision-making, making sure it contributes to the financial bottom line. The motivation is the business case; it is promoted if it is profitable (3) Caring CS/CSR consists of balancing economic, social and ecological concerns, initiatives go beyond legal compliance and profit considerations. The motivation is that human potential, social responsibility and care for the planet are important on their own (4) Synergistic CS/CSR consists of a search for well-balanced, functional solutions creating value in the economic, social and environmental realms of corporate performance in a synergistic, win-together approach with all the stakeholders. The motivation is the importance of CS/CSR in itself, recognized as being the inevitable direction progress takes (5) Holistic CS/CSR is fully integrated and embedded in every aspect of the company, aiming at contributing to the quality and continuation of life of every being and entity, now and in the future. Motivation is that sustainability is the only alternative since all being and phenomena are mutually interdependent and each person or organization has a universal responsibility towards all other beings.

Every company has the right to choose the level of CS/CSR they prefer while they can be structured into coherent institutional frameworks within the companies. However, these frameworks can be difficult thresholds preventing companies from adopting higher levels of corporate sustainability.

Additionally, companies implementing CSR, according to Porter and Kramer (2006) use four arguments to make the concept eligible and reaffirm it: sustainability, moral obligation, license to operate and reputation. Sustainability is reflected in the triple bottom line of economic, social and environmental performance. According to which companies should operate in ways that secure long-term economic performance by avoiding short-term behavior that is socially detrimental or environmentally wasteful. Moral obligation is understood as the duty of companies to be good citizens for example by operating within the law, honestly filing financial statements. License to operate is the most pragmatic approach out of the four mentioned. Companies identify and choose only those social issues that matter for their stakeholders and make decisions about them. Reputation argument is used by companies to justify CSR activities as those which will improve the company's image, strengthen its brand and enliven morale. These arguments stay in correspondence with the ambition level in reference to CS/CSR that was proposed by van Marrewijk (2003). The obstacle connected with implementation of CSR on the basis of those arguments is that they focus on the tensions between business and society rather than on their interdependence. As a result, companies are taking numerous actions connected with CSR concepts, but these are seldom rooted in the strategy and tied to operations of a specific company or the place it operates in.

3 Literature Review

In the field of assessment of impact of CSR on financial performance the vast majority of researches aim at showing immediate or long-term impact of performance related to sustainability on companies' financial results proving whether there exists a significant relationship between these two. The immediate effect is analyzed by event studies showing reaction of investors on the information disclosure about company being included (excluded) in (from) sustainability index (Consolandi et al. 2009; Cheung and Roca 2013; Daszyńska-Żygadło et al. 2014; Adamska and Dąbrowski 2016) The long-term effect (usually calculated as a yearly effect over a certain period of time) is analyzed on the basis of financial performance (accounting or market measures) regressed over sustainability (or CSR) performance that is measured by scoring or rating result (Wang and Berens 2015; Saeidi et al. 2015; Daszyńska-Żygadło et al. 2014).

The positive nature of the relationship between CSR and financial performance was found by several studies (Al-Tuwaijri et al. 2004; Burnett and Hansen 2008; Erhemjamts et al. 2013; Rodgers et al. 2013), as well as negative—by a number of studies (Baird et al. 2012; Peng and Yang 2014) (see also article of Witek-Crabb in this book) No relationship was found by Alexander and Buchholz (1978), Aupperle et al. (1985), Soana (2011), Sun et al. (2010), and McWilliams and Siegel (2000), while U-shaped and inverted U-shaped relationships were initially found by Bowman and Haire (1975) and then by Barnet and Salomon (2012).

Depending on the period of the analysis, proxies selected for financial performance and sustainability performance as well as a sample of companies and methodology of research assessed impact is either positive, negative or inconclusive and insignificant, Margolis et al. (2007) also confirm these findings based on their meta-analyses, concluding that a positive relationship is being found more often than other types (Margolis et al. 2007) Therefore, it still leaves a room for further studies. It is important to notice that empirical studies are mainly conducted on big samples from developed stock exchanges and developed markets while there is lack of evidence or limited number of studies (Daszyńska-Żygadło et al. 2014; Adamska and Dąbrowski 2016) that analyzed only short-term effects from less developed markets, especially central European countries. Additionally, it could be noticed that along with the development of CSR measures and disclosure practices more recent studies present conclusive and positive results, while those conducted back in 70s and 80s are equivocal or show no relationship.

Recently Galant and Cadez (2017) conducted research on operationalization and measurement approaches for the CSR and its relationship with corporate financial performance, they concluded that all approaches deployed in empirical literature suffer from weaknesses. They postulated that standardization of CSR reporting along with making CSR disclosure mandatory would be very beneficial for valid testing of the studied relationship from the perspective of academia, but also from the perspective of stakeholders as a support in their economic decision making.

Previous research results related to CSR performance of Polish companies published by Adamska and Dąbrowski (2016) as well as Daszyńska-Żygadło et al. (2014) confirmed that there exists significant reaction of investors for inclusion (exclusion) of companies in (from) the sustainability index RESPECT listed on Warsaw Stock Exchange, the methodology used was based on event study analysis. Conclusions drawn on the basis of these findings were consistent in the field of confirming positive effect of undertaking CSR activities by companies on the market value of their shares. Therefore, investors' reactions for that type of event were positive and significant. Another research based on event study methodology with buy-and-hold abnormal returns (BHAR) conducted on the basis of RESPECT index proved that investors care about reputational factors while making investment decisions and they perceive stocks of companies that were included in that index as more attractive (Blajer-Gołębiewska 2014).

Although research results of Zieliński (2015) do not show uniform relation between valuation multiples of companies from the RESPECT index with their sector peers, rejecting the hypothesis that companies from the sustainability index that implemented CSR are valuated higher that other companies from the same sectors. Chojnacka (2016) shows that CSR engagement of Polish publicly traded companies is not significantly related to their risk. Therefore, it is not possible to defend the thesis that along with incorporation of CSR goals risk of investing in those companies is decreasing.

Out of these findings we can conclude that results are mixed and not unequivocal, and even though we can find in literature confirmation that there is a positive relation between CSR activities and market value of companies included in the RESPECT index that confirmation relates only to short-term effects that are connected with reaction on the news and do not reflect long-term impact on financial performance. Abnormal stock returns at CSR index inclusion indicate that expectations of economic improvements underlie the market value of equities of firms being included in the RESPECT index. Therefore, there is still the room for further studies. To fill in the gap in existing literature the research question of this study has been defined as finding out whether CSR activities are beneficial or detrimental from a financial standpoint and if investors' expectations of economic improvements are ultimately met by the companies in a long-term perspective.

4 Sustainability on Warsaw Stock Exchange

The RESPECT index was initiated in November 2009 by the Warsaw Stock Exchange, evaluation and selection of companies to the index is done in cooperation with Deloitte. The index is a total return index including income from dividends and rights issue. Evaluation for inclusion is conducted in three stages, first of all, only companies that are included in WIG20 or mWIG40 or sWIG80 are acknowledged as liquid and could be entitled for further evaluation, second stage is the assessment of on-time and correct communication of companies with investors (current and periodical obligatory reports published on companies' websites) run by WSE and Association of Stock Exchange Issuers. The third stage is evaluation of sustainable performance of companies run by Deloitte based on verification of a questionnaire filled in by the companies. According to the current scheme assessment of companies and revision of the index is conducted once a year, but in years 2011–2013 there were two revisions of the index per year. It is an open index with variable number of companies, the purpose of revision is to add new or exclude companies. Final results of revisions are published, but the reasons for exclusions never are. There have been some exclusions of companies due to extraordinary circumstances, such as merger or bankruptcy, therefore not due to sustainability misconduct. Table 1 shows the history of participation of companies in the index in subsequent editions. One could observe persistent behavior of several companies. Although on the bottom of the table there are 11 companies that entered the index several years ago and disappeared from it after 1 or 2 years. Additionally, in the last, Xth edition there were four new entrants that are not included in table, these were: PCC Rokita SA, Trakcja PRKiI SA, Bank Pekao SA, Fabryki Mebli "FORTE" SA.

In regards to sustainability reporting practices of companies included in the RESPECT index in the final edition, only 9 out of 25 have published a report for 2016. In previous years two companies published some information about their CSR performance on the companies' websites, two have CSR section in their yearly report, four companies are publishing CSR reports every 2 years, and 17 are publishing a separate CSR report. It means only 68% of all companies from the RESPECT index are actually presenting their CSR data in separate reports. Presumably that results will improve up to 100% after 2017 when all the big companies traded at WSE will be obliged to publish reports with non-financial information.

5 Methodology and Data

The methodology applied in this research is based on event studies that employ accounting-based performance measures. This methodology is dedicated to assessment of long-term effect of corporate decisions on financial performance. In detail, it has been described by Barber and Lyon (1996) In the majority of cases it is used to analyze impact of strategic corporate decisions such as mergers, acquisitions or takeovers, as well as stock splits, dividend initiations or omissions (see Table 2). It has been widely used in research analyzing operating performance after mergers and acquisitions (Healy et al. 1992; Ghosh 2001; Linn and Switzer 2001; Perepeczo 2009).

That methodology has never been used in conjunction with sustainability. Keeping in mind that getting involved into sustainability activities is a strategic decision, while becoming a part of the sustainability index indicates the moment of positive acknowledgment done by an independent institution, one can assume that it can fall under the scope of significant events category. Additionally, selected studies confirm that financial performance improves for companies that implement CSR

Table 1 List of RESPECT participants according to their presence in the index in years 2009–2016

	2016 (X[a])	2015 (IX)	2014 (VIII)	2013 (VII)	2013 (VI)	2012 (V)	2012 (IV)	2011 (III)	2011 (II)	2009 (I)
Apator SA	v	v	v	v	v	v	v	v		v
Bank BPH SA		v	v	v	v	v	v			v
Bank Handlowy w Warszawie SA	v	v	v	v	v	v	v	v	v	v
Bank Millennium SA	v	v	v	v	v	v	v	v	v	
Bank Zachodni WBK SA	v	v	v	v					v	
Budimex SA	v	v	v	v	v	v	v	v	v	
Elektrobudowa SA	v	v	v	v	v	v	v	v	v	v
GPW SA	v	v	v	v						
Grupa Azoty SA[b]	v	v	v	v	v	v	v	v	v	v
Grupa LOTOS SA	v	v	v	v	v	v	v	v	v	v
ING Bank Śląski SA	v	v	v	v	v	v	v	v	v	v
JSW SA				v	v	v				
KGHM Polska Miedź SA	v	v	v	v	v	v	v	v	v	v
Netia SA			v	v	v	v	v	v		
Pelion SA	v	v	v	v	v					
PGE SA	v	v	v	v	v	v	v	v		
LW Bogdanka SA	v	v	v		v	v	v	v	v	
PKN Orlen SA	v	v	v	v	v	v	v	v	v	v
PGNiG SA	v	v	v	v	v	v	v	v	v	v
PZU SA	v	v	v	v	v	v	v			
Tauron PE SA	v	v	v	v						
Orange SA[c]	v	v	v	v	v	v	v	v	v	v
ZEW Kogeneracja SA	v	v	v	v	v	v	v			
Mondi Świecie SA							v	v	v	v
Bank Ochrony Środowiska SA		v								
Rawlplug SA			v							
Rafako SA				v						
Kredyt Bank SA						v				
Ciech SA							v			v
Fabryka Farb i Lakierów Śnieżka SA								v		
Grupa Żywiec SA										v
Barlinek SA								v	v	v
PBG SA							v	v		
DM IDM SA							v	v		
BRE Bank SA								v	v	

Source: Own elaboration
[a]Subsequent number of index edition
[b]Zakłady Azotowe w Tarnowie-Mościcach SA until 2013
[c]Telekomunikacja Polska SA until 2014

Table 2 Summary of studies analysing corporate performance following corporate events or decisions

Author(s)	Corporate event studied	Performance measure	Adjustment method
Healy and Palepu (1988)	Dividend initiation/ omission	Change in EPS/price	Industry
Asquith et al. (1989)	Stock splits	Change in EPS/price	Industry
Kaplan (1989)	Management buyouts	ROS; ROA (based on op. inc.) (Op. inc.—capex)/sales (Op. inc.—capex)/BV assets	Industry, size
Healy and Palepu (1990)	Seasoned equity offers	Change in EPS/price	Industry
Denis and Denis (1993)	Leveraged recapitalizations	ROA, Free cash flow/BV assets	Industry, size
Healy et al. (1992)	Takeovers	Op. cash flow/(MV equity + BV debt)	Industry
Ghosh (2001)	Acquisitions	Cash flows/assets	Performance, size
Linn and Switzer (2001)	Cash and stock acquisitions	Pretax cash flow return	Industry
Perepeczo (2009)	Mergers and acquisitions	Op. cash flow/BV assets	Not adjusted

Source: Own elaboration

(Al-Tuwaijri et al. 2004; Burnett and Hansen 2008; Erhemjamts et al. 2013; Rodgers et al. 2013). Therefore, findings of this study will contribute to the debate about relationship between CSR and financial performance.

The final sample consists of 22 companies that in the period of 2009–2014 were included in the sustainability index RESPECT on Warsaw Stock Exchange, these companies belong to the following sectors: banking, energy, oil and gas, construction, telecom, production of pharmaceutics, production of electric machinery, chemicals, mining, insurance. Analysis was conducted 3 years before and 3 years after the inclusion into the index in order to verify how operating performance has changed after inclusion in sustainability index. Results of the year of inclusion were treated as reflecting post inclusion performance. Therefore, it covers the period of 2006–2016. Based on the cited research and on the availability of data, it has been decided to use return on assets ratio (ROA) as a measurement of long-term operating financial performance, the ratio was calculated on the basis of operating income in relation to book value of total assets. An especially challenging task was to collect comparable data to calculate the benchmark ratio. Due to that fact only ROA could be chosen and other proposed measures were not taken into consideration. Benchmarks for each sector were calculated separately on the basis of selected peer groups for companies, chosen on the basis of the same code of major business operations (Polish business activity classification—PKD) as well as size of operations. Financial data was sourced from EMIS Intelligence database and Notoria Serwis database

Graph 1 Analysis of ROA differences before and after inclusions in RESPECT index. Source: Own calculations plotted in STATISTICA

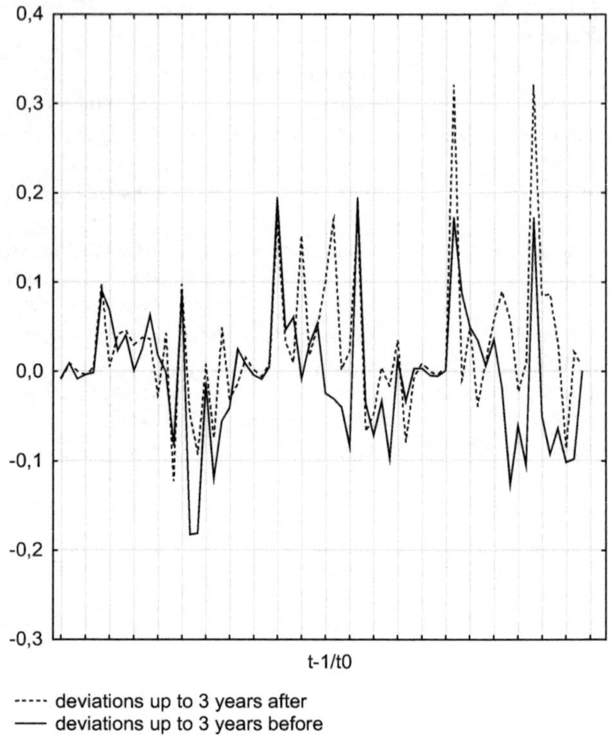

in order to calculate ROA for companies as well as for the peer groups. Additionally, sector averages available in reports published every year by the Polish Association of Accountants were used for the years before 2012.

On the basis of literature review, hypotheses were constructed. Hypothesis 0 was that before and after inclusion in the sustainability index operating performance of companies does not change, hypothesis 1 was that operating performance of companies changes positively after inclusion in sustainability index and the change is statistically significant. The analysis was conducted in STATISTICA software.

Graph 1 shows plotted differences between ROA of companies and benchmarks, the blue line indicates results of 3 years before the inclusion in RESPECT index while red line indicates differences after inclusion in the index within 3 years. On the basis of visual analysis one can notice that the majority of differences after inclusion in the index are higher than those before.

The data sample of abnormal returns (differences between ROA of a single company and its benchmark) was tested for normal distribution. The results are plotted on Graph 2.

One can observe that data sample in both subsamples are not normally distributed, therefore the recommended test to verify the significances of differences before and after the index inclusion is non-parametric test. As suggested in the

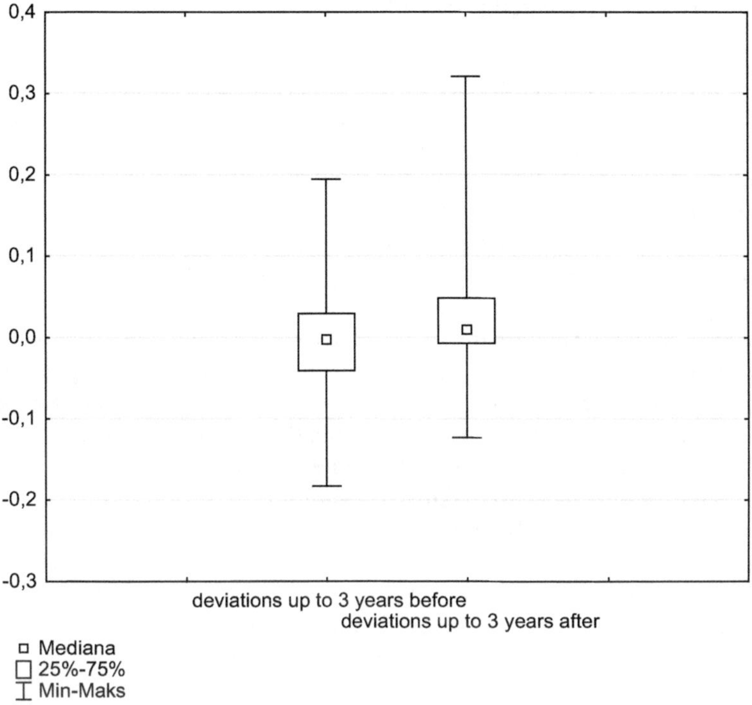

Graph 2 Analysis of distribution of data sample. Source: Own calculations plotted in STATISTICA

Table 3 Results of event study analysis for RESPECT index in 2 and 3 years' windows

Pairs of variables	N	T	Z	p (<0.05000)
Deviation up to 2Y before and 2Y after	44	324.000	1.995603	0.045978
Deviation up to 3Y before and 3Y after	66	568.000	3.433594	0.000596

Source: Own elaboration in STATISTICA

paper by Barber and Lyon (1996) Wilcoxon test was used. The results of analysis are presented in Table 3.

The obtained results show that there was a significant change in operating performance based on ROA relative to industry-average firms. The significance was also tested in a 1-year window, but it has not been achieved. These results indicate that the operating performance of companies from various industries changes significantly, level of significance is even higher in 3-years window (p value is lower for 3-years than for 2-years), following inclusion in the sustainability index. The test of signs indicated that in the sample of 2Y deviation there were 68.182% more negative differences before the inclusion than after the inclusion, while in the sample of 3Y deviation analogical result was 74.242%. Both results were statistically significant that indicates that financial performance of companies changes in positive way after inclusion in RESPECT index.

6 Discussion of Results and Conclusions

These results could lead to the conclusion that operating financial performance of companies being included in the RESPECT index improves after the decision of inclusion. Hence, the hypothesis of the research has been verified positively. In order to analyze what triggers that improvement additional study should be conducted, especially the direction of causality should be verified. The performed research does not answer the dilemma whether financial performance of a company is being improved thanks to introducing a CSR strategy, running CSR activities and getting verified by the RESPECT index procedure or maybe that the CSR strategy can be introduced due to financial performance improvement. Although, even without any further studies, it should be noticed that there is particular time lag between the verification procedure of the RESPECT index and CSR conduct in companies. In order to get qualified for the inclusion in the RESPECT index companies must show their CSR performance results. Thus, the moment of inclusion, which in this study is treated as the year of inclusion with post factum financial performance, represents the period in which effects of introducing CSR activities can be observed. While these effects indicate improvement of operating financial performance it can be stated that the hypothesis of this study was verified positively. Research results presented in this chapter stay in line with previous findings of Adamska and Dąbrowski (2016), Daszyńska-Żygadło et al. (2014) and Blajer-Gołębiewska (2014).

Additionally, findings of this chapter extend the previous findings by adding a long-term perspective. Cited research papers confirm that investors react positively on companies' inclusion into the RESPECT index and negatively on companies' exclusion from the RESPECT index, so they indicate that expectations of economic improvements underlie the market value of equities of firms, while this paper confirms that these expectations of investors are met in the long run, so they can achieve higher returns on assets even up to 3 years after company is being included in the RESPECT index.

Limitations of the study are connected with the small size of a sample and choice of proxies both for CSR activities and for operating performance. It would be recommended to verify the proposed way of analysis on a bigger sample. Although the major difficulty of this study lies in collecting data and finding proper benchmark within several years—industry average/median for comparisons with companies originating from different industries entering in different moments in time in the index which creates the need to find data 3 years prior and 3 years after the event in different time frames.

Acknowledgment Publication was prepared based on research financed from the scientific grant from Polish National Center of Science no 2011/03/B/HS4/05359 entitled "Wpływ społecznie odpowiedzialnych działań przedsiębiorstwa (CSR) na jego wartość".

References

Adamska, A., & Dąbrowski, T. (2016). Do investors appreciate information about corporate social responsibility? Evidence from the Polish equity market. *Inzinerine Ekonomika – Engineering Economics, 27*(4), 364–372.

Alexander, G. J., & Buchholz, R. A. (1978). Corporate social responsibility and stock market performance. *Academy of Management Journal, 21*, 479–486.

Al-Tuwaijri, S. A., Christensen, T. E., & Hughes II, K. E. (2004). The relations among environmental disclosure, environmental performance, and economic performance: A simultaneous equations approach. *Accounting, Organization and Society, 29*, 447–471.

Asquith, P., Healy, P., & Palepu, K. (1989). Earnings and stock splits. *The Accounting Review, 64*, 387–403.

Aupperle, K. E., Carroll, A. B., & Hatfield, J. D. (1985). An empirical examination of the relationship between corporate social responsibility and profitability. *Academy of Management Journal, 28*, 446–463.

Baird, P. L., Geylani, P. C., & Roberts, J. A. (2012). Corporate social and financial performance re-examined: Industry effects in a linear mixed model analysis. *Journal of Business Ethics, 109*, 367–388.

Barber, B. M., & Lyon, J. D. (1996). Detecting abnormal operating performance: The empirical power and specification of test statistics. *Journal of Financial Economics, 41*, 359–399.

Barnet, M. L., & Salomon, R. M. (2012). Does it pay to be really good? Addressing the shape of the relationship between social and financial performance. *Strategic Management Journal, 33*, 1304–1320.

Blajer-Gołębiewska, A. (2014). Do stock exchange indices based on reputational factors matter? *International Journal of Academic Research, 6*(4), 231–237.

Bowman, E. H., & Haire, M. (1975). A strategic posture toward corporate social responsibility. *California Management Review, 18*, 49–58.

Burnett, R., & Hansen, D. (2008). Ecoefficiency: Defining a role for environmental cost management. *Accounting, Organization and Society, 33*, 551–581.

Cheung, A., & Roca, E. (2013). The effect on price, liquidity and risk when stocks are added to and deleted from a sustainability index: Evidence from the Asia Pacific context. *Journal of Asian Economics, 24*, 51–65.

Chojnacka, E. (2016). The relationship between CSR engagement and firm risk in Polish public companies. In M. H. Bilgin, H. Danis, E. Demir, & U. Can (Eds.), *Business challenges in the changing economic landscape. Proceedings of the 14th Eurasia business and economics society* (Vol. 1, pp. 213–223).

Commission of the European Communities. (2001). *Promoting a European Framework for Corporate Social Responsibilities*, COM (2001) 366 final, Brussels.

Consolandi, C., Jaiswal-Dale, A., Poggiani, E., & Vercelli, A. (2009). Global standards and ethical stock indexes: The case of the Dow Jones sustainability stoxx index. *Journal of Business Ethics, 87*(1), 185–197.

Dahlsrud, A. (2008). How corporate social responsibility is defined: An analysis of 37 definitions. *Corporate Social Responsibility and Environmental Management, 15*, 1–13.

Daszyńska-Żygadło, K., Ryszawska, B., Słoński, T., & Zawadzki, B. M. (2014). Investors' reactions for sustainability index inclusion – Is CSR a good news? [w:] I. D. Czechowska, & R. Pastusiak (red.), Future of finance. Przyszłość finansów przedsiębiorstw I ubezpieczeń, "Acta Universitatis Lodziensis" Folia Oeconomica 2(300), Wydawnictwo Uniwersytetu Łódzkiego, Łódź 2014, s. 45–60.

Denis, D. J., & Denis, D. K. (1993). Managerial discretion, organization structure, and corporate performance: A study of leveraged recapitalizations. *Journal of Accounting and Economics, 16*, 209–236.

Erhemjamts, O., Li, Q., & Venkateswaran, A. (2013). Corporate social responsibility and its impact on firms' investment policy, organizational structure, and performance. *Journal of Business Ethics, 118*, 395–412.

Galant, A., & Cadez, S. (2017). Corporate social responsibility and financial performance relationship: A review of measurement approaches. *Economic Research-Ekonomska Istraživanja, 30*(1), 676–693.
Ghosh, A. (2001). Does operating performance really improve following corporate acquisitions? *Journal of Corporate Finance, 7*, 151–178.
Healy, P., & Palepu, K. (1988). Earnings information conveyed by dividend initiations and omissions. *Journal of Financial Economics, 21*, 149–176.
Healy, P., & Palepu, K. (1990). Earnings and risk changes surrounding primary stock offers. *Journal of Accounting Research, 28*(1), 25–48. https://doi.org/10.2307/2491216.
Healy, P. M., Palepu, K. G., & Ruback, R. S. (1992). Does corporate performance improve after mergers? *Journal of Financial Economics, 31*, 135–175.
Kaplan, S. (1989). The effect of management buyouts on operating performance and value. *Journal of Financial Economics, 24*, 217–254.
Linn, S. C., & Switzer, J. A. (2001). Are cash acquisitions associated with better postcombination operating performance than stock acquisitions? *Journal of Banking and Finance, 25*, 1113–1138.
Margolis, J. D., Elfenbein, H. A., & Walsh, J. P. (2007). Does it pay to be good? An analysis and redirection of research on the relationship between corporate social and financial performance. Working Paper, Harvard University.
McWilliams, A., & Siegel, D. (2000). Corporate social responsibility and financial performance: Correlation or misspecification? *Strategic Management Journal, 21*, 603–609.
Okoye, A. (2009). Theorizing corporate social responsibility as an essentially contested concept: Is a definition necessary? *Journal of Business Ethics, 89*, 613–627.
Peng, C. W., & Yang, M. L. (2014). The effect of corporate social performance on financial performance: The moderating effect of ownership concentration. *Journal of Business Ethics, 123*, 171–182.
Perepeczo, A. (2009). Fuzje i przejęcia – efektowność finansowa. Wyniki badań zagranicznych i polskich, Uniwersytet Szczeciński, Rozprawy i studia T (DCCCXX) 746, Szczecin.
Porter, M. E., & Kramer, M. R. (2006, December). Strategy and Society: The link between competitive advantage and corporate social responsibility. *Harvard Business Review, 84*, 78–92.
Rahman, S. (2011). Evaluation of definitions: Ten dimensions of corporate social responsibility. *World Review of Business Research, 1*(1), 166–176.
Rodgers, W., Choy, H. L., & Guiral, A. (2013). Do investors value a firm's commitment to social activities? *Journal of Business Ethics, 114*, 607–623.
Saadaoui, K., & Soobaroyen, T. (2017). An analysis of the methodologies adopted by CSR rating agencies. *Sustainability Accounting, Management and Policy Journal*. https://doi.org/10.1108/SAMPJ-06-2016-0031.
Saeidi, S. P., Sofian, S., Saeidi, P., Saeidi, S. P., & Saaeidi, S. A. (2015). How does corporate social responsibility contribute to firm financial performance? The mediating role of competitive advantage, reputation, and customer satisfaction. *Journal of Business Research, 68*, 341–350.
Scalet, S., & Kelly, T. F. (2010). CSR rating agencies: What is their global impact? *Journal of Business Ethics, 94*(1), 69–88.
Soana, M. G. (2011). The relationship between corporate social performance and corporate financial performance in the banking sector. *Journal of Business Ethics, 104*, 133–148.
Sun, N., Salama, A., Hussainey, K., & Habbash, M. (2010). Corporate environmental disclosure, corporate governance and earning management. *Managerial Auditing Journal, 25*, 679–700.
van Marrewijk, M. (2003). Concepts and definitions of CSR and corporate sustainability: Between agency and communion. *Journal of Business Ethics, 44*, 95–105.
Wang, Y., & Berens, G. (2015). The impact of four types of corporate social performance on reputation and finance performance. *Journal of Business Ethics, 131*, 337–359.
Zieliński, M. (2015). The impact of CSR concept on the valuation of stock companies. *Research Papers of Wrocław University of Economics, 401*, 642–650.

CSR of Banks in Poland

Łukasz Sułkowski and Justyna Fijałkowska

1 The Importance of Bank Sector in Poland

Poland's banking sector is the biggest banking market in the central and eastern European region. The banking sector plays a significant role in the economy. Banks have been an important pillar of the financial markets and the whole society, for centuries enabling them to function and develop. Generally, the financial services sector, with banks as its main actors, is viewed as a central element of modern capitalist economies (Merton 1995; Levine 2005). Banks act as financial intermediaries in society: they value financial assets, monitor loans, manage financial risks, organize payment systems (Greenbaum et al. 2015). In Poland, at the end of 2016, the number of banking entities covered 621 units, employing more than 170,000 people. At the end of 2015, banks' assets amounted to PLN 1.6 billion and the combined results were almost PLN 56 billion. The basic characteristics of the banking sector in Poland are presented in Table 1.

Emphasizing the size and importance of banks in the Polish economy, it can be noticed that there are 5 banks among the biggest 20 companies listed on the Warsaw Stock Exchange (WIG 20) and 7 among the biggest 30 (WIG 30). Financial institutions, and mainly banks, are among the most important pillars of the state budget; their tax for 2016 amounted in total to over PLN 5.1 billion (it accounts for more than 30% of the total tax paid to the budget). The total sales of the 20 largest

Ł. Sułkowski (✉)
Katedra Zarządzania Instytucjami Szkolnictwa Wyższego, Instytut Spraw Publicznych Uniwersytetu Jagiellońskiego, Kraków, Poland
e-mail: lsulkowski@san.edu.pl

J. Fijałkowska
Katedra Finansów i Rachunkowości, Społeczna Akademia Nauk, Łódź, Poland
e-mail: jfijalkowska@san.edu.pl

A. Długopolska-Mikonowicz et al. (eds.), *Corporate Social Responsibility in Poland*, CSR, Sustainability, Ethics & Governance,
https://doi.org/10.1007/978-3-030-00440-8_6

Table 1 Selected indicators of banking sector structure in Poland in 2011–2016

Characteristics	2011	2012	2013	2014	2015	2016
Number of entities conducting banking activity	642	642	640	631	623	621
– commercial banks	47	45	41	38	38	38
– credit institutions	21	25	28	28	27	26
– cooperative banks	547	572	571	565	561	560
Employment	176,658	175,071	174,321	172,659	170,935	172,803

Source: Own elaboration based on knf.gov.pl. (Access 30.04.2017); Financial system in Poland, (Warsaw 2017)

banks in Poland in 2016 amounted to PLN 71 billion, and their total net profit was almost PLN 14 billion (List of 500, 2017, edition No. 19, "Rzeczpospolita").

According to Pietraszkiewicz (2015) "after 25 years of the transformation and building of the Polish banking system it can now be said that the Polish sector is: stable, modern, effective, trustworthy, and backed by a solid infrastructure". Highlighting the role of the banking sector in the economy it is also worth mentioning that among the 330 most valuable brands in Poland there are 12 banks, in the 10 most valuable brands there are 2 banks, banks amount also to the one fourth of the 20 most valuable brands in Poland. The most precious brand in the banking sector is PKO Bank Polski—the value of this brand is estimated at over 3 billion PLN. In total, the 12 most valuable brands of banks account for more than 8 billion PLN (Ranking of the most valuable Polish brands 2016, Rzeczpospolita). According to Marcinkowska (2008) "the brand, its knowledge and market participants' ratings are the determinants of a bank's reputation—perhaps the most important factor determining its existence and stability. Taking care for credibility and brand strength is therefore an important element in building the image of a financial institution and helping to create value". Engagement in the CSR may become particularly useful and effective in this field.

2 CSR of Banks: General Concept

Corporate social responsibility (CSR) may be defined as "the process of communicating the social and environmental effects of organizations' economic actions to particular interest groups within society and to society at large" (Gray et al. 1987). Banks were never perceived as particularly harmful or dangerous for the environment. Before the crises banks were not extensively involved in CSR and in its disclosure (Novokmet and Rogošić 2016). Vigano and Nicolai (2006) indicate that European bankers have been slow in considering issues of sustainability (strictly related to CSR). However, taking into account both the functions performed by the banks and the size of the sector, it can be assumed that they have a significant impact on society and play an important role in economic development (Levine 2005;

Djalilov and Piesse 2011). Banks have a unique way of impacting on society, i.e. they can be engaged in CSR activities themselves and they can affect other businesses through financing the latter's projects with social, ethical and environmental issues (Scholtens 2006). Consequently, banks can also influence sustainable development. Korenik (2009) emphasizes that "banks with their actions and environmental impacts can help to limit or, on the contrary, they can enforce the socio-economic problems that plague modern societies". As it was underlined during the World Economic Forum in 2012 "finance plays a critical role for society at large, serving individuals, families, businesses, governments and civic institutions. The financial sector performs indispensable functions such as enabling saving and investment, providing protection from risks and supporting the creation of new jobs and enterprises. It is critical that the sector operates to provide these functions for society in a stable, sustainable way". It is also worth noticing that the impact of banks on the society may both be positive, but also negative, and in extreme cases it may be even destructive. The consequences of banks' mistaken decisions may have long-term impact and determine the lives of many people.

As Jurek (2015) underlines "CSR in financial institutions should not be seen only in terms of interactions between different stakeholders (shareholders, investors, depositors, borrowers, regulators) and financial institutions" in the fields of economic, legal, ethical and philanthropic responsibility, but "also in the context of their distinctive function of financial intermediaries". The main stakeholders, to which banks should be responsible, embrace the internal ones—board of directors and employees, as well as the external ones—clients, shareholders, government and its agencies, other financial institutions, financial analysts, competitors, suppliers, NGOs, media and society. It is worth mentioning that all these interest groups are important, however since the majority of bank assets come from depositors (i.e. society), not from shareholders, banks are required to provide feedback to the community more frequently than to other interested parties (Wu and Shen 2013).

The dimensions of CSR in financial institutions, of which banks are the main representatives, based on the Carrolls' (1991) framework are presented in the Table 2.

One of the main responsibilities of banks is the reliable information delivered to clients. The issue of the information asymmetry is highly important in this sector; both parts—financial institutions and client—should have an access to the full and transparent information on the services rendered by banks. It is unacceptable to include in the contracts small prints that are unfavorable to customers, not explaining the contract terms in full, do not transparently present the effects of conclusion or withdrawal from the contract with a bank. It is also unfair to divert all the risks associated with banking operations to an individual clients, as these clients, unlike corporations, do not have mechanisms that can mitigate the negative effects. Banks should not feed on customers, impose additional fees during the duration of the contract, and significantly increase the fees for their services. Whereas, banks should provide access to their offer to all market participants; they should not prefer only the strongest and most profitable actors, because it leads to inequalities in society and to discriminations. It is understandable that the goal of banks is profit and growth of

Table 2 Dimensions of CSR in financial institutions

Dimension of responsibility	Description
Economic	Financial institutions should improve the wealth of the owners, create new opportunities for risk management and offer the efficient transfer of funds
Legal	Legal responsibility is largely determined by legal regulations concerning for example the requirement of maintaining confidence, the limitations of risk taking, arbitrate relationship between the financial sector and society. Regulations constitute a pillar of financial institutions' responsibility and are supposed to rebuild trust and prevent from similar to 2007 and 2008 crises
Ethical	Ethical codes should be implemented in banks and financial institutions. Honesty and good faith, professional integrity, transparency and accountability are ethical values crucial for banks and financial institutions Regarding banks, that are the biggest players in the whole sector of financial institutions, Cowton (2002) indicates three aspects of ethics in banking: • Integrity—as it helps banks to generate trust that is vital for a financial system to flourish; • Responsibility—in the ways that banks have the responsibility to lend fairly (i.e. they have the responsibility not to exclude certain groups) and not being too hasty to foreclose when calling in loans; • Affinity—reflecting the ways in which depositors and borrowers can be brought closer together
Philanthropy	Philanthropy is concerned with altruistic acts not required by law that create social reputation and represent a social distribution of some proportion of profits. Philanthropy may be expressed also by the volunteering actions and charitable donations of the employees

Source: Fijałkowska et al. (2017)

value (Fijałkowska and Zyznarska-Dworczak 2017), but in view of their important social role, the goal should not be the maximum profit in the shortest possible time and at all costs.

3 Crises, Loss of Reputation and the CSR Remedy in Banking Sector

After the financial crises of 2007 and 2008 many financial sector institutions have lost their credibility in the eyes of consumers and investors. The economic crises originated from the financial markets and led to the lower trust in financial institutions, inferior investments and the recession. Therefore, public trust has become a pertinent concern of financial organizations (Norberg 2018). As Pietraszkiewicz (2015) underlines in Poland "banks with domestic capital played a very important and stabilizing role during the economic downturn between 2008 and 2013. First of all, they reduced lending to economic entities only slightly, and in many cases they took over the financing of entities for which banks with foreign capital could no longer provide loans". Moreover, in Poland, as in the rest of the world, CSR has been found to be a way for banks and other financial institutions to earn back their credibility (Cornett et al. 2014).

The services executed by banks and other financial institutions are characterized by information asymmetry and uncertainty which makes trust a necessary condition for doing business (Beck 2006; Schanz 2006). Trust is a central strategic issue for financial services firms (Llewellyn 2005; Olsen 2008). Decker and Sale (2009) note that “key CSR principles, such as accountability and transparency, are at the heart of regulatory efforts to ensure banks and bankers operate with trust”. Bossone (1999) and Olson (2006) argue that trust and confidence in banking will continue to be important and vital to society’s financial well-being. Today the credibility, accountability and reputation of financial sector is in its foundations and determines the stability of financial markets (Gaultier-Gaillard and Louisot 2006). Rossier (2003) highlights a number of reasons why bankers, in particular, have to consider issues of reputation. Bankers and other participants of financial sector trade money which represents other people’s security and well-being. They are also involved in a profession that has been held in contempt, since medieval times because of usury. As Stansfield (2006) states reputation in financial institutions concerns “financial performance and strength, client trust and confidence, client service, corporate social responsibility, corporate governance practices, corporate ethics, corporate disclosure practices, as well as relations with regulatory authorities and compliance in a broad sense”. The concept of CSR is closely related to trust and reputation and therefore it is an increasingly important issue in the international banking industry (Scholtens 2009) and in the financial sector as a whole. Banks perceive CSR as a way to restore trust, reputation, their image and credibility (Brammer and Pavelin 2005; Branco and Rodrigues 2006; Gössling and Vocht 2007; Minor and Morgan 2011; Hsu 2012; Schultz et al. 2013; Eberle et al. 2013; Hur et al. 2014). Many empirical results confirm that banks’ efforts to build a good reputation thanks to the CSR involvement brings concrete benefits and strengthens the banks’ performance (Forcadell and Aracil 2017). Banks got involved in CSR also because of the legitimization of the new banking process through social perceptions (Islam and Kokubu 2018).

The results of several studies (i.e. Pratihari and Uzma 2018) show that there is a significant impact of CSR components (economic, legal, ethical and philanthropic) on corporate branding to enhance customer brand loyalty in banks. The engagement of banks into CSR may also bring positive impact on their corporate financial performance (CFP) (Simpson and Kohers 2002; Wu and Shen 2013). However, the latest studies regarding the interrelation between CSR and CFP of banks in Poland (Fijałkowska et al. 2017) as well as those concerning generally the banking sector in Central and Eastern European countries (Fijałkowska et al. 2018) confirm that although banks got engaged in the CSR the response of their clients in this region as for now is slow and generally CSR involvement does not influence in a significant way the banks’ financial performance. This conclusions fall in line with the results of the other study concerning banks conducted by Matuszak and Różańska (2017b) and studies concerning the impact of CSR engagement on CSP e.g. in Italy (Soana 2011) and Turkey (Taskin 2015). The study of Fijałkowska, Zyznarska-Dworczak and Garsztka (Fijałkowska et al. 2017) indicates also that the bigger are the resources that are at the disposal of the bank the more it is engaged in the CSR. This is coherent with the latest work of Laguir et al. (2018).

4 CSR Engagement of Banks in Poland

According to report "CSR in Polish reality. Theory and Practice" (2015) banks go well when it comes to their interest and engagement in CSR. However, this opinion is worked out based mainly on the activeness of the biggest banks, listed in the Table 3. This table presents the engagement of the biggest commercial banks actively involved in CSR idea. All of those banks declare on their websites the interest concerning issues on CSR. All of them underline also the importance of human rights. 11 out of 13 banks analyzed below have codes of ethics, 12 banks announce that want to fight against corruption and the same number of banks declares that ecological aspects are very important in their operations. The policy on sustainable development has been developed by almost 77% of entities and almost 70% indicates that the whole supply chain is an important issue for them. In nine banks, there is a special person employed that is responsible for CSR issues. The reference to CSR in the annual report is present in all the banks. Banks are involved in the CSR activities and some of them try to disclose it in the CSR reporting. Table 3 shows the CSR involvement examples in the biggest banks in Poland.

It is however important to add, that the rest of banks, not included in this table, practically remains completely not engaged in CSR activities. The majority of smaller banks does not mention CSR on their website, usually the only announcement published on their websites concerns donation or sponsoring actions, rarely some separate activities regarding ecology, local society or volunteering actions. Therefore, in case of banks not included in the table, the information concerning CSR disclosed by them is occasional and cannot be understood as the manifestations of the regular CSR approach implemented into the strategy and practice of those institutions.

Still there are not many banks publishing CSR report as a special satellite document or a non-financial information document integrated with the financial report. The Table 4 shows the amount of banks publishing CSR report since 2004.

All together there are 42 CSR reports of banks in Poland published since 2004. The number of banks publishing annually is not high, out of 38 commercial banks operating in Poland in 2016 there were only 6 that decided to publish a separated CSR report. However, the result of all together 42 CSR reports place banking sector as the second most active in CSR reporting in Poland (after fuel industry). The Fig. 1 shows the number of banks reporting on CSR.

The study of Krasodomska (2015) highlights that banks in 2005–2011 tended to include CSR disclosures in the management commentary; they presented CSR information in a diverse manner, focusing mainly on community involvement.

Table 3 Engagement of commercial banks in Poland in CSR activities

		CSR tag onwebsite	Person responsible for CSR	CSR policy	Code of ethics	CSR report	GRI in CSR report	CSR in the annual report	Ecology	Supply chain	Human rights	Corruption
1.	Bank BGŻ BNP PARIBAS	+	+	+	+	+	+	+	+	+	+	+
2.	Bank Gospodarstwa Krajowego	+	+	+	+	–	n/a	+	+	–	+	+
3.	Bank Handlowy w Warszawie SA	+	–	–	+	–	n/a	+	+	–	+	+
4.	Bank Zachodni WBK SA	+	+	+	+	+	+	+	+	+	+	+
5.	Bank BPH SA	+	+	+	+	+	+	+	+	+	+	+
6.	Bank MILLENNIUM SA	+	+	+	+	+	+	+	+	+	+	+
7.	PEKAO SA	+	–	+	+	–	n/a	+	+	+	+	+
8.	CREDIT AGRICOLE Bank Polska SA	+	+	+	+	–	n/a	+	+	+	+	+
9.	GETIN NOBLE Bank SA	+	–	–	+	–	n/a	+	–	+	+	+
10.	ING Bank Śląski SA	+	+	+	+	+	+	+	+	+	+	+
11.	PKO BP SA	+	+	+	–	–	n/a	+	+	–	+	–
12.	mBank SA	+	+	+	–	–	n/a	+	+	–	+	+
13.	Raiffeisen Polbank	+	–	–	+	–	n/a	+	+	+	+	+

Source: Based on "CSR in Polish reality. Theory and Practice" (2015), p. 58

Table 4 The number of CSR reports published by banks in Poland

	Total	2004	2005	2006	2007	2008	2009	2010	2011	2012	2013	2014	2015	2016
PKO BP S.A.	1	0	0	0	0	0	0	0	0	0	0	0	1	0
BZ WBK S.A.	5	0	0	0	0	0	0	0	1	0	1	1	1	1
mBank S.A. (BRE)	6	0	0	1	1	1	1	0	0	0	0	0	1	1
ING BSK S.A.	3	0	0	0	0	0	0	0	0	1	0	1	0	1
Millennium S.A.	10	0	0	0	1	1	1	1	1	1	1	1	1	1
BGŻ BNP Paribas S.A.	6	0	0	0	0	0	0	0	1	1	1	1	1	1
Bank BPH S.A.	8	1	0	1	0	1	1	1	1	1	0	1	0	0
BOŚ S.A.	3	0	0	0	0	0	0	1	1	0	0	0	0	1
Total	42	1	0	2	2	3	3	3	5	4	3	5	5	6

Source: Own work based on the www.rejestraportow.pl database of CSRinfo

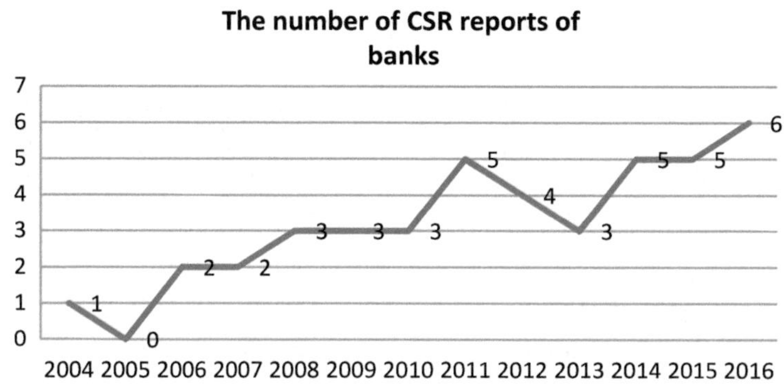

Fig. 1 Number of banks publishing on CSR. Source: own work

More recent research of Matuszak and Różańska (2017a) indicates that "the majority of Polish listed banks use more than one channel to communicate CSR activity but generally they attribute greater importance to annual reports comparing with the other channels", like CSR reports or websites. The latest research conducted by Fijałkowska and Zyznarska-Dworczak (2018) proves, that still the CSR disclosure level of bank reporting in Poland is quite low; in 2016 the intensity of disclosure was only slightly higher than in 2012. While analyzing the content of the CSR reports published by banks, Fijałkowska and Zyznarska-Dworczak (2018) observe that in 2012–2016 the most frequent disclosures included community involvement, corporate governance, risk management and consumer issues, while banks disclosure on fair operating practices and business model / value creation issues were at very low levels.

While analyzing the CSR involvement of banks in Poland, it is also worth mentioning that some of the banks make part of the Respect Index—an index of socially responsible companies quoted on the Warsaw Stock Exchange. The index comprises companies selected from among the largest public enterprises of WIG20, mWIG40 and sWIG80 indexes, operating in compliance with best corporate governance, information governance and investor relations standards as well as with consideration of environmental, social and economic factors. The selected companies undergo three-tier verification made by the Warsaw Stock Exchange, the Stock Issuers Association and an independent auditor. The index composition is reviewed once a year. The participation in the Respect Index confirms the responsible way of managing the business. In the last edition of the 19th December 2016 there were 25 companies admitted into this index, 5 of them were banks. Two banks (Handlowy and ING BS) were present in all ten editions of the Index and Millennium Bank was admitted nine times. The presence of banks into the Respect Index since 2009—the year of its first quotation—is presented in Table 5.

Table 5 The presence of banks into the Respect Index

Banks	Total number of presence	2009	2010	2010	2011	2012	2013	2014	2015	2016	2017
Edition		I	II	III	IV	V	VI	VII	VIII	IX	X
HANDLOWY	10	1	1	1	1	1	1	1	1	1	1
MILLENIUM	9	0	1	1	1	1	1	1	1	1	1
BANK BPH	7	1	0	0	1	1	1	1	1	1	0
INGBSK	10	1	1	1	1	1	1	1	1	1	1
BZWBK	4	0	1	0	0	0	0	1	1	1	0
mBank (BREBANK)	2	0	1	1	0	0	0	0	0	0	0
KREDYT BANK	4	0	0	1	1	1	0	0	0	0	1
PEKAO	1	0	0	0	0	0	0	0	0	0	1
BOŚ	1	0	0	0	0	0	0	0	0	1	0
Total	48	3	5	5	5	5	4	5	5	6	5

Source: Own work

J. B. Jorge,[1] Chairman of the Management Board of Millennium Bank, one of those banks, that are present in the Respect Index for the longest period of time, underlines: "*By acting in an ethical and responsible way we gain trust of our clients, counterparties and investors. We are also an appreciated employer both for current as well as potential employees—young people who see social responsibility of a company as one of the important selection criteria of their future workplace. We owe our long-term presence in Respect Index to number of factors: we continuously analyse the needs and expectations of our stakeholders and respond to changing environment; we pursue a transparent information policy; we are a customer service quality leader, (...) we continue to enrich our portfolio with more innovative products, intuitive in use and available to a broad range of clients, such as our mobile and internet banking*".

In the Tables 6 and 7 the main initiatives concerning the CSR engagement and the achievements or awards of banks in 2016 are presented.

5 Ethical Banking and Responsible Investment

Ethical baking is not a concept widely common in Poland. Ethical banks, also called alternatives, civic, or sustainable, finance only those ventures, projects, companies, and organizations that operate in accordance with certain basic rules. Attention is

[1]https://www.bankmillennium.pl/about-the-bank/press-centre/press-releases/-/news-info/16-12-2015-bank-millennium-po-raz-osmy-w-skladzie-respect-index (2015, access: 20.09.2017)

Table 6 The main initiatives concerning CSR in which banks in Poland are involved

Initiative	Description
Respect index	This includes Polish companies from the main exchange market securities in Warsaw, operating in accordance with the best management standards in corporate governance, information governance, and investor relations as well as accountability concerning environmental, social and labor factors
The responsible business ranking	The responsible business ranking has been built for 10 years by the business ethics Center of Kozminski Academy and "Dziennik Gazeta Prawna", the responsible business forum and Deloitte. The ranking presents the largest companies in Poland, evaluated in terms of the quality of management concerning corporate social responsibility. Deloitte's experts verify the implementation of socially responsible business policies in five areas: responsible leadership, dialogue with stakeholders, social engagement, responsible management and social innovation. Specific actions, operational procedures and policies and a holistic approach to management in all these areas are assessed
CSR leafs of Polityka magazine and Deloitte company	Classification of the best companies is prepared on the basis of international guidelines ISO 26000 social responsibility standards concerning seven areas: organizational order, human rights, work practices, environment, honest operational practices, consumer issues and social engagement. It concerns also chosen goals of sustainable development announced by the UN
Ranking of socially engaged companies	The ranking of responsible firms is prepared by Gazeta Bankowa and portal wGospodarce.pl. The ranking is a compilation of 40 most interesting projects created in companies in Poland from the area of CSR. The 2017 ranking shows the specific activities of large and small companies. Out of hundreds of such ventures, bold and original ideas have been collected and dubbed, which support the development of society in Poland as a whole, bring new quality to local communities, promote sustainable development and build the spirit of innovation.
Responsible business report in Poland in 2016	The report, published for 14 years by the forum of responsible business, presents initiatives that are socially and environmentally responsible implemented by companies in Poland
Golden banker ranking	Ranking organized by bankier.pl and "Puls Biznesu". Golden Bankier is the largest ranking of banking services and products in Poland. Since 2009 it has been setting the direction for the development of Polish banking sector. It promotes banks offering the highest quality of service. 2017 year's plebiscites provided for three categories: the best advertising spot of the bank, the CSR activity, and the social media activity

Source: Own work

Table 7 The main achievements and awards concerning CSR engagement of banks in Poland in 2016/2017

Initiative	Examples of banks' participation / admition / awards in 2016
Respect index	In 10 editions of the respect index banks appeared 48 times in. There are two banks present in the index from the very beginning of its creation, every year there are about 5 banks among about 20 companies accepted in this index
The responsible business ranking	In 2017 the ranking has been presented in a new release, broken down into three levels; the diamond level—that embraces the companies-leaders of excellence in CSR, platinum level—that includes well-advanced companies in CSR. The last stage involves companies from the crystal level, matured in CSR and active in this area enough to qualify for the listing. In the first level, out of 7 companies, there was Bank Zachodni WBK (on the 5th place). In the second level there was ING Bank Slaski SA (on the third place among 8 companies). The last level started with Bank BGŻ BNP Paribas in the first place, followed by Bank millennium (4th place), credit Agricole Bank Polska SA (6th place), Bank Handlowy w Warszawie SA together with Bank Gospodarstwa Krajowego (both at 7th place) and mBank SA (11th place)
CSR leafs of Polityka magazine and Deloitte company	In 2017 there were 111 companies participating. Golden leaf in CSR was awarded to Bank Zachodni WBK. For the second time Bank millennium was awarded silver leaf in CSR. Silver leaf was also granted to Bank BGŻ BNP Paribas. Bank Handlowy w Warszawie SA and Bank Ochrony Środowiska received white leafs
Ranking of socially engaged companies	Out of 40 initiatives, 5 awards went to banks. PEKAO S.A. took the first place among the projects having the best positive impact on the market. BGŻ BNP Paribas was honored for responsible funding. In the category of health and society third place was granted to PKO BP. The distinguished initiative was also granted to mBank in this category. In culture category Citibank was awarded the third place
Responsible business report in Poland in 2016	Among the 460 "Good Practices" described in the responsible business in Poland 2016 report, 58 referred to bank practices, including Bank Zachodni WBK (15), Bank BGŻ BNP Paribas (10), ING (6) BOŚ (5), mBank (5), credit Agricole Bank Polska (4), millennium (4), PKO Bank Polski (4), BGK (4), Citi Bank Handlowy (1)
Golden banker ranking	Golden banker title has been awarded to the mBank for the service quality, second place was assigned to Euro Bank, and the third place was given to Bank millennium and Getin Noble Bank. The golden banker statue for the most socially sensible bank was assigned to ING Bank Śląski

Source: Own work

drawn to the ecology, the social objectives pursued by the institution, the impact on the closest environment. Some products concerning responsible money saving are available in Poland. One example is affinity credit card. The first one to introduce it was the BISE Bank. Currently there are many banks such as BOŚ or BGŻ that offer those cards to their clients. When a customer makes an affinity credit card transaction, the percent commission, which is routinely collected, goes to assist the orphanage or to protect the lynx, black stork or for some other social or ecological purpose. In some banks this sum is covered by the bank itself. The more money the customer operates via credit card, the more is transferred by the bank for a noble goal, and the customer does not pay for it. Socially Responsible Investing is not well known and practiced in Poland. According to Adamska et al. (2015), while compared to 10 EU countries with different degrees of socially responsible investment (SRI) market development, Poland represented a low value of assets under management of ethical investment, but at the same time showed the highest growth rates of these assets. In addition, the segment of open ethical investment funds was considered underdeveloped, with moderate degree of heterogeneity in conjunction with the small size of the average fund. Adamska et al. (2015) underline also that "the low level of development of the domestic market of ethical investments is largely due to the existing system of institutional factors, which in contrast to the solutions found in countries with mature SRI market, does not create incentives for this type of investment". Banks in Poland should increase the access to ethical responsibility market investment, which is at the moment a marginal phenomenon in Poland while it is growing in a fast way in the global market of financial products. The rare example of the involvement in the responsibility investment are the initiatives of the Bank Gospodarstwa Krajowego that in case of some credits requires from the contract partners a statement that they are not involved in the operations concerning the violation of human rights, production of guns, pornography, gambling, drugs, ecc. It is worth noting that none of the Polish banks participates in any international initiative promoting responsible investment criteria (UN PRI,Equator Principles, Eurosif, etc.). Financial institutions should provide their individual clients with the data that helps in the funds allocations falling in line with their worldview and moral principles. This is still a rare practice in Poland.

6 Social Irresponsibility: Dark Side of the Banks in Poland

Banks make part of a particularly sensitive sector of business, turning large sums of money, the consequences of which may affect many areas of people's life. There is a risk of making "ethical" talks to create a smokescreen for unethical action while bothering only about a profit maximization. In Poland one of the biggest problems, revealing the dark side of the banks activities, concerned credit agreements in CHD that regards 500 thousand agreements. The Supreme Chamber of Control checked the franc agreements. It turned out that as much as 95% of cases contain illegal clauses. The Office of Competition and Consumer Protection also stated that the banks cannot themselves set a course, at which they provided loans and collected installments. Spread was indicated as a contradictory earnings of banks. Also the requirement of the additional securities was judged as illegal. Hence, the use of similar clauses is forbidden. Despite unfavorable sentences for banks, they generally do not react and still permanently use the forbidden practices. They cut off single sentences from the grounds of the court and interpret them in their own way. As it is described in the report of CentrumCSR Foundation (2015) concerning the situation of franc credits, banks were not willing to negotiate or make any concessions. When some proposals were worked out, banks did not agree with the clients on the sharing of co-responsibility. None of the banks joined the court's arrears. On contrary, in cases of clients having trouble with timely payment, in many cases banks offered another, but much more expensive cash loan. As a result, instead of getting out of debt some clients were falling into a loop from which was hard to get out. Another problem concerning Polish banks practices, indicated in the report of CentrumCSR Foundation (2015), is that banks often use the Bank Execution Title to run the eviction process in situation of just few instalments of the credit not paid. The time given to the client, compared to other European countries, is short and the appeal in the court does not halt the process of eviction. It may indicate that the business motives of banks are more important in that cases than the social ones.

7 Conclusions

The banking sector and its situation, in both financial and reputational aspects, strongly determine the quality and prospects of the financial services market, which has a direct impact on the whole economy. At the same time, the reputation of banks depends on many factors—to high degree on their social reception. It is increasingly common to see that banks are a symbol of crisis, speculation and injustice, and perceive some of their activities as one of the cause of poverty. In Poland, as in many parts of the world, banks try to rebuild their credibility and trust that were lost during the financial crises. One of the main ways is the engagement in the CSR activities. Trust, reputational and regulatory risks are of particular concerns in bankers' efforts to engage with CSR (Decker and Sale 2009). Many of banks strive to rebuild their image (Salehzadeh et al. 2018) and to get engaged in natural environment protection and socially responsible operations in many ways: by saving resources, financing pro-ecological organizations, offering special financial products, actively participating in local socially appreciated initiatives and programs, showing that they do care about their clients, employees, environment and local communities. Many big banks in Poland communicate on those engagements both on their website and in CSR reports. They are also actively involved in different programs and initiative concerning both the promotion and the evaluation of CSR engagement. This new trend is visible in Polish banks sector. The latest empirical research conducted in banks in Poland (Fijałkowska et al. 2017) proves however that the CSR engagement of these institutions is still not reflected in the bottom lines of their financial statement. Market and the public are reluctant in considering the CSR importance and the CSR engagement is not enough rewarded in bank sector. This undermines the role of CSR commitment in financial institutions. It may be assumed that CSR efforts are not focused properly, or they are still not communicated effectively.

Banks are a significant sector of economy in Poland. For many years banks in Poland enjoyed the status of "primacy of Europe". For a long time, there was no need to pay deposit guarantees from the Banking Indemnity Fund, the profitability of commercial banks allowed for a systematic increase of capital ratios, which strengthened their capital and their role in the Polish economy. Banks have played an active role in financing consumption and improving the quality of life in households, and the bank failures have not been heard for years. However, negative tendencies have started to be noticeable at least since 2014. The decline in yields was, however, mainly a result of the situation of 2008 and 2009 in the context of falling interest rates and the increase in capital requirements. The atmosphere after the financial crisis has also influenced the situation. Banks needed to adjust product prices to new market challenges, reduce costs, escape from low-margin products, and unprofitable business segments. Some of them got also engaged in CSR activities and their promotion.

Modern banking has faced a great task of rebuilding good name after the financial crisis. One of the solutions that can contribute to the development of this task is the involvement in CSR and its proper communication. In Poland, large banks are beginning to recognize this need and implement it into their practice. The smaller banks still lag behind.

References

Adamska, A., Dąbrowski, T., & Grygiel-Tomaszewska, A. (2015). Społecznie odpowiedzialne inwestowanie w Polsce na tle sytuacji w wybranych krajach Unii Europejskiej. *Nauk o Finans, 2*, 73–95. https://doi.org/10.15611/nof.2015.2.04.

Beck, T. (2006). *Creating an efficient financial system: Challenges in a global economy* (Policy Research Working Paper No 3856). World Bank.

Bossone, B. (1999). *What makes banks special? A study of banking, finance, and economic development* (Policy Research Working Paper). https://doi.org/10.1596/1813-9450-2408.

Brammer, S., & Pavelin, S. (2005). Corporate reputation and an insurance motivation for corporate social investment. *Journal of Corporate Citizenship, 20*, 39–51.

Branco, M. C., & Rodrigues, L. L. (2006). Corporate social responsibility and resource-based perspectives. *Journal of Business Ethics, 69*, 111–132. https://doi.org/10.1007/s10551-006-9071-z.

Carroll, A. (1991). The pyramid of corporate social responsibility: Toward the moral management of organizational stakeholders. *Business Horizons, 34*, 39–48.

Cornett, M., Erhemjamts, O., & Tehranian, H. (2014). Corporate social responsibility and its impact on financial performance: Investigation of US commercial banks. *Academy of Management Review, 14*(1), 65–107.

Cowton, C. J. (2002). Integrity, responsibility and affinity: Three aspects of ethics in banking. *Business Ethics: A European Review, 11*, 393–400. https://doi.org/10.1111/1467-8608.00299.

Decker, S., & Sale, C. (2009). An analysis of corporate social responsibility, trust and reputation in the banking profession. In *Professionals' perspectives of corporate social responsibility* (pp. 135–156). Berlin: Springer.

Djalilov, K., & Piesse, J. (2011). Financial development and growth in transition countries: A study of Central Asia. *Emerging Markets Finance and Trade, 47*, 4–23. https://doi.org/10.2753/REE1540-496X470601.

Eberle, D., Berens, G., & Li, T. (2013). The impact of interactive corporate social responsibility communication on corporate reputation. *Journal of Business Ethics, 118*, 731–746. https://doi.org/10.1007/s10551-013-1957-y.

Fijałkowska, J., & Zyznarska-Dworczak, B. (2017). Społeczna odpowiedzialność biznesu a tworzenie wartości banków. *Zarządzanie i Finans* (The Journal of Management and Finance), *15*, 107.

Fijałkowska, J., & Zyznarska-Dworczak, B. (2018). Sustainability reporting of Polish banks – A step towards greater accountability. *Entrepreneurship and Management, 19*(1), 47–61.

Fijałkowska, J., Zyznarska-Dworczak, B., & Garsztka, P. (2017). The relation between the CSR and the accounting information system data in Central and Eastern European (CEE) countries – The evidence of the Polish financial. *Accounting and Management Information Systems, 16*, 490–521. https://doi.org/10.24818/jamis.2017.04004.

Fijałkowska, J., Zyznarska-Dworczak, B., & Garsztka, P. (2018). Corporate social-environmental performance versus financial performance of banks in Central and Eastern European countries. *Sustainability, 10*, 772. https://doi.org/10.3390/su10030772.

Forcadell, F. J., & Aracil, E. (2017). European banks' reputation for corporate social responsibility. *Corporate Social Responsibility and Environmental Management, 24*, 1–14. https://doi.org/10.1002/csr.1402.

Gaultier-Gaillard, S., & Louisot, J.-P. (2006). Risks to reputation: A global approach. *The Geneva Papers on Risk and Insurance-Issues and Practice, 31*, 425–445. https://doi.org/10.1057/palgrave.gpp.2510090.

Gössling, T., & Vocht, C. (2007). Social role conceptions and CSR policy success. *Journal of Business Ethics, 74*, 363–372. https://doi.org/10.1007/s10551-007-9512-3.

Gray, R., Owen, D., & Maunders, K. (1987). *Corporate social reporting*. Englewood Cliffs, NJ: Prentice-Hall.

Greenbaum, S., Thakor, A., & Boot, A. (2015). *Contemporary financial intermediation*. London: Academic Press.

Hsu, K.-T. (2012). The advertising effects of corporate social responsibility on corporate reputation and brand equity: Evidence from the life insurance industry in Taiwan. *Journal of Business Ethics, 109*, 189–201. https://doi.org/10.1007/s10551-011-1118-0.

Hur, W.-M., Kim, H., & Woo, J. (2014). How CSR leads to corporate brand equity: Mediating mechanisms of corporate brand credibility and reputation. *Journal of Business Ethics, 125*, 75–86. https://doi.org/10.1007/s10551-013-1910-0.

Islam, M. T., & Kokubu, K. (2018). Corporate social reporting and legitimacy in banking: A longitudinal study in the developing country. *Social Responsibility Journal, 14*, 159–179. https://doi.org/10.1108/SRJ-11-2016-0202.

Jurek, M. (2015). CSR in financial institutions in the aftermath of the global financial crisis. *Kolegium Ekonomiczno-Społecznego Studia i Prace / Szk Główna Handlowa, 3*, 68–79.

Korenik, D. (2009). *Odpowiedzialność banku komercyjnego: próba syntezy*. Warsaw: Difin.

Krasodomska, J. (2015). CSR disclosures in the banking industry. Empirical evidence from Poland. *Social Responsibility Journal, 11*, 406–423. https://doi.org/10.1108/SRJ-02-2013-0019.

Laguir, I., Marais, M., El Baz, J., & Stekelorum, R. (2018). Reversing the business rationale for environmental commitment in banking. *Management Decision, 56*, 358–375. https://doi.org/10.1108/MD-12-2016-0890.

Levine, R. (2005). Finance and growth: Theory and evidence. *Handbook of Econometrics, 1*, 865–934. https://doi.org/10.1016/S1574-0684(05)01012-9.

Llewellyn, D. T. (2005). Trust and confidence in financial services: A strategic challenge. *Journal of Financial Regulation and Compliance, 13*, 333–346. https://doi.org/10.1108/EUM0000000007304.

Marcinkowska, M. (2008). Niematerialne zasoby banków jako źródło ich wartości. *Acta Univ Lodz Folia Oeconomica, 218*, 203–229.

Matuszak, Ł., & Różańska, E. (2017a). CSR disclosures in listed banks-current state and future perspective. *Pr Nauk Uniw Ekon we Wrocławiu, 478*, 311–323.

Matuszak, Ł., & Różańska, E. (2017b). An examination of the relationship between CSR disclosure and financial performance: The case of Polish banks. *Journal of Accounting and Management Information Systems, 16*, 522–533. https://doi.org/10.24818/jamis.2017.04005.

Merton, R. C. (1995). A functional perspective of financial intermediation. *Financial Management, 24*, 23–41.

Minor, D., & Morgan, J. (2011). CSR as reputation insurance: Primum non nocere. *California Management Review, 53*, 40–59. https://doi.org/10.1525/cmr.2011.53.3.40.

Norberg, P. (2018). Bankers bashing back: Amoral CSR justifications. *Journal of Business Ethics, 147*, 401–418. https://doi.org/10.1007/s10551-015-2965-x.

Novokmet, A., & Rogošić, A. (2016). Bank sustainability reporting within the GRI-G4 framework. *Zesz Teoretyczne Rachun, 88*(144), 109–123.

Olsen, R. A. (2008). Trust as risk and the foundation of investment value. *The Journal of Socio-Economics, 37*, 2189–2200.

Olson, M. (2006). *Are banks still special?* Washington, DC: Remarks at the Annual Washington Conference of the Institute of International Bankers.

Pietraszkiewicz, K. (2015). The Polish banking sector in the process of transformation of the Polish economy. *Argumenta Oeconomica, 1*, 5–32.

Piskalski, G. (2015). *CSR in Polish reality. Theory and practice*. Warsaw: CentrumCSR Foundation.

Pratihari, S. K., & Uzma, S. H. (2018). CSR and corporate branding effect on brand loyalty: A study on Indian banking industry. *Journal of Product and Brand Management, 27*, 57–78. https://doi.org/10.1108/JPBM-05-2016-1194.

Rossier, J. (2003). Ethics and money: What is required of the banking profession. *La lettre, Geneva Private Bankers Association, 23*, 1–8.

Salehzadeh, R., Khazaei Pool, J., & Jafari Najafabadi, A. H. (2018). Exploring the relationship between corporate social responsibility, brand image and brand equity in Iranian banking industry. *Journal of Islamic Accounting and Business Research, 9*, 106–118. https://doi.org/10.1108/JIABR-11-2014-0041.

Schanz, K.-U. (2006). Reputation and reputational risk management. *The Geneva Papers on Risk and Insurance – Issues and Practice, 31*, 377–381. https://doi.org/10.1057/palgrave.gpp.2510092.

Scholtens, B. (2006). Finance as a driver of corporate social responsibility. *Journal of Business Ethics, 68*, 19–33. https://doi.org/10.1007/s10551-006-9037-1.

Scholtens, B. (2009). Corporate social responsibility in the international banking industry. *Journal of Business Ethics, 86*, 159–175. https://doi.org/10.1007/s10551-008-9841-x.

Schultz, F., Castelló, I., & Morsing, M. (2013). The construction of corporate social responsibility in network societies: A communication view. *Journal of Business Ethics, 115*, 681–692. https://doi.org/10.1007/s10551-013-1826-8.

Simpson, W., & Kohers, T. (2002). The link between corporate social and financial performance: Evidence from the banking industry. *Journal of Business Ethics, 35*, 97–109.

Soana, M.-G. (2011). The relationship between corporate social performance and corporate financial performance in the banking sector. *Journal of Business Ethics, 104*, 133–148. https://doi.org/10.1007/s10551-011-0894-x.

Stansfield, G. (2006). Some thoughts on reputation and challenges for global financial institutions. *The Geneva Papers on Risk and Insurance – Issues and Practice, 31*, 470–479. https://doi.org/10.1057/palgrave.gpp.2510087.

Taskin, D. (2015). The relationship between CSR and banks' financial performance: Evidence from Turkey. *Journal of Yaşar University, 10*, 21–30.

Vigano, F., & Nicolai, D. (2006). CSR in the European banking sector: Evidence from a sector survey. In *Rhetoric and realities: Analysing corporate social responsibility in Europe (RARE)*. Cheltenham: Edward Elgar.

Warsaw. (2017). *Financial system in Poland 2015*. Warsaw: Narodowy Bank Polski.

Wu, M., & Shen, C. (2013). Corporate social responsibility in the banking industry: Motives and financial performance. *Journal of Banking & Finance, 37*, 3529–3547.

Part III
Legal, Financial, Environmental Aspects of CSR

Legal and Financial Aspects of Eco-efficiency Investments in Poland

Tomasz Słoński and Magdalena Ligus

1 Introduction

European Commission (2001) states that socially responsible actions of the enterprises means that, beyond legal constraints, firms commit on a voluntary basis to bear the cost of more ethical behavior in variety of practices, for example improving employment conditions and/or banning child labor in countries that do not respect human rights, protecting the environment and investing in abatement equipment to reduce the carbon footprint, developing partnerships with NGOs, or providing funds to charity, etc. For that reason implementing CSR is multi-faceted and triggers a multi-dimensional decision. Universally accepted approach (CFA Institute 2017; Financial Times Lexicon; PWC 2015) is to group those different dimensions into three main pillars: environmental, social and corporate governance (so-called ESG factors).

In the sense, it could be stated that CSR involves taking actions which reduce the extent of external costs. This approach reflected in OECD materials (OECD 2011) states the common aim of the governments to encourage the positive contributions that enterprises can make to economics, environmental and social progress and therefore be able to increase the social value added by corporate activity.

In previous findings of researchers a special attention was focused on an environmental performance and its impact on firm's value. Daszyńska-Żygadło et al. (2016) analyzed global data base and found environmental factor significant for Tobin's q in 8 macro sectors of economy but in each case the impact is negative (PE ratio decrease in case of 7 sectors). Derwall et al. (2005) and Semenova and

T. Słoński (✉) · M. Ligus
Katedra Finansów Przedsiębiorstw i Finansów Publicznych (KFP), Uniwersytet Ekonomiczny we Wrocławiu, Wrocław, Poland
e-mail: Tomasz.slonski@ue.wroc.pl; magdalena.ligus@ue.wroc.pl

A. Długopolska-Mikonowicz et al. (eds.), *Corporate Social Responsibility in Poland*, CSR, Sustainability, Ethics & Governance,
https://doi.org/10.1007/978-3-030-00440-8_7

Hassel (2008) stated that in the environmentally sensitive sectors the CSR's positive effects are difficult to achieve due to higher cost. The biggest reduction of assets' market-to-book were found in materials and in utilities sectors. Padgett and Galan (2010) found that manufacturer's environmental actions affect the firms' financial performance to a stronger degree. Since the findings did not confirm the positive effect of environmental actions on value creation one need to be aware of the need of environmental regulations imposed by the governments. Criscuolo and Menon (2015) and Kayser (2016) showed that renewable energy investments were mostly challenged by and highly sensitive to the risk of changing regulations.

The aim of this chapter is to analyze the impact of Polish government regulations on the future development of renewable energy sector which drew our attention for a few reasons: (1) to attract investors it receives special support from the government bodies in terms of financial means and regulations; (2) renewable energy sector has direct impact on the environment and for this reason it limits the cost to the society; and additionally since many companies report the renewable energy purchase/production as a part of their CSR policy (ThomsonReuters Asset4 Database 2017) renewable energy sector has (3) indirect impact on other sectors in terms of their CSR actions.

Authors as a research method applied a critical review of the legal regulations and strategic documents of the EU and at the national level and used the methodology of Regulation Impact Assessment Guidelines introduced by the EU Commission (EU SEC 791 2005) to analyze the consequences not only for sector participants but indirectly for economy and society as a whole.

From the micro-level perspective the analysis of government regulations gives the rare opportunity to: (1) measure growth prospects of renewable energy sector (RES), (2) find the preferable renewable energy technologies and (3) redefine the list of potential risks faced by the investors in the sector. The results of research are not narrowed down to renewable sector participants.

Our analysis of renewable energy sector accepts the view of the economy as a network of companies. Therefore, we reach far beyond the opinion stated by McWilliams and Siegel (2001) that "managers should treat decisions regarding CSR precisely as they treat all investment decisions". In our view, the analysis calls for contingent evaluation, and since environmental actions could negatively affect market capitalization of companies, they help to fulfill management's fiduciary responsibilities (Barnett 2007) in economy as a whole. For that reason the results show potential implications for the government bodies responsible for future development of the sector and the economy. Consequently, results will contribute to comparative studies on implementation of renewable energy solutions in emerging markets (Loper and Crittenden 2017).

The main group of risk factors discussed in this chapter are related to the revenue stream. These risk factors can be subdivided into two groups with respect to price and volume component. As a result of our analysis we suppose energy prices might increase in the auctioning system (adjusted to the reference prices), and this risk might grow over time. The other group of risk factors impacts the sales volume. Here the risks are tied to the auction mechanism. The auction characteristic (technological preferences, production volume, announcement date etc.) is in the hand of government bodies and plan for auctions in subsequent years is not scheduled. Furthermore, to fix the sales volume the investor needs to win an auction and carry operational risks. Risk affects not only revenues, since proposed legislation do not take into account the microeconomic considerations and criteria for making investment decisions, including the administrative costs (more broadly, the transaction costs).

2 European Regulations Shape the Renewable Energy Sector

The basic legal act specifying the situation of the renewable energy sector in Poland, and all European Union member states, is the Directive 2009/28/EC of 23 April 2009 on the promotion of the use of energy from renewable sources, which defines the obligatory goals with respect to the minimum volume of energy generated from renewable sources (see more in the article by Ryszawska in this book). Pursuant to its provisions, Poland is obliged to achieve a 15% share of energy production from renewable sources in 2020.

The current market conditions prevent the full competitiveness of alternative fuels and require dedicated support for the producers of energy from renewable sources (Daszyńska-Żygadło and Ligus 2017). By 1st July 2016, the incentive scheme addressed to the RES installations was primarily regulated in the Act of 10 April 1997—the Energy Law (Gov. Journ. of 2006 no. 89 it. 625 and no. 104 it. 708) and the transitional provisions of the Act of 20 February 2015 on renewable energy sources (Gov. Journ. of 2015 it. 478). It was based on tradable green certificate system since 2005. In green certificate system producers of energy from renewable sources were entitled to issue tradable green certificates (TGC), which constituted proprietary rights as the object of trade. At the same time, sellers of energy to the end-user were obliged to purchase the described certificates to surrender them. Otherwise, the seller was obliged to make a compensatory payment in the amount specified by the Energy Regulatory Office (ERO). The certificates might both be transferred in direct transactions and be the object of trade on the Polish Power Exchange, which is a reference point for the terms of direct transactions. Electricity producers must attain a minimum level of share of renewable energy from RES: 15% (Ordinance of the Minister of Economy of 18 October 2012) and 14.35% for the second half of 2016—pursuant to the amendment act of 22 July 2016 (Gov. Journ. of 2016, it. 925). The missing 0.65% is dedicated in the form of blue certificates to agricultural biogas. The share for 2017 is set at 15.4%—green certificates and 0.6%—blue certificates (Ordinance of the Minister of Energy). In 2018 the share is 17.5% for all renewable sources except for agricultural biogas and 0.5% for agriculture biogas. In 2019 it is 18.5% and 0.5%, respectively (Ordinance of the Minister of Energy). If they don't, they must fulfill the obligation by either making up the difference by purchasing green certificates on the market, or pay a compensation fee determined by the President of the ERO.

The methodology we used is based on the Impact Assessment Guidelines (EU Commission 2005) and accepted in Poland by the Ministry of Economy in 2006 (Ministerstwo Gospodarki 2006). It assumes assessment of institutions and their policy as well as the analysis of new regulation impact on society as a whole (Kasiewicz and Rogowski 2009). In our analysis we focus on the latter since government officials underline the significant overall cost of renewable energy sector. Referring to the Ministry of Finance guideline (Ministry of Finance 2003) we subdivide the costs into: direct financial costs, administrative barriers and adjustment costs carried exclusively by sector participants. For this reason, the analysis is performed mainly from the point of view of renewable energy sector.

The green certificate system had worked well for some time (Wood and Broom 2016). Due to the development of co-firing (co-combustion of coal and biomass in large coal-fired plants) created an enormous oversupply of green certificates in last several years. It caused the market value of green certificates drop significantly from around 280 PLN/MWh to even 30 PLN/MWh in 2016. The outcome was a reduction of profitability of co-firing plants. The RES Act also cut green certificates for co-firing by a half. This two factors made co-firing unprofitable. Unfortunately the oversupply also caused RES projects of different technologies unprofitable, among them wind on-shore—the second matured and developed renewable energy technology in Poland. The oversupply of green certificates persists. This is because the operating renewable energy plants usually work at high CAPEX and low OPEX and minimize the loss by working at full power. An exception is co-firing, as these installations were usually already fully depreciated.

Act of 20 July 2017 amending the Act on renewable energy sources (Gov. Journ. of 2017, it. 1593) changes the method of calculating the compensation fee set forth in art. 56.1. as 125% of the annual average price of property rights resulting from certificates of origin other than those issued for agricultural biogas electricity, but not more than 300.03 PLN/MWh (71.4 EUR/MWh). The amendment of the act will benefit the entities obliged to surrender property rights resulting from certificates of origin. Trading companies were establishing long-term contracts, mainly with wind energy producers for the sale of property rights, whose lowest guaranteed price was related to the amount of the substitution fee. It is estimated that as a result of the amendment, the revenues of energy producers from RES could fall by a third, which would require restructuring of loans or even lead to bankruptcy of companies.

3 The Auction System

Since the adoption of the 2009/28/EC Directive, several RES bills have appeared, but the legal risk—which in the case of the RES investment projects in Poland is the key risk source of each investment project—remained high. The act that was finally passed on the 20th of February 2015 introduced an auctioning system for the large providers (more than 40 kW). Choosing auction makes the investor ineligible for subvention, since two kinds of public support cannot be used together (or the auction price would be lowered). The laws from "CSR Versus Business Financial Sustainability of Polish Enterprises" of the RES Act, pertaining to the mechanisms and instruments of support for renewable energy sources, were to enter into force on the first of January 2016. On that day, the auctioning system for large providers was supposed to start functioning. Yet the Ministry of Energy has already passed two amendments. In December 2015 (Gov. Journ. of 2015, it. 2365)—postponing the entrance of "CSR Versus Business Financial Sustainability of Polish Enterprises" of the RES Act by 6 months and the most important on the 22th of June 2016 (Gov. Journ. of 2016, it. 925), introducing changes in the auction system in the direction of greater support for the production of energy from biomass and co-firing (as stable energy sources) over wind and PV which generate power intermittently. As a result, the auctioning system started functioning on the 1th of July 2016.

Auctions are divided into groups characterised by the reference to, in particular, the efficiency of an installation (instead of the type of technology). The Amending Act of 22 July 2016 introduces individual groups of RES installations for which the auctions will be held separately. Moreover, for each of the groups there will be separate auctions for installations with capacities below and above 1 MW. The aforementioned groups are as follows (Terlecka and Leśniak 2016):

1. installations where the total installed capacity level, regardless of the source of origin, exceeds 3504 MWh/MW/year;
2. installations using biodegradable waste to generate electricity;
3. installations emitting not more than 100 kg/MWh of CO2, with a total installed capacity level exceeding 3504 MWh/MW/year;
4. auctions for members of an energy cluster;
5. auctions for members of an energy cooperative;
6. installations using exclusively agricultural biogas for electricity generation;
7. other installations.

The first auctions were carried out by the Energy Regulatory Office on December 30, 2016. Auctions were carried out in four baskets:

1. for existing agricultural biogas plants up to 1 MW:7 bids were submitted, the minimum price was 502.23 PLN/MWh (119.6 EUR/MWh) and the maximum was 504.57 PLN/MWh (120.1 EUR/MWh);
2. for existing agricultural biogas plants with a capacity of more than 1 MW (minimum number of bids required i.e. three was not obtained);
3. for new installations so-called "others" with a power of no more than 1 MW: won 84 offers, mainly photovoltaic and several individual wind turbines, the minimum price was 253.5 PLN/MWh (60.4 EUR/MWh) and the maximum was 408.8 PLN/MWh (97.3 EUR/MWh);
4. for existing installations with a capacity of no more than 1 MW, meeting the criterion of the installed capacity of electric power above 3504 MWh/MW/year and of emissivity of no more than 100 kg/MWh, i.e. practically exclusively for hydro power plants: the minimum price was 30 PLN/MWh (7.1 EUR/MWh)—probably it was mistake of the bidder, and the maximum was 468 PLN/MWh (111.4 EUR/MWh).

Subsequent auctions were held on 29 and 30 June 2017. The first was targeted for new RES installations up to 1 MW. 472 bids were submitted in the auction, of which 352 were won (mostly PV projects). The second auction was targeted for existing hydro power plants up to 1 MW. 44 listings were submitted for this auction.

So most of the auctions were targeted for existing biogas and hydro power installations, giving them an opportunity to migrate from the certificate system to the auction system. Further auctions are scheduled on September 28, October 2, October 4 and October 6, 2017. Most of the auctions are again scheduled for existing biogas, biomass and hydro power installations as another opportunity to migrate from a certificate to auction system. It is not sure if there will be organized further auctions for new wind and solar installations in 2017.

4 Reference Prices

The Minister of Energy issues secondary legislation on reference prices for each RES technology and allocating total volumes and values of electricity that may be purchased in given year in each basket. The reference prices in force in 2017 are determined by the Regulation of the Minister of Energy of 16 March 2017 on the reference price of electricity from renewable energy sources in 2017 (Gov. Journ. of 2017, it. 634) for new renewable energy sources. They are shown in Table 1.

The fixed price for electricity agreed within the auction is guaranteed for 15 years regardless of market prices, save for yearly indexation of such price with annual average CPI. The fixed volume of purchased electricity agreed within the auction for each year will be verified after each 3-year settlement period. There is set the penalty for generation shortfall if generation from RES installation falls in any 3-year settlement period below 85% of the volume offered within the auction and accruing to such period. The amount of penalty is calculated as half the fixed auction price of the volume of undelivered energy.

Commissioning of the new RES installation covered by the auction should occur not later than within 48 months following the auction, such commitment to be secured with bank guarantee or money deposit in the amount of 30 PLN/kW (7.1 EUR/kW) of the installed capacity.

5 Major Shift in Support of Technology

Under the amendment act of 22 June 2016 subsidies for co-firing are re-instated as the reduction in support under the RES Act shall not apply to "dedicated biomass co-firing installations" where the share of electricity or heat produced from biomass, bio liquids, biogas or agricultural biogas is higher than 20% of the total amount of electricity or heat produced.

Table 1 Reference prices [EUR/MWh] of electricity from RES new installations in 2017

Installation	Reference price [EUR/MWh]
Wind on-shore < 1 MWe	320
Biomass [. . .] dedicated combustion installations	325
Wind on-shore >1 MWe	350
Other biogas	355
Biogas from sewage treatment plant	365
Waste incineration plants	385
Biogas from landfills	405
Hybrid systems >1 MWe	405
Biomass dedicated combustion installations or biomass hybrid systems ≤ 50 MWe	415
Photovoltaics >1 MWe	425
High-efficiency CHP dedicated biomass combustion installations or high-efficiency CHP biomass hybrid systems >50 MWe	435
High-efficiency CHP dedicated biomass combustion installations or high-efficiency CHP biomass hybrid systems ≤ 50 MWe	450
Photovoltaics <1 MWe	450
Hybrid systems <1 MWe	455
Geothermal installations	455
Wind off-shore	470
Bioliquids	475
Hydropower <1 MWe	480
Hydropower >1 MWe	480
Agriculture biogas <1 MWe	550
Agriculture biogas >1 MWe	550

Source: Based on [Ordinance of the Minister of Energy]

The Act on Windfarms of 20 May 2016 (Gov. Journ. of 2016, it. 961) (called also an anti-windfarms act) sets of new demands for wind power specifically aimed at restricting it's development (also significantly impedes or phase out the country's existing wind farms). The act sets out the minimum distance required between a wind power plant and residential buildings and forests. The distance is set at 10 times the height of a wind power plant (approximately 2 km) to prevent ice falls from rotor blades, which rules out 99% of land (Terlecka and Leśniak 2016). In addition, the new definition of a wind power plant will lead to an increase in the real estate tax imposed on the owners of the plants. Wind farm operators need to obtain an operation permit every 2 years from the Technical Supervision Office (TSO) with the cost capped at 1% value of a wind turbine. Operators also need to obtain an approval for any repair or modernization of technical fixtures of a wind turbine. This significantly increase operation costs and together with the drop in green certificate price make existing wind farms unprofitable, so massive bankruptcies are expected.

The government justifies the law with preference for stable sources of power and the dangers which would constitute windfarms eg. the impacts on human health or the imminent threat of ice falls from rotor blades and interference in the landscape. Wind power organisations suggest the government is instead seeking to protect lossmaking coal mines run by state.

The law significantly promotes agricultural biogas sector. The act of 22 June 2016 establishes separate certificates of origin (called blue certificates) awarded to agricultural biogas power plants with separate quotas for such certificates. If the price of blue certificates quoted at the Polish Power Exchange is lower than the substitute fee (set at 300.03 PLN/MWh which is 71.4 EUR/MWh) for the period of more than 1 month, there will be limited possibility to pay substitute fee as an alternative method of the fulfilment of the obligation to obtain and redeem certificates of origin.

The Amending Act of 22 June 2016 defines large installations as installations with a capacity equal to or exceeding 500 kW. According to the Amending Act, a mandatory energy purchase by an entity called an obliged seller is no longer applicable for such installations (except of all biogas plants). In the absence of an obligation to purchase energy from RES, many large installations may suffer from a reduction in their revenues and overall profitability.

Some of the aspects of the new RES regime are still to be determined by way of secondary legislation (including reference prices which should be published before 31 August each year). However, it appears that the Polish government intends to reduce support for the RES industry and will most likely treat biomass and biogas installations preferentially as opposed to windfarms and PV installations.

6 Discussion and Conclusions

Managers interested in launching eco-efficiency investments are aware of substantial risks present in renewable energy sector. The main group of risk factors discussed in this chapter are related to the revenue stream. These risk factors can be subdivided into two groups with respect to price and volume component.

Regarding the price, the administrative decision to set the price of the energy will depend on several factors. As a result of our analysis we suppose energy production costs (transaction costs) and prices might increase in the auctioning system, and this risk might grow over time, for two reasons. First, the auctioning system might limit the market of active support system participants in terms of technology and field even more (investors who are able to bear the risk and meet the criteria for access to auctions, including bank guarantees (7.1 EUR per each kW of capacity), connection allowances, and construction permits, and move through the unclear decision-making processes concerning the stability criterion); second, as the year 2020 approaches, the RES energy deficits in relation to the targets might be increasingly more visible (for instance, due to delays in the enforcement of the RES act and putting investments on hold, as well as the small supply of new projects due to the initial costs and risk). This might result in the necessity to increase the reference prices. In the auction system the prices are indexed with CPI index which is a good solution if the prices will fall in the future. If, however, the prices will increase in the future the opportunity cost could be substantial.

The other group of risk factors impacts the sales volume. Here the risks are tied to the auction mechanism and operational risks. The auction characteristic (technological preferences, production volume, announcement date etc.) is in the hand of government bodies and plan for auctions in subsequent years is not scheduled. Furthermore, to fix the sales volume the investor needs to win an auction and carry operational risks, which fall beyond the scope of our analysis (see i.e. ALTRAN, Arthur D. Little 2011; EIU 2011; Fraunhofer ISI 2016).

Risk affects not only revenues, since proposed legislation do not take into account the microeconomic considerations and criteria for making investment decisions, including the administrative costs (more broadly, the transaction costs).

The next step in the analysis from the microeconomic perspective is to design the valuation models which enables to grasp all important risks. The model should capture the revenue uncertainty as a value driver and allows for risk pricing. To price the risk in contingent valuation framework, i.e. real options approach to find the optimal executions of renewable energy projects.

Support system aims to guarantee the harmonious development of individual RES technologies. The existing support system, based on green certificates, has led to dominance in the RES market of co-firing technology and additionally wind energy. G. Wiśniewski estimates (on the basis of ERO data) that these two technologies accounted for almost 70% of the green energy market, and that the highest support in the form of green certificates in the amount of 9.8 billion PLN (2.33 billion EUR) between 2005 and 2015 was generated by co-firing biomass with coal and wind energy in the amount of 7.8 billion PLN (1.86 billion EUR) (Wiśniewski 2016). System shortages did not allow for the development of other RES technologies and the creation of a competitive RES market with a large group of investors, also diversified in terms of investment scale.

It is too early to evaluate of the auction system introduced recently. It seems, however, that the auction system in its present form is quite discretionary and centrally controlled rather than marketed. The Ministry of Energy, controlling the auction parameters, can decide which of the RES technologies will survive and develop and which will disappear from the market. The impact of the administration on the auction system can take place not only at the legislative stage (setting up the auction baskets) and at the executive level (the ERO president decision to announce or not announce the auction in selected baskets), but also at the stage of shaping reference prices. If the price is too low for LCOE (levelized cost of electricity), technology is virtually excluded from the system (even if there is a volume of energy in the basket). If the price is too high, the volume is large enough and the number of participants in the auction is small, there is a real risk of redundant support. The Institute for Renewable Energy (IRE) criticizes the reference price for a dedicated co-firing installations that is too high (325 PLN/MWh (77.4 EUR/MWh) in 2017) compared to the LCOE cost. According to IRE LCOE for co-firing is 240 PLN/MWh (57.1 EUR/MWh) (IRE 2016).

There are many doubts related, i.e. with the conditions of competitiveness and public aid to the auction system created by the Act on RES. The basic question concerns the shape of the energy mix and the market of RES that will emerge from the Polish power system.

The new auction system might restrict the access to new environmental solutions to limited number of big energy producers. The eco-efficiency solutions for small- and medium enterprises might fail to gain a substantial government support. The new means of government support can reduce the growth not only for RES but it might force to plot a different scenario for CSR-related, environmental actions within other sectors of economy. The increased risk level in the RES sector impacts the risk of different sectors—companies integrated in the supply chain, customers, banks and other financial intermediaries. Sustainable development demands for the reduction of business risk which supports positive relation between an enterprise and its stakeholders.

References

ALTRAN, Arthur D. Little. (2011). *Risk quantification and risk management in renewable energy projects*. International Energy Agency – Renewable Energy Technology Deployment. Retrieved from iea-retd.org

Barnett, M. L. (2007). Stakeholder influence capacity and the variability of financial returns to corporate social responsibility. *Academy of Management Review, 32*(3), 794–816.

CFA Institute. (2017). *Environmental, social and governance (ESG) survey*. Retrieved from https://www.cfainstitute.org/ESG

Criscuolo, C., & Menon, C. (2015). Environmental policies and risk finance in the green sector: Cross-country evidence. *Energy Policy, 83*, 38–56.

Daszyńska-Żygadło, K., & Ligus, M. (2017). Impact of risk factors on market value of companies from Polish renewable energy sector in changing regulatory environment. In S. Kapunek, & V. Krutilova (Eds.), *20th Annual International Conference "Enterprise and Competitive Environment"*, Conference proceedings (pp. 195–205). Brno. Available via DIALOG. Retrieved from https://ece.pefka.mendelu.cz/sites/default/files/imce/ECE2017_fin.pdf

Daszyńska-Żygadło, K., Słoński, T., & Zawadzki, B. (2016). The market value of CSR performance across sectors. *Inzinerine Ekonomika-Engineering Economics, 27*(2), 230–238.

Derwall, J., Guenster, N., Bauer, R., et al. (2005). The eco-efficiency premium puzzle. *Financial Analysts Journal, 61*(2), 51–63.

EIU. (2011). *Managing the risk in renewable energy*. The Economist Intelligent Unit. Available via DIALOG. Retrieved from http://digitalresearch.eiu.com/risksandrenewables/report

European Commission. (2001). *Green Paper: Promoting a European framework for corporate social responsibility*. Luxembourg: Office for Official Publications of the European Communities.

European Commission. (2005). Impact Assessment Guidelines SEC 791, Brussels.

Financial Times Lexicon. Retrieved from lexicon.ft.com

Fraunhofer ISI. (2016). The impact of risks in renewable energy investments and the role of smart policies. Final report WP3, DIACORE.

IRE President. (2016). *Energy auctions will support co-firing of biomass with coal*. This can lead to loss of energy independence (interview). Available via DIALOG. Retrieved from http://www.biznes.newseria.pl/news/prezes_ieo_aukcje_energii,p945083350

Kasiewicz, S., & Rogowski, W. (2009). *Hybrid investments – Towards new assessment of effectiveness* [in Polish]. Warsaw: WSE.
Kayser, D. (2016). Solar photovoltaic projects in China: High investment risks and the need for institutional response. *Applied Energy, 174*, 144–152.
Loper, T., & Crittenden, V. L. (2017). Energy security: Shaping the consumer decision making process in emerging economies. *Organization and Markets in Emerging Economies, 8*(1), 8–32.
McWilliams, A., & Siegel, D. (2001). Corporate social responsibility: Theory of the firm perspective. *Academy of Management Review, 26*(1), 117–127.
OECD. (2011). *Guidelines for multinational corporation.* Retrieved from http://mneguidelines.oecd.org/guidelines
Padgett, R. C., & Galan, J. I. (2010). The effect of R&D intensity on corporate social responsibility. *Journal of Business Ethics, 93*, 407–418., Available via DIALOG. https://doi.org/10.1007/s10551-009-0230-x.
PwC. (2015). *Bridging the gap.* Retrieved from https://www.pwc.com/gx/en/services/sustainability/publications/responsible-investment.html
Semenova, N., & Hassel, L. (2008). *Industry risk moderates the relation between environmental and financial performance.* Sustainable Investment Research Platform (Working Papers 08–02). Available via DIALOG. Retrieved from http://web.abo.fi/fak/esf/fei/redovisa/kurser/Lasse/Mistra%20wp/sirp%20wp 2008-02%20Semenova%20and%20Hassel.pdf
Terlecka, K., & Leśniak, K. (2016, July 4). Act on windfarms and amendments to the act on renewable energy sources. Mondaq Business Briefing.
The Ministry of Economy. (2006, October). Guidelines to Regulatory Impact Assassment [in Polish] Warsaw.
The Ministry of Finance. (2003). Administration barriers – The analysis of the problem [in Polish], Haga.
ThomsonReuters Asset4 Database. (2017). Retrieved from https://financial.thomsonreuters.com/en/products/data-analytics/company-data/esg-research-data.html
Wiśniewski, G. (2016). Unclear role of co-firing biomass with coal. *Clean Energy, 12*(185), 22–28.
Wood, I., & Broom, R. (2016, August 26). *Reforming Poland's renewable industry, renewable energy focus.* Retrieved from http://www.renewableenergyfocus.com/view/44664/reforming-poland-s-renewable-industry/

Legal Acts

Ordinance of the Minister of Economy of 18 October 2012 regarding detailed obligations to obtain and redeem certificates of origin, substitution fee payment, purchase of electricity and heat 2012, it. 1229.
Ordinance of the Minister of Energy of 11 August 2017 on the change in the volume of the share of the total amount of electricity resulting from the redeemed certificates of origin confirming the generation of electricity from renewable energy sources in 2018-2019. Gov. Journal of 2017, it. 1559.
Ordinance of the Minister of Energy of 16 March 2017 on the reference price of electricity from renewable energy sources in 2017 and the periods of obligatory generators who won the auctions in 2017. Gov. Journal of 2017, it. 634.
Ordinance of the Minister of Energy of 17 October 2016 on the change in the volume of the share of the total amount of electricity resulting from the redeemed certificates of origin confirming the generation of electricity from renewable energy sources in 2017. Gov. Journal of 2016, it. 1753.
The Act of 10 April 1997 on energy law. Gov. Journal of 2006 no. 89 it. 625 and no. 104 it. 708.
The Act of 20 February 2015 on renewable energy sources. Gov. Journal of 2015 it. 478.

The Act of 20 July 2017 amending the Act on renewable energy sources. Gov. Journal of 2017, it. 1593.
The Act of 20 May 2016 on investments in wind power plants. Gov. Journal of 2016, it. 961.
The Act of 22 June 2016 amending the Act on renewable energy sources and certain other acts. Gov. Journal of 2016, it. 925.
The Act of 29 December 2015 amending the Act on renewable energy sources and the Act on energy law. Gov. Journal of 2015, it. 2365.

The Role of CSR in the Transition to a Green Economy

Bożena Ryszawska

1 Introduction

The green economy concept has emerged in recent years in the mainstream public debate. The reason for such a widespread interest in green economy has been the multidimensional economic, ecological and social crisis, as well as initiatives launched by global and regional organizations seeking recovery from the crisis. The crisis disclosed in 2008 has been referred to as the most severe one since the 1930s. The diagnosis of the situation, which was made after the crisis had finished, revealed an imbalance which had been increasing for a number of years on the financial market and, in broader terms, in the economy (Jessop 2011), as well as pressure on the natural environment, which puts the quality of people's life at risk (Jackson 2009; Rockström et al. 2009), and, in the social dimension, global inequalities with regard to income distribution and consumption level (Dowbor 2012). Describing imbalances in so many dimensions made politicians and decision-makers realize systemic relations between the contemporary problems. The analysis of the causes and effects of the global financial crisis showed that the features of the current economic system, which is based on economic growth (measured by the GDP), generate imbalance. An important point in the international debate was putting forward programs for recovery from crisis and new development strategies. It was agreed that economic, ecological and social balance would be restored by making use of the opportunities offered by the concept of sustainable development that leads to sustainability transition (Geels 2013). In this paper, attention is given to the present transformation (2008–2017) of the economic and social model towards a

B. Ryszawska (✉)
Katedra Finansów Przedsiębiorstw i Finansów Publicznych, Uniwersytet Ekonomiczny we Wrocławiu, Wrocław, Poland
e-mail: bozena.ryszawska@ue.wroc.pl

A. Długopolska-Mikonowicz et al. (eds.), *Corporate Social Responsibility in Poland*, CSR, Sustainability, Ethics & Governance,
https://doi.org/10.1007/978-3-030-00440-8_8

model that is more sustainable (see more in the article of Rok in this book) in terms of the environment and socially equitable which is reflected in transition to green economy. The purpose of this study is to emphasise the role of corporate social responsibility (CSR) in the transition to green economy. The main thesis is: current economic model and traditional approach to CSR have failed. The present scope and methods used in the implementation of corporate social responsibility in business are insufficient according to radical green economy transition. The green economy needs new business models and a new concept of CSR called radical CSR or CSR 2.0. This paper is based on broad desk studies, systemic multilevel approach and actor analysis. The analysis confirmed that radical CSR supports the radical transition to green economy.

2 The Green Economy Concept as a Research Issue

2.1 Definitions of Green Economy

The term "green economy" first appeared in 1989, in the report *Blueprint for a Green Economy* for the British government. It was prepared by three famous economists: D. Pearce, A. Markandya and E.B. Barbier. The report was supposed to support the British government in implementing the idea of sustainable development (Pearce et al. 1989).

The present process of defining and researching green economy began in 2008 with establishing the Green Economy Initiative by UNEP (United Nations Environmental Programme). The first measure taken within the Initiative was Barbier's report titled *Global Green New Deal* (Barbier 2010). The next step was taken after the climate conference in Copenhagen in 2009 when UNEP put forward the concept of green economy as a method to address the most important environmental, social and economic challenges and as an opportunity for a development that would be favorable to all the states. UNEP started preparations for the international conference on sustainable development which was supposed to take place in 2012, again in Rio de Janeiro (the so-called RIO + 20). One of the flagship reports produced in partnership with research centers and business entities as part of the preparations was the Green Economy Report (UNEP 2011). Simultaneously, other reports and studies were being prepared by international organisations, associations and partnerships, based either on the green economy concept or the green growth idea, which is an object of interest in OECD studies. Green economy is presented in this chapter from two perspectives:

- as a political concept defined in strategic documents provided by international organisations, which is a priority and a guideline for economic, agricultural, transport, fiscal, financial and even social policy in a number of states worldwide;
- as a growing sector of the real economy (production of renewable energy sources, organic agriculture, sustainable transport, organic food and environmentally friendly services).

There are numerous perceptions of green economy. It is predominantly seen as:

- a direction for transforming the economy and the entire economy–society–environment system (ethical, responsible, based on respect for the planet and the human being) (Jackson 2009),
- a process of civilisation changes, which is necessary due to exceeding the planet's limits (Zovanyi 2013; Rockström et al. 2009),
- the key element and objective of strategies and programmes for recovering from crisis, the driving force of the economy that creates jobs (Frérot 2011; UNEP 2011),
- a model desirable in the sustainable development process, which combines economic, environmental and social aspects (Kosoy et al. 2012),
- an alternative to the brown economy (UNEP 2011),
- the process and the way to achieve the goal (greening the economy, transformation towards green economy) (Bigg 2011; EEA 2010),
- a specific objective of the current environmental policy and ecological modernisation (Jänicke 2011),
- a real, growing sector of the economy that reflects structural changes in the economy (Ocampo 2011)

The common element in the way of thinking presented by the major entities of the global economy and politics is the belief that green development, green growth, green economy (Table 1) may become the driving force recovering the economy from the post-crisis stagnation, creating new jobs and an added value in the economy, attracting direct and indirect investments, creating innovations and technologies which reduce the environmental pressure. Simultaneously, they may function in synergy and thus contribute to solving numerous environmental and social problems (inequalities and poverty) which determine the quality of human life.

The process of transformation that can be witnessed at present is—in its essence—a manifestation of structural changes. Research results and statistical data confirm the growing impact of the green economy upon the GDP and employment rates, thus elevating its significance as a source of competitive advantage, a mark of sectoral development and a potent attractor for public and private investors. This is a new and very distinct trend (Barbier 2016; Wyszkowska 2016). Observation and interpretation of structural changes associated with such concepts as the green economy, green growth, green entrepreneurship and green employment are an important and current topic for in-depth studies. This is particularly evident in those areas where science meets politics. On the one hand, politicians voice their demand for broad and accurate analyses of the complex system of economy-society-environment. On the other hand, the scientific community requires political support and political experiences to facilitate the implementation of practical solutions and remedies.

Analysis of numerous definitions of green economy (Table 1) distinguished the most important keywords used in the definitions. Clusters of words create the most important areas: Society (inclusion, inequalities, poverty, wellbeing, social integration)

Table 1 Selected definitions of green economy

The introducing entity	Characteristics and definitions of green economy
UNEP 2011	Green economy is one that has a positive influence on people's Well-being and social equity while reducing environmental risk and consumption of natural resources
ICC 2012	Green economy is the economy where the economic growth is connected with ecological responsibility, which strengthens each other in the process of supporting the social progress
European Environmental Agency 2013	Green economy is the economy where environmental, economic and social policies and innovations support societies in the effective use of resources and while at the same time improving human Well-being, accentuating social integration and protection of natural systems which sustain life on the earth
European Commission 2011	Green economy is a low-carbon and a resource-efficient economy which provides growth creates jobs and eliminates poverty by investing in and protecting the natural capital, on which the survival of out planets depends on the long-term
AASA 2011	Green growth means growth that is smart, innovative, cooperative, inclusive and low-carbon
European Union	The economy which is resource-efficient, low-carbon and competitive, one which protects the environment and strengthens social and territorial cohesion

Source: Own worked based on (AASA 2011; European Commission 2011; ICC 2012; UNEP 2011).

Economy (competitiveness, growth, jobs), Environment (pollution and waste, natural capital, consumption of natural resources).

2.2 *Green Economy: The New Model*

There is a growing recognition that achieving sustainability rests almost entirely on getting the economy right. It also emphasizes the crucial point that economic growth and environmental stewardship can be complementary strategies. The concept of green economy should be seen as consistent with the broader and older concept of sustainable development, which has been mainstreamed into the work of the United Nations. The specificity of the broader concept is associated with its both its *holistic* character, as it encompasses the three pillars of development—economic, social and environmental- and its particular focus on *inter*-generational equity (Ocampo 2011).

As has already been noted, what is essential to the concepts of green economy is the understanding that the benefits of environmental sustainability outweigh the costs of investing in and protecting the ecosystems, so that it is possible to have a win-win or "double dividend" strategy of growth with environmental sustainability, and even win-win-win or "triple dividend" strategy that also includes poverty eradication and broader improvements in social equity (Ocampo 2011). This is interesting from the point of view of policy makers because they had to deal with many dimensions of the crisis and the idea of sustainable development is offering the path to transform economy but also to support citizens and environment (See more in the article of Magala in this book).

There is a growing debate among scholars about the importance of green economy concept for future development policy. Below author presents few of them: Mol, Sonnenfeld, Spaargaren say: "*The concept of a green economy is crucial for guiding policies for sustainable development since it goes to the heart of the matter: organize economies in ways that fit local and global ecological prerequisites and long-term dynamics.*" (Fulai et al. 2011). Pirgmaier mentions: "*A 'green economy' concept is an essential step towards the transition to a resilient and sustainable economy. However, it does not go far enough as long as it does not address the dilemma of economic growth: every expansion of economic activity implies more stress on the planet's ecosystems*" (Fulai et al. 2011).

In the opinion of Downs, "*The 'green economy' concept is useful to the extent that it engages policymakers, economists, and businesses in critical dialogue with other stakeholders to compare alternative pathways for development. The comparison should then consider economic criteria alongside social, political, cultural and ecological criteria, and how "business-as-usual" development and more sustainable alternatives trade-off*" (Fulai et al. 2011). By recognizing that the economy is a subsystem of human society which is itself a subsystem of the environment, the green economy concept helps to prioritize sustainable development and resource management on decision makers agendas, influencing their thinking and giving them a mandate to ensure that our economic activities do not exceed environmental tipping points sad Polzin and Kostka (Fulai et al. 2011).

2.3 The Enterprises and Other Actors of Transition to a Green Economy

The transition we discussed in this paper is a conglomerate of processes occurring in parallel in the spheres of economy, technology, politics, and civil societies. The above processes stimulate the involvement of many actors—those can be segregated into the following groups: state authorities (governments), decision makers, and local self-governmental bodies; enterprises; consumers and non-governmental organisations (civil society).

Each of the above groups plays their specific role in the transition processes, using their specific sets of resources and other instruments, as adequate for the task at hand (Røpke 2013).

It seems that state authorities (governments) play the most decisive role in the processes at hand, at least in the initiation of changes. After all, the political decision-makers, in cooperation with scientific communities, were the ones to initiate the process of diagnosing the present crisis situation. Governments and international organisations are the obvious proponents of any strategies, policies and remedial measures designed to address the challenges associated with the systemic crises. The state and other public bodies take up a new role in the task of addressing the complex tangle of economic, social and environmental problems: they are expected to formulate strategic objectives and institutional frameworks in support of the transition and serve as intermediaries for cooperation between governments, business entities, and societies (see more in the article of Kacprzak in this book).

The enterprise sector, in turn, is involved in the realization of the concept of social and environmental responsibility, designed to reduce the negative impact of the sector upon the natural environment and the society at large. These actors are involved in the introduction of new, clean technologies and Eco-Management and Audit Schemes (EMAS) in support of environmental innovations. Recent years have brought an increased interest in the concept of *Corporate Sustainability* to broaden the scope of the existing Corporate Social Responsibility and to better address the challenges of the green economy transition, as discussed herein.

The last group of actors—consumers and non-governmental organizations—is involved in the task of changing the existing cognitive and behavioral patterns through education, information campaigns, and grassroots pressure upon decision makers. The growing awareness of the consumer base will result in increased demand for environmental goods and services, to the effect of stimulating greater involvement of business entities in the task of changing the existing production and supply of these products.

3 The Failure of the Brown Economy and CSR 1.0

3.1 From Brown to Green Economy

The brown economy is based on unlimited economic growth and represents business as usual which is criticized in public debate after the crisis in 2008. It is focused on fossil fuels, high consumption of natural resources, CO_2 emissions, reduction of biodiversity, overconsumption, finally creates global social inequalities. The crisis showed lack of responsibility and transparency of financial institutions, banks, and global corporations. As a result, lack of social trust appeared. The transition is a term used to describe conversion (evolution) from the existing model of the brown economy towards one based on increased social and environmental responsibility. The new model (green economy) should emphasize such elements as: low emission, resource efficiency, clean technologies, responsible consumption, social justice and equality (both inter- and intra-generational). Transition to green, efficient, low carbon economy (called *sustainability transition*) in conventional reception is associated with the process of multi-level transformation of the economic and socio-technological system intended to increase its environmental sustainability and social fairness.

The transition to a green economy has reached an unprecedented scale, with the active involvement of a broad spectrum of stakeholders (business actors, the government, the consumers. One of its manifestations is the sizeable contribution on the part of international organisations and individual countries—their activities, undertaken in response to global crisis, come as part of much broader strategies of development and new policies, strongly rooted in the concepts of sustainable growth, green economy and clean production as the most effective remedies with potential to revitalise and sanitise the economy. Business entities gradually depart from the *business as usual* model of operation, looking for new business models (and thus introducing the idea of *corporate sustainability*). The above process can also be identified among consumers, as attested by the growing interest in such ideas as sustainable consumption and collaborative consumption (See article of Karwacka and article of Szumniak-Samolej).

To sum up, it may be observed that some of the processes which, up till recently, were only registered at the niche segments and peripheral areas of the economy, are gradually seeping to the mainstream, turning the brown economy of overabundance into a green economy of moderation and restraint (Table 2). Some of the manifestations of this process include such phenomena as: the growing share of energy from renewable sources in total energy production, waste recycling, reduced greenhouse gas emission, of modern products and technologies with improved energy efficiency, sustainable transportation, sustainable supply chains, sustainable consumption, collaborative consumption, circular economy, corporate social and environmental responsibility, corporate sustainability.

In any document related to the development framework for Poland there is no direct indication that the main goal is to create a green economy. However as a member of EU Poland is implementing targets of strategy *Europe 2020*. The central part of the strategy is sustainable growth which means low carbon, resource efficient, green economy. The government accepted National Plan of Low-emission Economy in 2015 focused on development of clean technologies, renewable energy sources, improving energy and material efficiency, changing the consumption model and production for a more balanced, integrated product policy, green public procurement, green jobs, or ecological fiscal reform.

The newest national document constituting the response to Agenda 2030 (United Nation Sustainable Development Goals) is the Strategy for Responsible

Table 2 Transition from brown to green economy (author's own elaboration)

Brown economy	Green economy
Unlimited growth	Decoupling economic growth from natural resources consumption
Based on fossil fuels	Renewable energy
High consumption of natural resources	Energy and resource efficiency
CO_2 emission	Clean production
Reduction of biodiversity	Biodiversity
Global social inequalities	Sustainable consumption
Over consumption	Intergenerational and interregional justice
Business as usual	Socially responsible firms and investors
Lack of responsibility	Social trust, partnership, transparency
Lack of social trust	

Development (Strategy for Responsible Development 2017). The strategy presents a new development model—responsible development, i.e. one that builds competitive strength using new development factors, ensures participation and benefits for all social groups inhabiting various places in the country. At the same time, in the new model, the needs of the present generation will not jeopardize the opportunities of future generations. The impact of the business activity for the environment, the stability of ecosystems, biodiversity and climate change create a space for green economy development.

Actions in the field of green economy are also determined by the Partnership Agreement with the EU for the years 2014–2020 regarding European funds. The main priorities of this agreement concern financing low-emission economy, energy efficiency, development of renewable energy as well as mitigation and adaptation to climate change. According to this financial framework more than 1100 small municipalities created Plans of Low Carbon Economy and are making steps towards green economy.

3.2 CSR 1.0

The issue of corporate social responsibility (CSR) is now widely discussed in the literature. CSR is defined in a narrow and a broad way. According to the first understanding CSR takes into account social, environmental and ethical risks in order to protect and increase the value of the company to the shareholders (Davis and Blomstom 1975; Certo and Certo 2006). The author of the wider definition is Carroll (1979). According to him, *the corporate social responsibility covers the economic, legal and philanthropic society's expectations to the company*. Particularly useful for the paper are the definitions of the CSR based on the concept of stakeholders. Every company coexists in the local environment with other participants of social life and is facing some expectation especially after the last financial crisis. B. Rok writes: "Facing challenges of the severe recent economic crisis leaders of largest companies, as well as representatives of leading research centers, search ways of further development which includes the expectations of the various groups of stakeholders in the process of creating value in the rapidly changing environment" (Rok 2010).

To understand the current issues around CSR we will focus on the failure of CSR after the financial crisis. Existing concept of the CSR is criticized in the context of the global financial crisis (Fig. 1) W. Visser writes: *CSR, as a business, governance and ethics system, has failed. This assumes that success or failure is measured in terms of the net impact (positive or negative) of business on society and the environment* (Visser 2012).

Most companies listed in the rankings of CSR have played a leading role in extremely unethical actions by creating a risky financial instruments and selling them to clients in an aggressive way, common practice was also the privatization of profits and shifting cost of their activities to the public sector (eg. negative

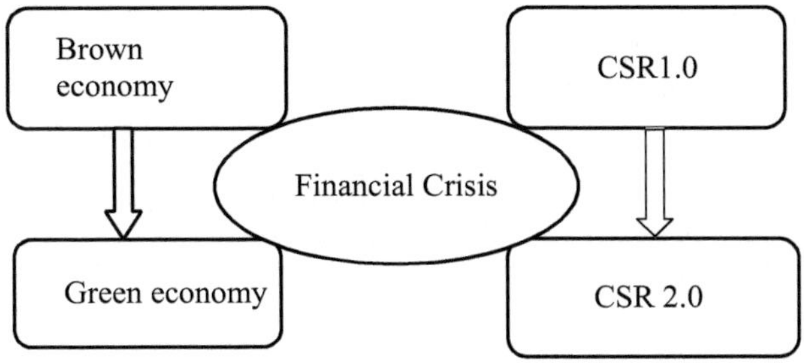

Fig. 1 Transition from brown to green economy and from CSR 1.0 to CSR 2.0 after financial crisis 2008. (Author's own collaboration)

externalities). The failure of the idea of corporate social responsibility in companies have several reasons: CSR has remained largely restricted to the largest companies, and mostly confined to PR, or other departments, rather than being integrated across the business (see article of Witek-Crabb and article of Dancewicz); CSR has adopted the quality management model, which results in incremental improvements that do not match the scale and urgency of the problems CSR does not always make economic sense, as the short-term markets still reward companies that externalize their costs to society.

4 The Future of CSR 2.0 and Green Economy

W. Visser proposes a new concept of the responsibility of companies, which is called CSR 2.0 or Corporate Sustainability and Responsibility. This new approach acknowledges that 'sustainability' (with roots in the environmental movement) and 'responsibility' (with roots in the social activist movement) are really the two main games in town (Visser 2012). This is the adequate approach in the "Age of Responsibility", and is searching for a new, sustainable and responsible development. CSR 2.0 means moving away from philanthropic activities, sponsorship of events, integration meetings and build the image of the company, and the move to create a strategy for limiting the negative impact on the environment and the social environment, to invest in responsible projects, for example. "clean" technologies and energy-saving products (Table 3).

Visser believes that CSR 2.0—or Systemic CSR (sometimes calls Radical CSR or Holistic CSR)—represents a new model of CSR (see Table 4).

According to Visser the essence of the CSR 2.0 DNA model is the four DNA Responsibility Bases: Value creation, Good governance, Societal contribution and environmental integrity. This approach harmonizes with priorities and values of the green economy which represents a new paradigm of development. Implementation of CSR 2.0 leads to designing and adopting an inherently sustainable and responsible business model. The process is supported by a reformed financial and economic system called green economy (Table 5). EU countries also Poland take actions towards low carbon, resource, and energy efficient economy. Polish government introduced National Program of Low Carbon Economy and Strategic Plan of

Table 3 CSR 1.0 to CSR 2.0—meta-level ontological shifts (Visser 2012)

CSR 1.0	CSR 2.0
Philanthropic	Collaborative
Risk-based	Reward-based
Image-driven	Performance-driven
Specialized	Integrated
Standardized	Diversified
Marginal	Scalable
Western	Global

Table 4 CSR 1.0 to CSR 2.0—micro-level methodological shifts (Visser 2015)

CSR 1.0	CSR 2.0
CSR premium	The base of the pyramid
Charity projects	Social enterprise
CSR indexes	CSR ratings
CSR departments	CSR incentives
Product liability	Choice editing
Ethical consumerism	Service agreements
CSR reporting cycles	CSR data stream
Stakeholder group	Social networks
Process standards	Performance standards

Adaptation to Climate Change to 2020 (Szyja 2016). More than 60 bln PLN is dedicated to energy efficiency, renewable energy and low carbon economy till 2020 (NPRGN 2015). The main source of finance for low carbon economy in Poland are European structural funds which supports development of renewable energy, mitigation and adaptation to climate change, protection of environment and biodiversity, reduction of CO_2 emissions. Polish regions and municipalities by investing the European funds into infrastructure, energy efficient buildings. SMEs and agriculture become low carbon, green regions till 2020 (Wyszkowska 2016; Saikku et al. 2015).

The next important aspect which influences future of CSR is the European strategy for the transformation of the smart, efficient, low carbon economy. Strategy emphasizes the role of CSR as an institutional project, which seeks to regulate the business impacts on stakeholders. The European Union has adopted in October 2011, a new strategy in the field of corporate social responsibility (EU strategy 2011–2014 for Corporate Social Responsibility 2011). This document highlights the impact of CSR on six key factors in the competitiveness of businesses: reduce costs, human resource management, customer relationships, innovation, risk management, and financial results. The recommendations of the strategy say that implementation of the principles of CSR requires the collection and disclosure of information on socially and environmentally responsible business activities. Member states implement Directive 2014/95/EU (called CSR directive) of The European Parliament and of The Council. It regards to disclosure of non-financial and diversity information. They must address "policies, risks and results" in relation to "social, environmental and human rights impact, diversity and anti-corruption policies" in their annual reports. The European Union is a world leader in the disclosure of non-financial information.

Table 5 CSR 2.0 supports the transition to green economy (Visser 2015; Ryszawska 2013)

Radical holistic CSR 2.0			Radical green economy
DNA code	Strategic goals	Key indicators	Areas and values
Value creation	Economic development	Capital investment (financial, manufacturing, social, human and natural capital), beneficial products (sustainable and responsible goods and services, inclusive business (wealth distribution, the bottom of the pyramid markets)	New economic model sustainable development, Sustainable production and consumption, Sharing economy Green growth, Wealth distribution
Good governance	Institutional effectiveness	Leadership (strategic commitment to sustainability and responsibility) transparency (sustainability and responsibility reporting, government payments) ethical practices (bribery and corruption prevention, values in business)	Responsibility, transparency, social trust, partnership, long-term perspective
Societal contribution	Stakeholder orientation	Philanthropy (charitable donations, provision of public goods and services) fair labor practices (working conditions, employee rights, health and safety) supply chain integrity (SME empowerment, labor and environmental standards)	Social justice Poverty eradication Stakeholder orientation Intergenerational justice, reduction of inequalities
Environmental integrity	Sustainable ecosystem	Ecosystem protection (biodiversity conservation and ecosystem restoration) renewable resources (tackling climate change, renewable energy and materials) zero waste production (cradle-to-cradle processes, waste elimination)	Low carbon emission, renewable energy, resource efficiency, biodiversity conservation, tackling climate change, zero waste production (cradle-to-cradle)

CSR is moving from charitable actions supporting various social and environmental events through donations and sponsorships, to effective management strategy of the enterprise which contributes to increasing the competitiveness of the company, build a reputation and at the same time shaping favorable conditions for social and economic development (see also article of Rok in this book). This process is supported by ISO standard 26,000 establishing new areas of corporate social responsibility activities: *Organizational governance*—practicing accountability and transparency at all levels of your organization; *Human rights*—treating all individuals with respect; making special efforts to help people from vulnerable groups; *Labor practices*—providing just, safe and healthy conditions for workers; *Environment*—identifying and improving environmental impacts of your operations; *Fair operating practices*—respecting the law; practicing accountability and fairness in your dealings with other businesses; *Consumer issues*—providing healthy and safe products, promoting sustainable consumption; *Community involvement and development*—getting involved in the betterment of the local communities that your organization operates in. ISO 26000 proposes, a broader definition of CSR, which means the responsibility of the organization for the impact of its decisions and activities on society and the environment. The commitments of CSR companies must take into account the expectations of stakeholders, must align with the law and international codes of conduct, and be integrated with the activities of the organization (International Organization for Standardization 2011).

Main trends presented above point out the new significant role of CSR in the ongoing transition to green economy.

5 Conclusion

The transition to a green economy is a complex problem. Strategy Europe 2020 is shaping most of regulation, policies and financial frameworks of EU. Flagship programs of the strategy emphasize smart growth, innovations and knowledge-based economy, also resource efficient, renewable energy, low actions in industry and agriculture. To understand the ongoing transition to green economy better we analyzed green economy concept and CSR 2.0. In this context, CSR 2.0 is a part of the process, together with public regulation and policies also awareness and action taken by consumers and civil society. We can observe growing importance of sustainability issues in business. According to Visser: *sustainability and responsibility can be thought of as different, yet complementary elements of CSR*. Hence, sustainability can be conceived as the destination—the challenges, vision, strategy and goals, i.e. what we are aiming for; while responsibility is more about the journey our solutions, responses, management and actions, i.e. how we get there (Visser 2015).

Poland implemented in 2017 Directive 2014/95/ of EU on non-financial reporting. This tool is able to transform our CSR performance. Large entities, which are public interest entities, exceeding the criterion of average number of 500 employees have to disclose non-financial information at least concerning environmental, social and labor issues, respect for human rights, counteracting corruption and bribery. EU or international standards or guidelines. About 300 firms will mandatory report starting from the fiscal year 2017. The areas of reporting cover core aspects of business activities and gradually depart companies from CSR 1.0 to CSR 2.0.

Polish economy thanks to European strategic priorities and founds is slowly moving from brown to green economy. The process is specially intense in small regions, municipalities, cities (smart cities) but also in companies which discover advantages of green economy in their future development.

More empirically grounded research could explore in more depth how ongoing green economy transition and their actors, specially enterprises implementing CSR

2.0, might influence dynamics of the process. Based upon the analytical framework chosen major challenges of multilevel green economy transition has been recognized. The green economy transition process has reached an unprecedented scale (planetary, global, local) and involved many actors (international organizations, EU, national and local governments, enterprises, civil society, universities) (Barbier 2016). Financial crisis 2008 was a turning point for EU strategy, recovery programs and policies. The crisis made us aware that "business as usual" is not possible anymore.

Business entities gradually depart from the *business as usual* model of operation, looking for new business models (and thus introducing the idea of *corporate sustainability*). Corporation that adopt significant aspect of CSR 2.0 also standards of non-financial reporting (GRI 4 or according to ISO 26000) can align with changes in niches (innovation, new technologies) and landscape level (climate change, low carbon green economy). State authorities play significant role as a rule setter, agent of change and leverage for action taken by business. For example the influence of regulatory policy of EU about corporate responsibility seems be effective.

To conclude I use words written by Mike Townsend, in his newest book The Quiet Revolution: *We are in a new era. The systems, rules and behaviours that led to business success in the twenty-first century are no longer working. Since the onset of the longest and deepest financial crisis in living memory, capitalism is still suffering a crisis of liquidity, reliability and confidence. There is a huge question mark over whether our economic system will allow us to make the necessary transition to a more sustainable world or whether we are locked into a fatal collision course* (Townsend 2016).

Acknowledgment Publication was prepared based on research financed from the scientific grant from Polish National Center of Science no 2011/03/B/HS4/05359 entitled "Wpływ społecznie odpowiedzialnych działań przedsiębiorstwa (CSR) na jego wartość".

References

AASA. (2011). *Towards a sustainable Asia: Green transition and innovation.* Heidelberg: Springer.

Barbier, E. B. (2010). *A global green new deal – Rethinking the economic recovery.* Cambridge: Cambridge University Press.

Barbier, E. B. (2016, November). Building the green economy. *Canadian Public Policy* [Analyse de politiques]. https://doi.org/10.3138/cpp.2015-017.

Bigg, T. (2011). Development governance and the green economy: A matter of life and death? *Review of Policy Research, 28*(5), 459–465.

Carroll, A. (1979). A tree-dimensional conceptual model of corporate social performance. *Academy of Management Review, 4*(4), 497–505.

Certo, S. I., & Certo, S. (2006). *Modern management.* Singapore: Pearson Prentice Hall.

Davis, K. I., & Blomstom, R. (1975). *Business and society: Environment and responsibility.* New York: McGraw-Hill.

Dowbor, L. (2012). Szalony świat wymyka się z rąk, "Gazeta Wyborcza", 11–12 sierpnia 2012.

EEA. (2010). Środowisko Europy 2010 – stan i prognozy. Synteza, EEA, Kopenhaga.
EEA. (2013). *Towards a green economy in Europe*. Copenhagen: European Environmental Agency.
EU strategy. (2011). *2011–2014 for corporate social responsibility*. http://eur-lex.europa.eu/LexUriServ/LexUriServ.do?uri=COM:2011:0681:FIN:en:PDF
Europe 2020. (2010). *A strategy for smart, sustainable and inclusive growth*. European Union. Accessed August 15, 2017, from www.europa.eu
European Commission, A Roadmap for moving to a competitive low carbon economy in 2050, COM (2011) 112 final, Brussels 2011.
Frérot, A. (2011). Unia Europejska a wyzwanie stworzenia zielonej gospodarki, "Kwestie Europejskie" 23 maja, nr 206.
Fulai, S., Flomenhoft, G., Downs, T. J., et al. (2011). Is the concept of a green economy a useful way of framing policy discussions and policy making to promote sustainable development? *Natural Resources Forum, 35*, 63–72.
Geels, F. (2013). The impact of the financial-economic crisis on sustainability transitions: Financial investment, governance and public discourse. *Environmental Innovation and Societal Transition, 6*, 67–95.
ICC, Green Economy Roadmap, International Chamber of Commerce. (2012). http://www.iccwbo.org
International Organization for Standardization, ISO 26000. (2011). https://www.iso.org/files/live/sites/isoorg/files/archive/pdf/en/discovering_iso_26000.pdf
Jackson, T. (2009). *Prosperity without growth: Economics for finite planet*. London: Earthscan.
Jänicke, M. (2011). *Green growth: From a growing eco-industry to a sustainable economy*. Berlin: Freie Universität Berlin.
Jessop, B. (2011). *The global economic crisis, the green new deal, and the no-growth economy*. Conference of UNRISD.
Kosoy, N., Brown, P. G., Bosselmann, K., Duraiappah, A., Mackey, B., Martinez-Alier, J., et al. (2012). Pillars for a flourishing earth: Planetary boundaries, economic growth delusion and green economy. *Current Opinion in Environmental Sustainability, 4*, 74–79.
Narodowy Plan Rozwoju Gospodarki Niskoemisyjnej, Warszawa. (2015). http://www.me.gov.pl/files/upload/24672/NPRGN_konsultacje%20i%20uzgodnienia%20zewn%C4%99trzne.pdf
Ocampo, J. A. (2011). The macroeconomics of the green economy. *The transition to a green economy: Benefits, challenges and risks from a sustainable development perspective*. UNEP. Accessed December10, 2012, from http://www.unep.org/greeneconomy/Portals/88/documents/research
Pearce, D., Markandya, A., & Barbier, E. (1989). *Blueprint for green economy*. London: Routledge.
Rockström, J., et al. (2009). A safe operating space for humanity. *Nature, 461*(24), 472–475.
Rok, B. (2010). Instytucjonalizacja społecznej odpowiedzialności przedsiębiorstw. Perspektywa teoretyczna i strategia badawcza. W. W. Pańków, B. Rok i M. Strumińska-Kutra, Oblicza społecznej odpowiedzialności przedsiębiorstw, red. W. Pańków, B. Rok, M. Strumińska. Warszawa: Scholar.
Røpke I. (2013). Sustainability transitions in the perspective of ecological macroeconomics, The 10th Biennial Conference of the European Society for Ecological Economics, 18–21 June, Lille, France.
Ryszawska, B. (2013). *Zielona gospodarka - teoretyczne podstawy koncepcji i pomiar jej wdrażania w Unii Europejskiej*. Wrocław: Wydawnictwo Uniwersytetu Ekonomicznego we Wrocławiu.
Saikku, L., Antikainen, R., Droste, N. et al. (2015). *Implementing the green economy in a European context: Lessons learned from theories, concepts and case studies*, Helsinki. Accessed September 15, 2017, from https://www.researchgate.net/publication/301349044
Strategy for Responsible Development until 2020 (with a view until 2030), Resolution No. 8 of the Council of Ministers of February 14. (2017). Accessed February 20, 2018, from http://www.monitorpolski.gov.pl/mp/2017/260/1

Szyja, P. (2016). Transition to a low carbon economy at the level of local government research papers of Wrocław University of Economics Nr 437. https://doi.org/10.15611/Pn.2016.437.43.
Townsend, M. (2016). *The quiet revolution: Towards a sustainable economy*. London: Greenleaf Publishing.
UNEP. (2011). *Towards a green economy: Pathways to sustainable development and poverty eradication*. Accessed August 3, 2015, from www.unep.org/greeneconomy
Visser, W. (2012). *The future of CSR: Towards transformative CSR, or CSR 2.0*. Kaleidoscope Futures Paper Series No. 1.
Visser, W. (2015). *Sustainable frontiers: Unlocking change through bussiness, leadership and innovation*. Sheffield: Greenleaf Publishing.
Wyszkowska, D. (Ed.). (2016). Wskaźniki zielonej gospodarki w Polsce, Urząd Statystyczny w Bialymstoku. ISBN 978-83-89643-62-9.
Zovanyi, G. (2013). *The no-growth imperative: Creating sustainable communities under ecological limits to growth*. London: Routledge.

Part IV
Practical Aspects of CSR

CSR Level of Enterprises in Poland: Before and After Transition

Magdalena Rojek-Nowosielska

1 Introduction

The discussion about the issue of corporate social responsibility in Poland implies referring to the situation which Poland faced after the second world war. It had a significant effect not only on the scope and mode of performing community-oriented activities, but also on the ability to launch and regulate such initiatives by enterprises. Due to the fact that after the Second World War Poland found itself in the area of Soviet influence, the state authorities embarked on shaping the reality in line with the socialist ideology, with an enterprise being assigned a critical role in the performance of socially-oriented activities. The emphasis on the social aspects in the economic policy adopted at that time by the Polish state was also reflected in the Constitution of the People's Republic of Poland, which was then in place.

The socioeconomic and political background of post-war Poland had an impact on social unrest, which later led to the systemic transition. The year 1989 is taken as the starting date for the transition (Crane 1991; Giffin and Ellington 1995). That breakthrough had also an essential influence on shaping a new social order, which involved social responsibility of enterprises. Yet, a question arises, as to whether there are differences in the degree of social responsibility between enterprises established before and after transition in Poland. Since providing a precise answer to this question requires a more profound analysis, the question became the research problem addressed in this paper. In order to answer this question, the purpose of the study was set to present and describe the pre-transition situation in Poland and to discuss the level of CSR in the post-transition reality of enterprises. The

M. Rojek-Nowosielska (✉)
Department of Sociology, Wroclaw University of Economics, Wrocław, Poland

Social Policy Department, Wroclaw University of Economics, Wrocław, Poland
e-mail: Magdalena.Rojek-Nowosielska@ue.wroc.pl

A. Długopolska-Mikonowicz et al. (eds.), *Corporate Social Responsibility in Poland*, CSR, Sustainability, Ethics & Governance,
https://doi.org/10.1007/978-3-030-00440-8_9

accomplishment of this purpose requires determining a specific methodology of research, which includes the methods of literature sources analysis and a diagnostic poll with a survey questionnaire as a research tool. A research-specific CSR continuum model was used to determine the level of social responsibility.

2 Pre-transition Situation in Poland

The scope of socially-oriented measures taken by the pre-transition Polish state was shifted on enterprises who, as stakeholders in social policy, were obliged to perform them under the then applicable provisions of law. These activities were part of so-called non-productive functions of an enterprise, i.e. functions which did not result from the basic substance-focused tasks, for whose performance a given workplace was established (e.g. production, administration, services, training etc.) (see also: Laurentiu 2016; Basu et al. 2005; Johnson and Loveman 1995; Robinson 2004).

The focal issues of concern within the activity of an enterprise included among others the policy of rational employment, ensuring safe and healthy working conditions, maintaining company social infrastructure facilities (as kindergartens, holiday resort buildings) or assistance in meeting employee housing needs.

Despite the declared concern about social issues, the policy of the socialist government focused mainly on the development of heavy and defense industries; (as L. Balcerowicz wrote: *this system was dominated by the state sector, characterized by heavy industrial concentration, equipped with distorted prices due to massive price subsidies and controls, geared to import substitution, and deprived of both internal and external competition. (Balcerowicz* 1994*)*); for that reason farming, transport, building and construction, as well as trade and commerce were neglected, which resulted in a deteriorating overall standard of living. Additionally, the elimination of the economic calculation from the enterprise management strategies and the introduction of the planning system, led to a disturbance in the process of setting the equilibrium on the employment, production, investment, supply and wage markets. This situation caused numerous political and socioeconomic tensions, which reached their peak in 1989 (with the effect of system transition) (see also: Crane 1991; Burrell 2011; Keane and Prasad 2002; Blanchard 1994; Rondinelli and Yurkiewicz 1996).

3 Post-transition Reality

The processes initiated by the system transition contributed to many further changes, not only for entrepreneurs. It is worth noting that they were a consequence of top-down reforms which started from changes in the state's social policy. In the initial period of creating the new system, the area of social policy found itself in a systemic vacuum. It resulted from the collapse of the previous system's structure, with no new forces to have yet appeared to create a new system of social policy. The state, following its own declaration, financial difficulties and the ongoing privatization processes, had already withdrawn from the social area (i.e. nurseries, kindergartens, health care, company vocational schools, scholarships, home-building-related funding, workers' holiday) or drastically limited its participation in it. On the other hand, the new social groups, capable of creating new structures and mechanisms to fill this gap under the new rules, had not yet appeared.

Additionally, the social security guarantees, which were previously in place, ceased to apply. They included:

- full employment (which was replaced by a single to double-digit unemployment rate),
- price subsidies in the case of nearly all basic goods and services (see also: Kowalik 2012; Księżopolski 2013; Golinowska 2013; Sterhenz 1993).

Furthermore, many social services, previously available without any additional payments, were now subject to fees and charges, either newly imposed or increased.

The implementation of the free market mechanism, along with the competition resulting from it, forced entrepreneurs to control costs. That, on the other hand, led the employers to seek all possible means to effectively reduce financial expenditures. In this way, financial constraints became the reason of a radical decrease in employee welfare benefits; even the guarantees offered by the Employee Benefit Fund (pl. *Zakładowy Fundusz Świadczeń Socjalnych*) were not always covered (see also: Stroinski 1998; Milanovic 1993; Kramer 1995). It is hard to blame business owners and managers for this situation; although many of them would oppose such a conclusion, a statement can be made that these were the costs which the society needed to bear in order to achieve 'a better standard' later.

The simple summary presented in Table 1 allows to see significant changes in determining and management of socially-oriented activities in enterprises in pre-and post-transition economy in Poland. Yet, it seems that the key (however basic) objective of both socially-oriented activities of an enterprise and corporate social responsibility was the same. It consisted in supporting and securing adequate working and existential conditions of "workplace team members" (now: employees) and their closest environment (now: local community, customers, suppliers—i.e. stakeholders). Comparing both periods and their inherent concepts, it is possible to attempt making a statement that the character of the community-oriented activities of an enterprise was reactive, whereas CSR is proactive. In other words, responding to current problems, concern about the fundamental working and existential conditions, together with some opportunities to participate in cultural events offered in the pre-transition era, are the reactive features of the abovementioned concept. Nowadays, the social responsibility of enterprises reaches much further, treating the abovementioned aspects as the basic minimum. The proactivity of CSR is reflected in continuous development, as well as forecasting and creating the future (and not passively surrendering to it).

The unquestionable difference consists also in the implementation process, available resources and tools, and, above all, in the origin of the initiative itself, or the

Table 1 The comparison of enterprise's social role before and after transition

Differentiating criterion	Social activities of a workplace	Corporate social responsibility
Degree of assumed responsibility	Usually towards employees and their families	Usually towards employees, but also towards local community, natural environment, customers and other stakeholders
Initiative	– imposed (rather non-voluntary) – top-down (coming from the state-level)	– usually voluntary – bottom-up—often comes from employees themselves, but also results from the will and ideas of company management
Degree of formalization	High—the degree of activity precisely defined by adequate provisions	Different, depending on the stage of development of CSR in a given enterprise; a requirement of 'legal liability' is in place, which enforces respecting the applicable provisions as a minimum
Enterprise's main goal	Guaranteeing production which responds to social needs (in ideological terms; in reality it consisted in fulfilling central plans)	Economic responsibility—(as the basis for operation and—in consequence—the capability of achieving specific objectives) this goal is expressed in a more or less official and/or formal way

Source: Own elaboration

degree of autonomy both in taking decisions about performing community-oriented activities and in the possibility to shape the directions of its development. Imposed solutions, which have not been discussed with the persons whom they directly concern, and which these persons will have to use, are usually implemented with massive resistance and resentment. Furthermore, lack of understanding may occur, which in extreme cases goes together with a strong resistance against agreed lines of action (Rojek-Nowosielska 2016).

3.1 *The Description of CSR Continuum Model*

The main point of reference for the CSR continuum model is the concept of maturity models.[1] The concept of maturity (as a tool for analysis and measurement) was introduced by Crosby (1979). Another significant article which was published the same year is the Nolan's paper describing six stages that have to be achieved to reach maturity (Nolan 1979). Many of published papers also correspond with the Capability Maturity Model which was developed for software organizations to provide them with guidance on how to gain control over their processes for development and

[1]A part of description of the model was originally printed in (Rojek-Nowosielska 2014).

arriving at excellent achievements. By focusing on a limited set of activities and working aggressively to achieve them, an organization can steadily improve its organization-wide software development process to enable continuous and lasting benefits in software process capability (Paulk et al. 1996).

Maturity models are applied in a large number of scientific studies, in different scientific fields. In this regard, it is worth to mention fields such as environmental sustainability (Babin and Nicholson 2011; Hynds et al. 2014), corporate sustainability (Robinson et al. 2006; Golinska and Kuebler 2014; Edgeman and Eskildsen 2014; Amini and Bienstock 2014), supply chain (Okongwu et al. 2013; Srai et al. 2013; Reefke et al. 2014; Kurnia et al. 2014) and also e-government (Andersen and Henriksen 2006), business process management (De Bruin and Rosemann 2005) or IT outsourcing (Gottschalk and Solli-Sæther 2006).

Talking about the maturity models, there are in the literature two general points of view—a life cycle perspective and a potential performance perspective (McBride 2010). The life cycle perspective indicates that an organization usually develops and expands over time and that is why (automatically) has to go through all stages of growth (due to the fact of improvement). It is important to mention, that the life cycle perspective models have well defined the final stage of maturity. In contrast, the potential performance perspective, is a "picture" of possible development way that correspond with the consecutive levels of maturity. The user's decision is whether proceed the next stage or not (Crosby 1979; Nolan 1979). These models not only show the path of development, but also focus on the potential improvements which take place by moving along (Wendler 2012).

The corporate social responsibility continuum model correspond with the models of potential performance perspective. It is a multilevel hierarchy of criteria, that allows to place the declared business actions on the "ladder" of social responsibility, which consist of three main levels (Fig. 1).

The first stage—**C (Common)**—was divided into two sub-stages (C+ and C++). This allows for a more precise determination of the entity's involvement degree in achieving the objectives of CSR. At this stage, the main determinant is the lack of formalization and spontaneous action. The C+ sub-stage includes those entities

Fig. 1 Levels of corporate social responsibility model. Source: Own elaboration

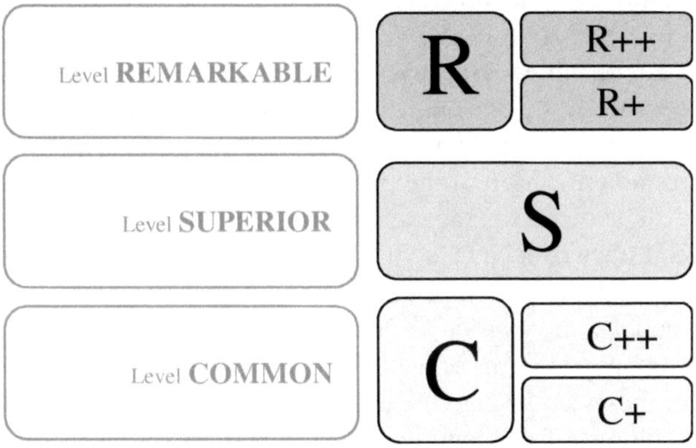

which, while integrating the concept of CSR, do not benefit from the experience, and their actions are directed by temporary inspiration. The C++ sub-stage, however, includes those entities which, despite the absence of formal memorandum or documents, use the previously gained experience, and so achieve faster task implementation, greater efficiency, and effectiveness.

We can therefore make an attempt to establish that there was a transition from chaotic action (C+ sub-stage) to more deliberate and planned behaviour (C++ sub-stage), so that the activities performed may show repeatability features, and the managers can take decisions based on previous experience.

The second stage—**S (Superior)** is the level where the actions identified at the previous stages are not only repeatable but, more importantly, formalised and described in appropriate documents and procedures. These documents (given that they are not just another type of read-only reports), indicate a thorough consideration of the described processes and choosing the solutions, which, having considered the existing conditions, constitute the best and the most optimal solutions. One of the most interesting aspects of running a business entity is the ever-changing character of the factors influencing the company, which leaves little space for routine actions. However, the same factors which speak in favour of a given phenomenon, can, under different conditions, speak against it. This is exactly the case with the turbulences of business environment. Given this reality, the documents created at the Superior stage should be verified and updated systematically. Within the scope of these assumptions, the last stage—**R** (**Remarkable**) has been developed. In order to allow for the changes taking place in the enterprise, the model organisation which can be qualified as being at the highest stage assumes two sub-stages, R+ and R++. An entity that is at the R+ sub-level may be characterized by the task repetitions that are included in the relevant documents, which provides order and improves the daily functioning. Additionally, (at R++ sub-level), continuous observation and analysis of the performed activities are made in order to modify, improve and upgrade the existing solutions, adequately to changes in the market place and stakeholders' expectations (Table 2).

The fundamental elements constituting the social responsibility continuum model are the following parameters: the CSR **area** and the CSR **institutional level**. The multitude of aspects related to the CSR subject requires the introduction of order and systematisation of problems which may occur. That is why five general areas have been identified: **employees, customers, suppliers, natural environment and local community**. Each area has been characterised by a set of crucial aspects, which have been incorporated into questions and put into the questionnaire to constitute the basis for a verification of the progress of an assessed company.

The second parameter, which makes it possible to more accurately establish the end stage of social responsibility development in a given company, is the level of the institutional development of the given CSR area. In reference to the CMM-type models, the particular institutional levels characterise the inspected areas of a company in terms of the orderliness level of the undertaken actions, the repeatability of these actions and the ability to verify and improve the implemented changes and solutions. The identified institutional levels also point to the stages of transition

Table 2 Description of corporate social responsibility continuum model levels

	CSR areas				
CSR level	Employees	Customers	Suppliers	Natural environment	Local community
R++	Responsibilities towards different stakeholder groups are implemented in a repeatable manner, are described in formal documents and there are improvements implemented aiming at the improvement of the existing situation				
R+	Responsibilities towards different stakeholder groups are implemented in a repeatable manner and are described in formal documents and undergo assessment (for their effectiveness)				
S	Responsibilities towards different stakeholder groups are implemented in a repeatable manner and are described in formal documents and procedures				
C++	Responsibilities towards different stakeholder groups are implemented in a repeatable manner but are not formally described in documents, the company bases its conduct on previously acquired experience				
C+	Responsibilities towards different stakeholder groups are implemented in an unsystematic and chaotic manner and do not result from established plans and procedures				

Source: Own elaboration

between the consecutive levels of entrepreneurs' awareness related to corporate social responsibility (Rojek-Nowosielska 2014).

4 CSR Level—Research Results

The concept of the CSR continuum model, which was put into practice in this study, enabled to carry out empirical research allowing to verify the CSR level in enterprises founded before and after transition. According to the Kumar's classification, the research based on quantitative variables, and can be classify as cross-sectional, retrospective, non-experimental (Kumar 2012). The leading method was the diagnostic survey method, which uses questionnaire survey as a research tool. The research was conducted in 2014, between January and April.

The analysis of the obtained results shows that even in general terms there are visible differences between the assessed entities with reference to the achieved CSR level. The enterprises created before the transition process reached a lower level (C+ +) than younger entities (level S). A more detailed characteristics of the CSR level in the analysed entities will enable to determine differences existing in selected areas. The only field in which the assessed enterprises achieved the same level is the area referring to employees. Both groups of respondents reached the superior level (S).

The field in which the enterprises achieve the highest level is the area of customers, although there are differences between entities established before and after transition. The solid majority of firms created before the year 1989 reached the highest levels, i.e. R+ and R++. The indications within the group of younger enterprises, created after the onset of free market economy, achieved in vast majority the level R+ (Table 3).

Table 3 Enterprises classified as representing particular CSR levels (in general terms and in selected areas of operation) [in %]

	General		Employees		Customers	
CSR level	Before 1989	After 1989	Before 1989	After 1989	Before 1989	After 1989
R++	–	0.91	2.13	1.82	31.91	19.39
R+	19.15	8.18	17.02	12.42	31.91	44.55
S	31.91	50.00	53.19	65.15	27.66	21.82
C++	46.81	39.39	21.28	16.97	4.26	11.21
C+	2.13	1.52	6.38	3.64	4.26	3.03
Total	100	100	100	100	100	100

Source: Own elaboration

Table 4 Enterprises classified as representing particular CSR levels (in selected areas of operation) [in %]

	Suppliers		Local community		Environment	
CSR level	Before 1989	After 1989	Before 1989	After 1989	Before 1989	After 1989
R++	2.13	0.91	2.13	0.91	23.49	5.15
R+	4.26	4.24	2.13	1.52	4.26	8.18
S	25.53	24.24	34.04	19.70	8.51	18.18
C++	27.66	47.27	34.04	45.45	27.66	38.79
C+	40.43	23.33	27.66	32.42	36.17	29.70
Total	100	100	100	100	100	100

Source: Own elaboration

The area of suppliers is by no means a difficult and demanding field. On the one hand, thanks to them a given entity is able to operate, and as long as the quality of the supplied good or service has a direct impact on the quality of final good or service, selecting an appropriate supplier will remain essential for a given entity. On the other hand, when choosing a supplier in line with the CSR concept, it is essential to consider not only the factors which have a direct impact on our operation, but also e.g. the way in which our supplier treats their employees, natural environment, their own suppliers etc.; this turns out to be a far more difficult task, with the effects of expenditures being visible only in a longer term. All these potential difficulties to overcome are reflected in the results of empirical research. The entities established under the centrally planned economy system reached the lowest of all possible levels, which is the sub-level C+; the entities created after the system transition achieved a slightly higher sub-level of C++ (Table 4).

Results similar to those referring to the area of suppliers are observed within the field related to the natural environment. The awareness of environmental responsibility and the ability to implement it (within the period of the empirical research) in entities established before transition remained on the lowest possible level. In younger enterprises the dominant level was C++.

In the case of so called 'silent stakeholder' (e.g. natural environment), who will rather not defend themselves, the extent of responsibility measures being taken may result either from the awareness of the given entity (i.e. its management), or relevant institutional regulations, whose non-observance may entail adequate sanctions.

In the area of local community better results were achieved by pre-transition entities. The highest number of indications within this group of respondents is observed at the levels S and C++. The level C++ was also reached by the majority of post-transition entities.

5 Conclusion

The analysis of results obtained in the empirical research enabled to attempt answering the research question as to whether there are differences in the level of social responsibility between enterprises created before and after systemic transition in Poland. In conclusion, a statement can be made that (1) it was reasonable to select entities created before and after the transition process within the analysed group of respondents; (2) the obtained results, whose analysis was conducted with regard to different areas of responsibility, do not indicate a clear advantage of one selected group of respondents; (3a) by formulation of generalized comparisons, it can be observed that both groups of analysed entities achieve the same level of responsibility (S) only in the employee-related area; (3b) entities created under free market economy achieved a better level within the CSR continuum model than entities whose origins date back to the centrally planned economy only in the areas of suppliers, natural environment, and in the general comparison considering all levels; (3c) entities established before the year 1989 came in higher with regard to the customer- and local community-related areas.

References

Amini, M., & Bienstock, C. (2014). Corporate sustainability: An integrative definition and framework to evaluate corporate practice and guide academic research. *Journal of Cleaner Production, 76*, 12–19.

Andersen, K. V., & Henriksen, H. Z. (2006). E-government maturity models: Extension of the Layne and Lee model. *Government Information Quarterly, 23*(2), 236–248.

Babin, R., & Nicholson, B. (2011). How green is my outsourcer? Measuring sustainability in global IT outsourcing. *Strategic Outsourcing: An International Journal, 4*(1), 47–66.

Balcerowicz, L. (1994). Transition to the market economy: Poland, 1989–93 in comparative perspective. *Economic Policy, 9*, 73.

Basu, S., Estrin, S., & Svejnar, J. (2005). Employment determination in enterprises under communism and in transition: Evidence form central Europe. *Industrial and Labor Relations Review, 58*(3), 353–369.

Blanchard, O. J. (1994). Transition in Poland. *The Economic Journal, 104*(426), 1169–1177.

Burrell, K. (2011). Opportunity and uncertainty: Young people's narratives of 'double transition' in post-socialist Poland. *AREA, 43*(4), 413–419.

Crane, K. (1991). Institutional legacies and the economic, social, and political environment for transition in Hungary and Poland. *American Economic Review, 81*(2), 318–322.

Crosby, P. B. (1979). *Quality is free: The art of making quality certain*. New York: McGraw Hill.

De Bruin, T., & Rosemann, M. (2005). Towards a business process management maturity model. In D. Bartmann, F. Rajola, J. Kallinikos, D. Avison, R. Winter, P. Ein-Dor, J. Becker, F. Bodendorf, & C. Weinhardt (Eds.), *Faculty of science and technology*. London: Verlag and the London School of Economics.

Edgeman, R., & Eskildsen, J. (2014). Modeling and assessing sustainable enterprise excellence. *Business Strategy and the Environment, 23*(3), 173–187.

Giffin, P. E., & Ellington, L. (1995). The origins of capitalist markets: Transition in Poland with comparisons to East Asian capitalism. *Journal of Economic Issues, XXIX*(2), 585–590.

Golinowska, S. (2013). The Polish welfare state birth and twilight. In P. Michoń, J. Orczyk, & M. Żukowski (Eds.), *Facing the challenges. Social policy in Poland after 1990*. Poznań: Poznań University of Economics Press.

Golinska, P., & Kuebler, F. (2014, June 18–20). The method for assessment of the sustainability maturity in remanufacturing companies. In *Proceedings of the 21st CIRP conference on life cycle engineering (CIR LCE)* (Vol. 15). Trondheim, Norway.

Gottschalk, P., & Solli-Sæther, H. (2006). Maturity model for IT outsourcing relationships. *Industrial Management and Data Systems, 106*(2), 200–212.

Hynds, E. J., Brandt, V., Burek, S., Jager, W., Knox, P., Parker, J. P., & Zietlow, M. (2014). A maturity model for sustainability in new product development. *Research – Technology Management, 57*(1), 50–57.

Johnson, S., & Loveman, G. (1995). Starting over: Poland after communism. *Harvard Business Review, 73*(2), 44–54.

Keane, P. M., & Prasad, E. S. (2002). Inequality, transfers, and growth: New evidence from the economic transition in Poland. *The Review of Economics and Statistics, 82*(2), 324–341.

Kowalik, T. (2012). *From solidarity to sellout. The restoration of capitalism in Poland*. New York: Monthly Review Press.

Kramer, M. (1995). Polish workers and the post-communist transition, 1989–1993. *Communist and Post-Communist Studies, 28*(1), 71–114.

Księżopolski, M. (2013). In search of a new post-socialist social policy model in Poland. In P. Michoń, J. Orczyk, & M. Żukowski (Eds.), *Facing the challenges. Social policy in Poland after 1990*. Poznań: Poznań University of Economics Press.

Kumar, R. (2012). *Research methodology. A step-by-step guide for beginners*. Los Angeles: Sage.

Kurnia, S., Rahim, M. M., Samson, D., & Singh, P. (2014, June 9–11). Sustainable supply chain management capability maturity: Framework development and initial evaluation. In *Proceedings of the 22nd European Conference on Information Systems, ECIS*. Tel Aviv, Israel.

Laurentiu, I. (2016). *The social and economic evolution of Poland and Romania for the last 100 years*. Annals of the „Constantin Brâncuşi" University of Târgu Jiu, Economy Series, Issue 2.

McBride, T. (2010). Organisational theory perspective on process capability measurement scales. *Journal of Software Maintenance and Evolution: Research and Practice, 22*(4), 243–254.

Milanovic, B. (1993, August). *Social costs of transition to capitalism: Poland, 1990–1993*, (Policy research working paper series, no. 871). Washington, DC.: The World Bank.

Nolan, R. L. (1979). Managing the crisis in data processing. *Harvard Business Review, 57*(2), 115–126.

Okongwu, U., Morimoto, R., & Lauras, M. (2013). The maturity of supply chain sustainability disclosure from a continuous improvement perspective. *International Journal of Productivity and Performance Management, 62*(8), 827–855.

Paulk, M. C., Curtis, B., Chrissis, M. B., & Weber, C. V. (1996). *Capability maturity model^{SM} for software*. Pittsburgh: Carnegie Mellon University.

Reefke, H., Ahmed, M. D., & Sundaram, D. (2014). Sustainable supply chain management – Decision making and support: The SSCM maturity model and system. *Global Business Review, 15*(4), 1S–12S.

Robinson, I. (2004). Enterprise restructuring in transition: Case study evidence from Poland. *Journal of East-West Business, 10*(2), 45–64.

Robinson, H. S., Anumba, C. J., Carrillo, P. M., & Al-Ghassani, A. M. (2006). Steps: A knowledge management maturity roadmap for corporate sustainability. *Business Process Management Journal, 12*(6), 793–808.

Rojek-Nowosielska, M. (2014). Corporate social responsibility level – Theoretical approach. *Management, 18*(1), 38–40.

Rojek-Nowosielska, M. (2016). Question of terminology or the change of meaning? On the corporate social responsibility. In O. Kowalczyk, A. Kubów, & W. Wątroba (Eds.), *Społeczeństwo i zmiana* (pp. 284–285). Wrocław: Publishing House of Wrocław University of Economics.

Rondinelli, D. A., & Yurkiewicz, J. (1996). Privatization and economic restructuring in Poland: An assessment of transitions policies. *American Journal of Economics and Sociology, 55*(2), 145–160.

Srai, J. S., Alinaghian, L. S., & Kirkwood, D. A. (2013). Understanding sustainable supply network capabilities of multinationals: A capability maturity model approach. *Proceedings of the Institution of Mechanical Engineers, Part B: Journal of Engineering Manufacture, 227*(4), 595–615.

Sterhenz, J. (1993). Modern development of polish unemployment law. *Comparative Labor Law Journal, 14*(4), 163.

Stroinski, K. (1998). Poland: The reform of the pension system. *Economic Affairs, 18*(1), 29–33.

Wendler, R. (2012). The maturity of maturity model research: A systematic mapping study. *Information and Software Technology, 54*, 1317–1339.

Social Responsibility Management in a Small Enterprise: Selected Problems and Good Practices

Agnieszka Sokołowska-Durkalec

1 Introduction

The state and directions of research into social responsibility of a small enterprise are intensely differentiated in content, methodically and spatially. Polish achievements in this area, in the context of world achievements, seem to be disproportionately modest. Exploration of the problems related to socially responsible activities of small enterprises appears quite rarely. Firstly, the issue is mainly dealt with in large and medium-sized enterprises or small and medium-sized ones from a sectoral perspective, without clear reflection on the differentiation between these two groups. Secondly, NGOs (non-governmental, public) are more concerned with the issue of social responsibility, promoting it, carrying out their research at a very diversified level of content and methodology. In addition, it can be noticed that in the modest research output, the inside diverse issue of social responsibility is discussed rather briefly and superficially. However, the terminological ambiguity associated with the concept of social responsibility results in a different understanding of the category. Few Polish research initiatives in only a small (to a minimum) extent reduce the knowledge deficit on the issue of social responsibility of small enterprises. There is not only a deficit of research on the specificity of management or systematic approach to the issue, but above all we lack the complex exploration dealing with the economic, social, institutional, organizational and individual context of the problem. Fragmentary research usually has a local or regional scope. In general, Polish literary and empirical achievements need filling in their deficits practically in all dimensions of the concept and possibilities of applying its assumptions in small enterprises. The more so, the main assumptions of the concept of social

A. Sokołowska-Durkalec (✉)
Uniwersytet Ekonomiczny we Wrocławiu, Wrocław, Poland
e-mail: agnieszka.sokolowska-durkalec@ue.wroc.pl

A. Długopolska-Mikonowicz et al. (eds.), *Corporate Social Responsibility in Poland*, CSR, Sustainability, Ethics & Governance,
https://doi.org/10.1007/978-3-030-00440-8_10

responsibility of a small enterprise in its ideological layer are still being explained, and one may get impression that it "stuck" at the level of argument presentation for its application, not sufficiently discerning its tool dimension. There are many reasons for this, but one of the key assumptions is the one of the socially insufficiently responsible activities of this group of enterprises. The conclusions presented were formulated on the basis of a detailed analysis of the condition and direction of social responsibility research of a small enterprise worldwide and in Poland, presented in the author's monograph and selected articles (Sokołowska 2013, 2016; Sokołowska-Durkalec 2017). They contain, among others, a detailed description of research areas in the field of corporate social responsibility and their main representatives arranged as: main areas of research, selected issues, representatives and the most important features of the selected research on social responsibility of small and medium enterprises in the following order: author/authors, research object, place of research, subject/purpose of research, research sample, methodology, kind and type of study (using more than 250 subject literature items).

Moreover, a detailed description of individual, selected results and conclusions from empirical research is presented, that identifies many important gaps required to be filled in, to which the author's literature and empirical research aspire. When reviewing various (mainly due to the area and place of research) research projects (both foreign and Polish), it can be stated that research to date on social responsibility of small and medium-sized enterprises has allowed, inter alia, to: partially answer questions concerning specific features of this group of enterprises, distinguishing them from large companies; learn about (also partially) possibilities of adapting the CSR concept principles to SMEs and involving companies in this area (demonstrating ambiguous possibilities of the concept principles realising in small enterprise functioning conditions, mainly due to their corporate nature); identify and compare specific features of large, medium-sized and small enterprises from the perspective of strategic orientation related to social responsibility (learning about differences between formal and informal CSR strategies); determine small enterprise impact on the natural environment; assess the scope of small enterprise involvement in local community activities; identify selected issues regarding the ethical area of small enterprise activities (comparison of ethics institutionalization in the form of strategies and instruments of social responsibility and recognition of the ways of integrating CSR elements with business ethics); assess the use of quality management systems and corporate social commitment; indicate the impact of the prosperity level in a given country on enterprise willingness to socially responsible activities; identify categories related to social capital measurement, etc.

Meanwhile, the author's research findings (Sokołowska 2013, 2016) show numerous indications, and in some cases, targeted, methodical and even procedurally sanctioned actions related to social responsibility in a small enterprise. It turns out that the benefits of cultivating long-lasting and universal values and principles, and of sustainable activities based on social dialogue are more often recognized by entrepreneurs, their employees and many external stakeholders. Social phenomena and processes are inherently long-lasting, difficult to capture and create, and their development requires involvement of many entities and means, as well as adaptation of an organization. That is why questions arise here: *what key issues/problems can be identified in the realisation of specific management functions of social responsibility in a small enterprise? and what good practices can be identified and in what areas of the concept realisation in a small enterprise?* The considerations presented in this article are a continuation of multidimensional literature and empirical research on the identification, evaluation and improvement directions of social responsibility of a small enterprise from the perspective of managing its areas. They touch on the dimension of social responsibility improvement and report on the further issue exploration (empirical research results), and more precisely they observe assumption enforcement and application of the selected concept instruments in small Polish enterprises.

The aim of this article is to present selected problems and good practices in the management of social responsibility in a small enterprise, and thus to approximate the tool dimension of the concept of social responsibility. Due to the limited scope of the article, only selected issues related to some sort of dysfunctions of social responsibility management and best practices are presented. As a result, the description of the extended research process and the presentation of the detailed empirical research results are excluded in favour of synthetic, so far unpublished in such a form and perspective, conclusions concerning the subject matter.

2 Assumptions for Social Responsibility Management from the Perspective of a Small Enterprise: Original Concept

The multidisciplinary and internally diversified category of social responsibility of enterprise has been defined in a number of ways. An analysis of the broad spectrum of definitions and approaches to this issue allows us to adopt an author's comprehensive approach incorporating all key perspectives: economic (e.g. Friedman 1970), ethic-philosophical (e.g. Kung 1999), highlighting the social response of an enterprise in the context of relations with stakeholders (e.g. Carroll 1999; Freeman and Phillips 2002), normative with patterns and mechanisms facilitating socially responsible actions (e.g. Garriga and Mele 2004; Madueño et al. 2016), in the scope of which it can be stated *that social responsibility is economic, legal, ethical and philanthropic commitment of an enterprise to internal and external social groups (and individuals), and that it can be the object of expedient, rational and institutionalized action, which can become a source of competitive advantage* (Sokołowska 2013).

A review of the most important literature and empirical research results on social responsibility (including a small enterprise) in domestic and foreign literature entitles to formulate the statement that the continuum of the discussion over social responsibility aims at systematizing it and capturing its complex social "matter" and consolidating its place in the collection of important issues for modern enterprise management. The concept of social responsibility consists of its ideological layer with a set of assumptions constituting it and its tool dimension, rich in methods, techniques and management tools. In the ideological layer, selected principles/ standards developed by international institutions can be adopted. These are the Caux Round Table, Global Compact and Sullivan principles that define norms of conduct concerning building confidence in business. They recommend fair and lawful behaviour, respect for the dignity of the human person and the environment. Fundamental moral values such as honesty, truthfulness, respect, honesty, keeping the word are the underpinning and they facilitate implementation of activities in accordance with the norms of decency and the law. Whereas, the tool framework of the concept requires a decision about formalized management (using and adapting guidelines for standardization and evaluation according to the international standards such as SA 8000, AA1000, ISO 26000) or not formalized using methods and tools borrowed from related concepts such as: human resources management, marketing management, logistics, etc. The management process can be described on a set of sequential stages such as: identification of areas (economic, legal, ethical and philanthropic) and social responsibility entities (internal entities: owner/owners, manager/managers, employees and external entities: clients, suppliers, local community, etc.), planning (e.g. value statement, locating social goals in the bundle of enterprise objectives, development or choice of methods of measuring, reporting and developing social responsibility, etc.), organizing (i.e. providing appropriate conditions for the implementation of the concept: organizational structure, organizational culture, strategic solutions, integration within the existing management system, etc.), developing/shaping (specific socially responsible actions within the areas dedicated to the entities, using supportive concepts, methods, ways and management tools) and controlling social responsibility (measurement/monitoring of social responsibility (possibly reporting), analysis and evaluation of effects, corrective actions and implementing improvements) (cf. Sokolowska 2013; Sokołowska-Durkalec 2017).[1] The attitude of the owners and management towards social responsibility is also very important, i.e. their knowledge and resulting from it behavior that influence all enterprise activities dedicated to these issues. The result of the realisation of the process are notable achievements coming from the responsible functioning of an enterprise resulting from its responsible management. The benefits of implementing the principles of social responsibility are related to the cohesion of its realisation and controlling. In this context, the relationships between the set of principles and their transposition into the language of practice and identification of the accompanying conditions are essential.[2]

[1]A detailed description of the entire process of social responsibility management together with the assumptions and instruments embedded in the situational context of a small enterprise can be found in the author's monograph (Sokołowska 2013). Due to the limited page size of the study, the decision was made to present the original concept synthetically.

[2]Discussions regarding the issues can also be found in the studies e.g.: (Stawicka 2017; Hąbek 2017; Aluchna 2016; Wójcik-Jurkiewicz 2017 etc.)

Characteristic features of a small enterprise can influence the realisation of the concept of social responsibility. By recognizing the essence and importance of a responsible and transparent functioning of an enterprise, a small company can gain many benefits by gaining the trust of its key players such as: employees, customers, local communities, etc. The central figure and the decision maker (the combination of owner and manager's functions[3]) in an enterprise is the owner/owners; and success of the realisation of the concept depends on his knowledge, behaviour and beliefs about the rightness of making changes; he shares the values with the stakeholders and institutionalizes them. Small enterprise owners should take advantage of the fact that social process courses in this enterprise group are often determined by personal relationships that can result in relationships based on trust and creative collaboration. "Agility" (flexibility, dynamism, and openness) of a small enterprise creates the conditions for recognizing the importance of social processes and related to them: information flow, knowledge, intellectual and social capital; and also for developing a sustainable organizational culture based on trust and dialogue.

The proximity of the market on which a small enterprise operates creates an opportunity for very close contacts with external partners. Collaboration with customers, suppliers, etc. is often based on personal contacts that define their relationships as often going far beyond formal contacts. A small enterprise can gain dominance over its rivals through responsible and transparent operations in all dimensions, competing not only with companies from its size group, but also with those large ones with established positions. Specific characteristic features resulting from the nature of a small size of economic activity also define characteristics of areas (scope of the areas: economic, legal, ethical and philanthropic), and entities of social responsibility (in terms of their objectives, expectations, scope of the relationship with an enterprise), attitudes of the owners and management towards social responsibility, realisation of the social responsibility management process and others internal and external conditions which can be different than those in large enterprises.

3 Research Methodology

The conclusions presented in the paper, resulting from reference sources and empirical studies refer to the research process no. N N115 55 0138 entitled: *Social responsibility management in a small enterprise. Identification—evaluation—improvement directions*, financed by Polish Ministry of Science and Higher Education in the period of 2010–2012. The research process consisted of several stages:

[3]In "larger" small enterprises the separation of these functions becomes more natural with the development of the business, then their importance gain—in the context of social responsibility—two-way relationships existing on the line: owner-manager, manager-employee.

preliminary research (2008–2009) and three phases of relevant research, at which 187 small Polish enterprises from 12 voivodeships and 133 small companies from Ireland, Great Britain, Germany, the Czech Republic, Norway, Spain, Italy were researched through questionnaires, interviews, observations, analysis of organizational documentation, case-based comparative analysis. Altogether 320 companies were surveyed. The research sample is characterized by quite large variations. The enterprises researched were divided into (according to size criterion—the number of employees): micro-enterprises employing 1–10 employees and other small enterprises employing 11–50 persons ("larger" small enterprises). In addition, the division of the enterprises (according to the core activity) was made: production, service, commercial, and also (according to the criterion of specific characteristics) into family and non-family. Finally, the last division refers to the time of operation on the market for the enterprises surveyed, and so there were distinguished: new enterprises operating on the market for 1 year (very young), enterprises operating from 1 to 3 years (young) and enterprises older than 3 years (older).

The main aim of the research was to assess social responsibility of a small enterprise by analysing entrepreneurial awareness, identifying determinants, indications and ways of managing social responsibility (including reporting), and benefits/ costs of social responsibility (from the perspective of identifying social responsibility areas and entities and planning, organizing, developing and controlling of social responsibility) and to analyse formal and informal management, as well as identify which enterprises and when can be considered as socially responsible. In the theoretical part, the method of descriptive analysis was used on the basis of extensive studies of Polish and foreign literature of the subject. The empirical research was based on a model theoretical concept based on the modified A.B models of Carroll, Y.Ch. Kang and D.J. Wood. K. Davis and R. Blomstrom, S.L.Wartick and P.L. Corchan and on the theory of stakeholders.

The research is idiographic in its character, hence the conclusions drawn from it refer only to the population studied. It seems that any generalizations formulated on the basis of the obtained results, mainly due to the absence of sampling procedure application, are unauthorized. On the other hand, the interpretation of the obtained results allows to develop knowledge about social responsibility and management of its areas in Poland and partly—very circumspectly—in selected European countries.

After more than 4 years (2015–2017) from the completion of the relevant surveys, the state of realisation of the concept of social responsibility has been re-examined. Modified research tools (survey questionnaires for the owner, employee, customer and owner interview questionnaires) were used. Extensive and in-depth empirical research is being carried out on purposefully selected 48 small Polish companies from 12 voivodships. The targeted selection of enterprises was directly linked to the results of the empirical research from the second stage of the 2011 study, whereby enterprises aspiring to be considered socially responsible were selected because they identified features that emphasized the socially responsible nature of the business. The research process is ongoing.

4 Problems Related to the Management of Social Responsibility in a Small Enterprise: Selected Conclusions from Empirical Research

The results of the empirical research[4] show that the state of social responsibility management in small enterprises can be defined as quite substantially not in line with the model assumptions presented in the synthetic form in the previous sections of the article. It turns out that the clash of the concept of social responsibility, specific characteristics of small enterprises and situational context results in many described below problems which were identified during the research process.

The owners' level of expertise about the social concept is in opposition to the actual verification of the individual aspects of the concept in practice, and thus the awareness of the importance of social responsibility in business management. Therefore, it is not sufficient to consciously build the competitive advantage of an enterprise on this competence (entrepreneurs do not mention it among the competitive factors) and aspire to create a socially responsible enterprise. The most popular source of knowledge about social responsibility is the mass media.

Most of the respondents (70% small Polish enterprises) do not know other companies that manage social responsibility.

The entrepreneurs most often identify social responsibility with economic and ethical responsibility, only few associate social responsibility with the law and philanthropy. Most respondents see social responsibility more in terms of compulsion than in volunteering, perceiving more problems and costs than benefits (78% small Polish enterprises). Funds for social purposes are most often passed on to activities in the economic area of social responsibility, the least of which is dedicated to charities.

The indicator analysis conducted in the surveyed enterprises reveals a small, financial commitment to the declared socially responsible activities. Marketing activities appear to be the most financially provided area, including professional customer service. The measure has shown insufficient funding for employee affairs, environmental protection and charity work.

Passive attitude of the entrepreneurs towards social responsibility dominates in most of the surveyed enterprises (94%). In this type of enterprises, social responsibility occurs objectively without the consciousness of its creation in a very diverse range, and many manifestations of responsible actions can be found within the framework of current day-to-day operations.

The individual elements within the framework of social responsibility management function are realised without the entrepreneurs' knowledge about possibilities to categorize them, and in principle no group of the enterprises distinguished themselves with anything unique within the framework of this management function.

[4]Presented, selected results of the empirical research apply to only small Polish enterprises.

The identification of social responsibility areas and entities takes place in an informal manner in most of the surveyed companies (88%). The entrepreneurs do not place particular emphasis on a systematic analysis of the objectives, expectations and manifestations of stakeholders' power. External stakeholders are of the greatest importance to the entrepreneurs, in the form of customers, financial institutions and regulators, specifically representatives of government and self-government administration. Employees occupy quite a high position among the groups interested in the activities of the surveyed enterprises. Relatively little recognition is given to the environment.

Social responsibility planning is an incidental phenomenon, and it only comes down to individual actions in this area. Most of the surveyed companies (73%) do not realise at all or only partially realise elements of the strategic management process. Among the goals dominate those of an economic nature, few enterprises have missions, visions and strategic objectives (often interchangeably) in which traces of social responsibility can be found.

The entrepreneurs do not know the international rules of social responsibility (e.g. Global Compact, Sullivan rules, or the Caux Round Table rules), from which they could draw inspiration for their business. The entrepreneurs practically—apart a few exceptions—do not use methods of strategic analysis, and the strategy has mostly informal character and exists more often in the owner's head than is written anywhere. No records concerning social responsibility were identified in companies that operate on the basis of a formal strategy (most often not fragmented according to a functional order).

Organizing social responsibility is mainly related to the organizational structure. In none of the small Polish enterprises the concept of social responsibility was located in the organizational structure, in the form of a separate position or organizational unit. In many companies, social responsibility issues are dealt with by posts or organizational units related to employee affairs, marketing, logistics or finance. Most often, the enterprises can be observed in the initial stage of maturity of the organizational concept, where social responsibility issues are dealt with "by chance" as part of additional activities of other departments (in "larger" small businesses).

Most entrepreneurs (77%) recognize the existence of organizational culture in their enterprises, but they do not declare that it is based on universal ethical values, but rather more often on the utilitarian and selfish assumptions originating from their attachment to the economic domain of business.

More than half of the enterprises can't boast about ethical institutionalization in the form of an ethical code.

The vast majority of the entrepreneurs (73%) fail to recognize the need to implement the concept of social responsibility and integrate it into the management system.

Almost completely unappreciated is developing/shaping, although it is the most developed function of social responsibility management. The essential issue, of not universal character, is the rule of reciprocity in cooperation with stakeholders who do not always act economically rationally, ethically and legally (not to mention any form of philanthropy).

Usually low budgets of small enterprises exclude the use of certain tools, especially in the area of employee-directed economy (e.g. extended incentive schemes) or philanthropy towards the local community (e.g. charity support, cultural patronage).

The least developed social responsibility function is controlling. The analysis and assessment of social responsibility is alien to the overwhelming majority of the surveyed enterprises. The entrepreneurs have never heard of the international standards for standardization and reporting of social responsibility. No measurement of social responsibility is carried out in any of the enterprises.

There was not found any enterprise with a comprehensive management program, including accomplishment of all the functions.

The interesting and surprising conclusion is that problems, which were identified while planning, organizing, developing/shaping and controlling social responsibility, did not have any particular characteristics in terms of size, nature of activity, age, family spirit and origin of the studied enterprises, but these determinants influenced: knowledge and attitude of the entrepreneurs towards social responsibility and identification of social responsibility areas and entities.

5 Good Practices in Social Responsibility Management in the Surveyed Small Enterprises

The emerging from the empirical research image of social responsibility management points to the problems and significant gaps in both indications and also deliberate impacts on this still under-valued area of economic activity. However, in the group of the surveyed enterprises one can find those in which selected elements of the concept are more or less intentionally accomplished (and even implemented).

Few entrepreneurs (only 6%) are characterized by an active attitude towards social responsibility. Among their enterprises a varied level of social responsibility management can be noticed: from symptoms of individual elements of management to conscious realization of them. The enterprises surveyed are also at different stages of designing and implementing social responsibility principles. It can be estimated that most of these are at the initial and incremental stages of the implementation of the rules.

Most respondents (91%) acknowledge social responsibility for a duty that involves bearing consequences. Quite a large group (34%), however, associates responsibility with compulsion.

In spite of the identified knowledge deficiencies, mainly in the tool dimension of the concept, the entrepreneurs recognize many knowledge asymmetries related to social responsibility, particularly in the scope of different ways of social responsibility management, their knowledge of this subject matter and possibilities to meet the stakeholders' expectations. Larger, family, service enterprises, operating on the market for over 3 years, demonstrated the most positive characteristics in the area of the owners' knowledge on social responsibility issue and its management.

The vast majority of the respondents perceive the specific characteristics of their enterprises as useful in socially responsible activities.

Most of the surveyed entrepreneurs (62%) declare informal management of social responsibility, although a large group of them (23%) does not recognize the particular need, especially in terms of benefits, of affecting this dimension of their activities.

Few enterprises (only four enterprises) were diagnosed with missions, visions, and strategic goals, in which references to ethics and social responsibility can be found. Individual cases consciously focus their strategic values around declarations of responsibility and reliability in action.

It is comforting that in many strategic, tactical and operational plans (generally not distinguished formally), social responsibility entities are taken into account.

The most developed function of social responsibility management is developing/shaping. Many actions regarding employees (including managers), customers, suppliers, competitors, local communities, financial and state institutions, and so forth, and even not sufficiently recognised in identifying the areas and entities—the environment, assume very extended dimensions and scope. The entrepreneurs declare many actions that can be considered socially responsible, practically in all the areas of social responsibility towards many entities. These activities are largely unequated with social responsibility.[5]

The entrepreneurs use a variety of management practices, many of which can support socially responsible business.

There are single cases in which the entrepreneurs show a very active attitude towards the recognition of the importance of social processes in enterprise management. They are cognitively open-minded, creative and willing to learn. They participate in seminars and conferences on the issue (acquired knowledge is shared with employees). They even set up bookshelves where employees can read publications and brochures about new trends in management, social responsibility, business ethics. In addition, the owners organize cyclical meetings with employees, at which they not only discuss current enterprise functioning issues, but also problems concerning ethical, responsible and highest quality standard activities in all dimensions (usually this way management procedures and ethical codes are established). The teams work to improve physical and behavioral artefacts, as well as norms and simple procedures that reflect values in organizational culture.

It happens that in some enterprises their owners minimize reluctance of workers to changes, for example, by engaging them in the process of formulating the mission and strategic goals, as well as creating tools facilitating the identification of customer needs and contacts with suppliers. Employees jointly work out the main objectives of the concept implementation (inter alia: maintaining and improving the quality of

---

[5]Descriptions of the specific examples of actions for each of the social responsibility entities in the surveyed small businesses can be found in the author's publications.

service, work and cooperation) and appoint a person to coordinate activities and report on the progress towards accomplishment of the goals.

The entrepreneurs prepare their enterprises not only socially but also organizationally (for example, introducing changes in the organizational structure, procedures, job descriptions, etc.) and financially (e.g. finding funds in their budgets, etc.).

In some of the surveyed enterprises areas and entities of social responsibility are cyclically identified. The entrepreneurs use stakeholder expectation and goal monitoring tools by interviewing employees, carrying out short surveys or interviews with clients, surveys for suppliers and the local community. In addition, they develop detailed guidelines for planning, organizing, developing/shaping and controlling social responsibility, etc.[6]

6 Recommendations

The dysfunctional issues and best practices presented in the preceding paragraphs lead to the presentation of some recommendations.

Implementing the assumptions of the concept of social responsibility requires time and commitment, which can stand in opposition to entrepreneurs' expectations, acting under short-term pressure, quick effects and costs.

Entrepreneurs' knowledge (and consequently employees') and the attitude associated with it require to orient enterprise activities to social aspects, with particular emphasis on social responsibility. So called ethical training with an emphasis on case studies and simulations of ethically dubious situations are recommended here.

Identification of the areas and entities should take the form of more precise stakeholder relation management, it is required to systematically analyse them paying special attention to expectations and impact power on enterprise activities. Optimal choice and hierarchy of stakeholder expectations will ensure greater balance of economic and social goals.

In planning, it is important to accept and define the conditions for adherence to values (the existing international principles of social responsibility may be helpful). In addition, it is necessary to gradually formulate a general strategy with a module on social responsibility and, ultimately, along with the development of the enterprise potential, also functional.

In organizing it is important to gradually, methodically introduce elements of the concept, with which close integration within the organic functions is linked, while identifying specific conditions.

The proposal of extended stages of the social responsibility management process should be replaced by "tailor-made", personalized programs dedicated to specific enterprises or groups.

---

[6]Sample activities were initiated during the empirical research as a result of the author's cooperation with the small enterprise owners.

Developing/shaping social responsibility should be rooted in the use of a wide spectrum of domain management practices so that individual areas are correlated with the expectations of entities.

Functionality and usefulness of the proposed management methods are essential. Social responsibility measurement causes many difficulties (and thus it is discouraging); controlling social responsibility requires adaptation of the available measures and normalization standards to the conditions of a small enterprise functioning. Initially, social responsibility can be assessed instead of precisely measured, and later—after developing available options and convincing the entrepreneur about the rightness of the action associated with social responsibility—corporate social involvement of the enterprise can be systematically monitored. It is also important to take corrective measures in case of significant deviations, etc.

7 Conclusion

The researched enterprises are characterized by varied levels of preparation for being a socially responsible company, it is indicated by the list of selected problems and good practices being still recognized by the author in the research process and synthetically presented in the article. This specification does not exhaust the topic. The general conclusion is that although formal management of this dimension of activity was not found in any of the surveyed enterprises, practically each of them had a specific characteristic element worth noting, noticing, and even imitating. The processes of implementing and accomplishing the assumptions of the concept are intermingled and complementary, and the attitude and current actions of the entrepreneurs augur well when it comes to the possibility of their full realization. Also, the recommendations presented in the article should initiate formulation of recommendations for small enterprise owners whose specific characteristics require specific solutions. It must also be borne in mind that even when modelling the reality of small enterprises, the situational context makes this group of companies highly heterogeneous and difficult to identify unequivocally.

References

Aluchna, M. (2016). From ideas to impacts: CSR in the eyes of top managers in Poland. In S. O. Idowu & S. Vertigans (Eds.), *Stages of corporate social responsibility: From ideas to impacts* (pp. 47–64). Cham: Springer.

Carroll, A. B. (1999). Corporate social responsibility: Evolution of a definitional construct. *Business & Society, 38*(3), 268–295.

Freeman, R. E., & Phillips, R. A. (2002). Stakeholder theory: Libertarian defense. *Business Ethics Quarterly, 12*(3), 331–349.

Friedman, M. (1970, September 13). The social responsibility of business is to increase its increase its profits. *New York Times Magazine, 32–33*, 122–124.

Garriga, E., & Mele, D. (2004). Corporate social responsibility theories: Mapping the territory. *Journal of Business Ethics, 8*(53), 51–71.

Hąbek, P. (2017). CSR reporting practices in Visegrad Group countries and the quality of disclosure. *Sustainability, 9*(12), 2322. https://doi.org/10.3390/su9122322.

Kung, H. (1999). Pokój na świecie – światowe religie – światowy etos. In J. Sekuła (Ed.), *Idea etyczności globalnej* (pp. 386–398). Siedlce: Wydawnictwo Seculum.

Madueño, H., et al. (2016). Development of corporate social responsibility in small and medium-sized enterprises and its nexus with quality management. *Cogent Business & Management, 3*, 1–21.

Sokołowska, A. (2013). Społeczna odpowiedzialność małego przedsiębiorstwa. Identyfikacja-ocena-kierunki doskonalenia. Seria: Monografie i Opracowania nr 232, Wydawnictwo Uniwersytetu Ekonomicznego we Wrocławiu, Wrocław.

Sokołowska, A. (2016). Uwarunkowania implementacji założeń koncepcji społecznej odpowiedzialności w małym przedsiębiorstwie. Zarządzanie i Finanse. *Journal of Management and Finance, 14*(2), 351–360.

Sokołowska-Durkalec, A. (2017). Identyfikacja warunków, przejawów i problemów badawczych dotyczących reguły wzajemności w relacjach małego przedsiębiorstwa z interesariuszami w procesie zarządzania społeczną odpowiedzialnością. Marketing i Rynek, Polskie Wydawnictwo Ekonomiczne, Warszawa, Rok XXIV, Listopad, Nr 11/2017: 553–566.

Stawicka, E. (2017, October). Social responsibility model of the sustainable development in enterprises in Poland. *The Journal of Human and Work, 4*(2), 143–149.

Wójcik-Jurkiewicz, M. (2017). Role of CSR reporting. Evidence from Poland. Zeszyty Teoretyczne Rachunkowości. *Stowarzyszenie Księgowych, 94*(150), 173–188.

The Influence of MNEs on CSR in Poland

Bożena Dancewicz and Franziska Struve

1 Introduction

The objective of this chapter is to diagnose the condition of CSR in Polish branches of multinational corporations. However Poland does not belong to countries commonly referred to as poor, it neither belongs to the world's economic leaders. According to a study of the Global Finance Magazine Poland is the 48th wealthiest of 189 surveyed countries. The study is based on Gross Domestic Product (GDP) in relation to the Purchasing Power Parity (PPP). Thereby the ranking indicates a comparison of living standards and economic strengths across countries (see Global Finance Magazine 2017).

Polish society is at the stage of building its civic awareness. Also in this aspect it cannot be classified among the world's leaders. Therefore an attempt to answer the question what has to happen in order for the branches of multinational corporations to be willing to act in a socially responsible way in given circumstances may contribute to filling in the research gap. A diagnosis of the status of CSR in Polish branches of multinational corporations was presented along with own research results and a summary of the work. This research contributes to better understanding of the factors that led multinational corporations to prosocial behaviours.

B. Dancewicz (✉)
Nowy Dwór Mazowiecki, Poland

F. Struve
German Graduate School of Management & Law (GGS), Heilbronn, Germany
e-mail: franziska.struve@ggs.de

A. Długopolska-Mikonowicz et al. (eds.), *Corporate Social Responsibility in Poland*, CSR, Sustainability, Ethics & Governance,
https://doi.org/10.1007/978-3-030-00440-8_11

2 CSR in Poland: Current Status and Historical Development

Poland became a subject of world's economic activity relatively late, which is why Polish branches of multinational companies develop in a slightly different way than their parent organisations. This is also due to the national culture and the sum of specific historical events and applies especially to the concept of corporate social responsibility (CSR). The period after 1989 was characterised by deep transformation processes and rapid catching up on economic and social lags. Together with corporations, business knowledge and experience came along, which created a different economic reality and changes in people's mentality. The level of development of the civic society constitutes one of the conditions of conducting business in a socially responsible manner.

Based on the global Study of Gallup Inc. on Civic Engagement one of the most important challenges for global society will be the personal engagement beyond law of the citizens of each country. The study examines the citizen's willingness to invest time, money and assistance based on 1000 interviews conducted in 2015. Based on these three categories Poland reached a total score of 25 (scale of 0–100) and is therefore comparable with countries such as Simbabwe and Saudi Arabia (see Gallup 2016: 13). Societies with higher standards of civic awareness monitor, assess and disapprove to a greater extent all kinds of professional misconduct. The expansion of international corporations and globalisation in the second half of twentieth century has given social responsibility a new meaning (Dancewicz 2016: 19–23).

When in developed Western countries civic ideas and ecological movements were maturing, Poland was occupied with regaining independence and struggling with basic existential issues. That influenced how Poles were setting their priorities. Traces of these economic and especially mental transformations are visible until now.

In Poland multinational corporations appeared relatively late, for historical reasons, which influenced their role in the local reality, including their approach towards CSR and its execution. The concept of CSR is relatively new—especially in vicegrad countries—but is rapidly spreading, based on the influence of MNEs as foreign investors and their integration to the European Union (EU) (Hąbek 2017). Because CSR in Poland is mainly driven by the influence of foreign investors it's influence on CSR disclosure was examined. "In fact, only foreign capital share has turned out to be important in the context of the environmental disclosure (Dyduch and Krasodomska 2017: 18)".

2.1 The Role of MNEs

Progressive globalisation is an undeniable fact with Multinational Enterprises (MNEs/MEs) as an inseparable element of this process worldwide. "They (MNEs) usually comprise companies or other entities established in more than one country and so linked that they may coordinate their operations in various ways. While one or more of these entities may be able to exercise a significant influence over the activities of others, their degree of autonomy within the enterprise may vary widely from one multinational enterprise to another" (Organisation for Economic Co-operation (OECD) 2011, p. 17).

The power multinational enterprises have sometimes makes them appear as "a global institution in terms of its power, authority, and relative autonomy (Ruggie 2017: 13)." This creates a concern of their uncontrolled further expansion.

Most laws consider only the separate entities of MNEs whereas only few consider the MNE as a whole construct. Therefor a difference between the legal regulations and the daily economic business practice appears. This difference is one of the key characteristics of MNEs and leads to a global governance gap (Ruggie 2017: 13; Piasecki and Gudowski 2017: 152). The behaviour of some multinational enterprises generates distrust, because MNEs often use their power to ask for exceptions or weaker regulations, especially in developing countries. In many cases developing countries have natural resources and cheap manpower, which attract multinationals to take advantage of. Which lead to the assumption that MNEs impose conditions that negatively impact the national sovereignty of developing countries (Piasecki and Gudowski 2017: 152).

Most research in CSR focuses on the theoretical exploration of the topic, of particular segments of socially responsible behaviours of corporations, such as protection of the environment, human rights and employee rights protection. There is relatively little attention given to examining factors that might prompt MNEs to realize socially responsible programs. Research conducted in less developed areas is rather directed at issues of breach of CSR rules (Dancewicz 2016: 104–114).

2.2 *Evolution of CSR in Poland: Example of the Retail Market*

Poland is based on its comprehensibly transition process into a market economy an interesting case for researching the evolution of CSR in a certain country without the usually already existing boundaries. Especially the branch of retailing seems to be interesting as they were very welcome on the one hand and criticized on the other hand.

"D(d)ue to the fact that the retail market was not saturated at the beginning, consumers were only slightly protected by the law, and there were no institutions promoting the implementation of social responsibility standards by companies (Smigielska and Oczkowska 2017: 1)."

The early stage foreign retails in Poland sometimes lacked ethical standards which would be considered today as "immoral management". This might be based on incomplete legal frameworks in the country as well as the faced challenge to meet specific quality standards which appeared to be higher in Poland than abroad. Therefor CSR in Poland at least regarding MNEs applied on a voluntary basis might be a myth (see Smigielska and Oczkowska 2017: 7–11).

In this case three major developments can be highlighted:

- In 1991 the government set up The Foundation of the Polish Promotional Programme "Now Poland" with the aim to promote domestic production.
- In 1999 the Responsible Business Forum was set up. Its mission is promoting CSR activities and sustainable development. Since 2002, it publishes annual reports to highlight good practices of companies operating in Poland
- In 2000 the Office of Competition and Consumer Protection elaborated the Act on competition and consumer protection which defined the principles of the entire system of consumer protection.

It was observed that, when the retail market was not saturated, and there was a lack of legal regulations, only few non-governmental organizations as well as CSR-unaware consumers, multinational retailers did not take care of CSR, even though they were applying CSR practices in their host markets (see Smigielska and Oczkowska 2017: 11).

The fact that multinational retailers started to implement CSR standards when facing growing competition suggests that their attitude should be classified as a business case for CSR. This raises the question of whether activities should not rather be considered as a part of societal marketing. Societal marketing is defined as "delivering to the target consumers superior value in a way that maintains or improves the consumer's and the society's well-being" (Kotler and Armstrong 1996: 19).

In 2004 a study including Danone and Bayer examined the extent to which MNEs build social capital in Poland. Their conclusions included, but where not limited to the following findings: "Compared to the number of projects undertaken by overseas governments, the number of projects undertaken by European multinationals is small (around 7% of the government figure). Though MNC figure as a percentage of currently ongoing projects may be relatively higher. Many European MNCs in Poland demonstrate little or no social engagement. Of those firms that are socially engaged, their engagement tends to be at the national and local levels, mainly dealing with environmental and education issues (Jones et al. 2004: 25)".

2.3 *Corporate Social Responsibility*

In contrast the concept of corporate social responsibility (CSR) is somehow a response to the lack of possibilities to use institutional fines internationally and the lack of self-regulatory activity of corporations. Literary sources provide various definitions and explanations of CSR (Dancewicz 2016: 46–48). For the purpose of this work, a definition proposed by B. Rok has been used, according to which CSR is 'a long-term and strategic approach, based on principles of social dialogue and search for solutions beneficial to everyone' (Rok 2012: 427). Although the assumptions of CSR have characteristics of soft legal regulations, they increasingly force enterprises to adhere to the rules of social coexistence. This is influenced, apart from the pressure from public opinion in Poland, by international institutions, non-governmental organisations, and corporations which are CSR leaders, such as Solidarity (polish trade union), Caritas Polska (see The Guardian 2011).

2.4 *Stakeholder Approach to Tackle New CSR Challenges*

Within Poland's transition process towards a market economy new problems emerged, such as non-effective law in the global dimension, uncontrolled, degenerating exploitation of resources. At the same time, social awareness increased, that resulted in disapproval of lack of responsibility from corporations' side. Mutual dependence of corporations and stakeholders marks out the area of benefits and losses resulting from this co-dependency (Freeman and Reed 2010).

As an example, there are the practices of CSR reporting within the Energy and Mining sector of polish companies. "CSR reports of Polish companies have similar structures at a very high level of analysis, but a more detailed study reveals much diversity in the approaches to the report's content. Even fairly similar companies may devote significantly different amounts of space to the same issue. The number of similar stakeholders varies, irrespectively of the company's size. Considerable diversity of reporting patterns results from the nature of CSR reporting, because it concerns highly entity-specific issues (Szczepankiewicz and Mućko 2016: 15)."

Showing the connection between stakeholders' expectations and the possibility to achieve its objectives by organisations is the reason that the stakeholder theory became the basis for formulating and developing ideas connected to socially responsible business (Damali 2008). Identifying stakeholders, their expectations and effectively managing them can be key factors for building the enterprise's value in the long run (Domagała-Korona and Herman 2006: 11). This becomes particularly meaningful in moments of crisis, where relying on company's reputational capital proves to be essential.

2.5 *Role of NGOs in Polands CSR Process*

A growing importance for monitoring the ethics of corporate actions is seen of - watch-dog organisations (Batko-Tołuć and Izdebski 2012) and spontaneous initiatives organised by social media users that publicise antisocial behaviours of corporations. Fierce actions of corporations trying to neutralise the effects of such initiatives indicate how damaging for the company's reputation they can be and how can they influence their economic results causing boycotts (Wyborcza 2017).

Apart from the company's key stakeholders there is the role of international and social organisations that monitor the way companies conduct business and that set out norms of conduct. As the development of new technologies advances, it becomes more and more difficult for companies to hide actions that are socially harmful (Dancewicz 2016: 239–255).

Non-governmental organisations (NGOs) are an important factor that influences the behaviour of multinational corporations. In Poland they have a long tradition (Szustek 2008). Nevertheless, due to the aforementioned circumstances, until 1989 they focused mainly on independence (Solidarność) and challenging economic conditions. The diffusion in civic values and initiatives from the Euro-Atlantic region impacted the formation of Poles' mind-sets. Poland joining the European Union was a milestone in NGOs' development in the country.

At that time also, non-governmental organisations whose interest was in corporate social responsibility started to emerge. Somewhat earlier, in 2000, one of the most active and opinion-forming organisation of this kind was created under the name of the Responsible Business Forum (RBF). The state of knowledge and interest in CSR at that time is reflected in the words of the Forum's founders: In year 2000 not many people in Poland heard about CSR and business ethics was associated with an academic discipline. However: Responsible Business Forum was not created in a void, but was a result of ideas coming from the countries of the West and changes in perceiving the tasks that business in Poland is faced with (Forum Odpowiedzialnego Biznesu 2015).

Responsible Business Forum (RBF) is responsible for shaping, analysing and promoting CSR in cooperation with academia, business, public administration and non-governmental organisations. Based on CSR reports they contribute to promoting good practice, and a high position in RBF's ranking means prestige for a corporation, which equals good reputation. Organisations like RBF set new standards and shape the attitudes towards CSR in branches of multinational corporations in Poland. Apart from publishing good practice it also condemns the malpractice that can negatively influence enterprise's reputation.

Since 2002 it publishes a report called 'Responsible Business in Poland', which is the biggest review of socially responsible business practice. Although each year the number of firms applying for their actions to be published in the report raises, they still make a small percentage of the whole number of business entities (FOB 2016). In 2016 total of 180 companies applied, whereas the biggest foreign investors in Poland amount to ca 2400, which means that only about 7.5% of major foreign investors are actively engaged in the communication of their local CSR activities (Polska Agencja Inwestycji i Handlu 2016).

2.6 Reasons to Engage in CSR: MNEs in Poland

A recent study on a sample of 60 reports from Polish companies listed on the Warsaw Stock Exchange examines (Dyduch and Krasodomska 2017).

The profit-maximizing rule of business does not explain the behaviour of many companies nowadays. Business executives recognize the importance of non-economic purposes. CSR disclosure is a response to the changes that have occurred in business operations in recent years and which consist in the growing belief that companies' social and environmental performance has an impact on their general performance (Dyduch and Krasodomska 2017; Piasecki and Gudowski 2017: 152).

Factors that affect the CSR disclosure among companies listed on theWarsaw Stock Exchange are "the company's size cannot be perceived unequivocally as a determinant of CSR disclosure (Dyduch and Krasodomska 2017: 18)" only the turnover has a direct positive impact on the disclosure of CSR. "They care for the maintenance of their good corporate image in order to acquire and retain the business partners' and customers' loyalty. They are also able to absorb extra costs of CSR disclosure (Dyduch and Krasodomska 2017: 18)".

The matter of responsibility in general does bring a number of dilemmas of ontological nature. This concept, according to literary sources, is recognised as an attribute of humanity (Krzysztofek 2015). Since enterprises are not human being, they do not possess moral sensitivity (Jędraszewski 1984: 370), It therefore does make sense to pose a question: what can lead corporations to execute the postulates of conducting business in accordance with social interest?

One of the factors that motivate enterprises to socially responsible actions is the fear of losing good reputation (Dancewicz 2016). The source literature shows the important role that reputation plays in increasing the value of the enterprise (Roszyk-Kowalska 2006; Dyduch and Krasodomska 2017: 18). Thus enterprises which in the past had reputational losses are more willing to engage in pro-social actions. They tend to more often consider social interests and expectations. They pay special attention to the declarations being in accordance with the execution of socially responsible programs. A clear gap between declarations and the execution of CSR policy can be one of the main reasons of reputation loss (Dancewicz 2016: 270–271).

Legitimacy theory seems to be the most relevant one for explaining CSR disclosure practices of Polish listed firms. "Polish firms report on CSR activities in order to respond to public pressures and build or sustain corporate legitimacy. Companies whose operations are harmful to the environment use CSR disclosures as a means to publicize their image and legitimize their corporate initiatives.

In addition, companies with higher turnover are motivated to devote more financial resources to initiatives promoting a positive impression of their performance. Corporate legitimacy is also related to public or social visibility. The positive influence of the inclusion in the RESPECT Index on CSR disclosures is proof that Polish firms that act in a socially responsible manner are capable of providing high-quality CSR disclosure. Companies under investigation report on CSR activities mainly to act and be seen as acting within the bounds of what is considered acceptable according to the expectations of society about how their operations should be conducted (Dyduch and Krasodomska 2017: 18)".

"One conclusion seems inescapable: in light of the multinational's power, authority, and relative autonomy, the time-worn mandatory/voluntary dichotomy (of CSR and regulations) inhibits rather than advances our coming to grips with the challenges posed by corporate globalization (Ruggie 2017: 14)".

Even though the concept of CSR is sort of a soft law, the consequences of breaching it can be as severe as if the institutional law was broken. The benefits of respecting the CSR guidelines and the disadvantages of not adhering to them are noticed by the business community. Also, more and more often the latter support fulfilling the civic duties by corporations. Thus, they not only set out standards of conduct, but also act as a guarantor of establishing a new order of rules for conducting business.

3 Examining the Engagement of MARS and Avon

3.1 Applied Methodology

The main objective of this work is to diagnose the situation of CSR in Polish branches of multinational corporations.

Therefore, the subject of the research is the way the CSR policy is implemented in Polish branches of multinational corporations. Hence the research problem has been formulated as: CSR in Polish branches of multinational corporations. The research has explorative character, because its objective is to deepen the knowledge required to understand the scope of the global CSR programs' implementation in local branches.

Because of applied methodology and scope of research (two enterprises, ten interviewee's sample) the research results can only be treated as an illustration of the assumptions used and it cannot serve as material for making any general conclusions.

The undertaken literature studies and the secondary sources analysis lets one formulate detailed objectives of the work. It is also possible to identify differences between the global and local approaches to CSR policies, which confirmed the validity of the undertaken research objective and influenced the choice of Polish branches of multicultural corporations for further exploration. Two Polish branches of MNEs have been examined, which helped to address the stated detailed research problems. Further qualitative research was conducted in the form of partially structured in-depth interviews. The research respondents were chosen based on a non-probabilistic sampling, i.e. on a purposeful one. The respondents were managers, external co-workers and representatives of local communities related to the enterprises being examined. The interviews were conducted in person by the author in July and August 2017. The main research problem, detailed research problems and questions are shown in Table 1.

Table 1 Main research problem, detailed research problems, research questions

Main research problem CSR in polish branches of multinational corporations	
Detailed research problems	Research questions
Setting scope of global CSR programme realised locally	Does the company have ethical code? Where is it published? Is there a department or person in the company responsible for CSR? How much autonomy does the polish branch have when creating and implementing the local CSR policy? Is a local CSR strategy being created?
Identification of factors that can encourage polish braches of MCCs to realise socially responsible programs	How can implementation of CSR policy influence firm's operation?
Identification of lack of consequence in case of not implementing CSR programs	Can lack of CSR policy influence firm's operation?

Source: Own elaboration

3.2 *MARS Case*

In Forbes' ranking of America's Largest Private Companies in 2016 Mars came 6th (Forbes 2017). This American corporation established in 1911 is a producer of sweets, beverages, ready meals, chewing gum and animal feed. It operates in 74 countries in the world, employing around 80,000 people. It prides itself on a 100-year tradition of family firm, for which the relationship with stakeholders is a priority and which constitutes the basis for mutual benefits. The asset of a family firm lays in its independence from stakeholders and stock market fluctuations, which favours running the business in accordance with chosen hierarchy of values. Mars indicates five rules as its basis of organisational culture and approach to business: quality, responsibility, reciprocity, efficiency, liberty. These rules sound laudable and correspond with the rules of socially responsible business. On the company's official website there is a brief description on what stands behind the slogans (Mars 2018a).

On the website of their Polish branch not even a small note regarding the topic not be found. The only information the Polish customer gets is the mention of the values of the company. What exactly they mean and how they are put into practice remains a mystery, unless one takes as an explanation the following brief note: 'The customer is our boss, quality is our work and a favourable price is our objective. Our firm's domain is the highest quality in all aspects of work. Quality is an uncompromising standard of our actions and derives from our passions and pride of being part of the Mars community. High quality work, which is a result of our personal engagement, is the first factor of high quality brands and the source of our reputation because of high standards which we comply with (Mars 2018b).

It is worth noting that Mars has been present in Poland since 1992, in the form of three partnerships: Mars Poland, Wrigley Poland and Royal Canin Poland.

The company employs around 2500 people, and its products made in Poland are distributed in several dozen countries of the world.

In 2011 the company issued a document called ‘Our principles in action’, where the firm’s objectives, the scope of their realisation and future commitments are presented. The document shows areas where Mars undertakes socially responsible initiatives globally (Mars 2018c).

Respondents who took part in the study (two managers of higher rank and three representatives of local communities) expressed their absolutely positive opinion about the firm. They highlighted the firm’s ethical and socially responsible attitude. None of the statements indicated possible antisocial behaviours of the firm.

The local branch has a responsible person who takes care e.g. of the identification of needs of local communities and coordinating initiatives to fulfil them. Long term activities, like e.g. taking care of animal shelters are coordinated by people who take up these responsibilities through the employer’s voluntary service (own research).

According to the respondents the Polish branch has a big autonomy in creating and implementing CSR projects. They independently create strategic plans in this area based on previously identified local needs. Apart from the plan for the whole department (One Mars), every factory that is part of the Mars Group creates its own CSR program. Also a margin is left in the form of budget for interim actions. ‘We are proud that our company saves rain forests or sponsors the ecological farming of cocoa trees. Above all, however, we would like to create friendly work conditions and to be a *good neighbour’*, explains one of the respondents. She talks with passion about the employees’ engagement in the animal care: ‘animals are defenceless. We feed dogs and cats in shelters and we also support guide dogs. We are aware that maintaining such a dog is costly and often a person in need cannot afford it’. ‘We sponsor dog therapy for those who really need it’, she adds. ‘We want our factories not to bother people living nearby. We try our best to make sure our neighbours do not live in polluted environment. We also try to help build local infrastructure which would serve the locals’ (quote from the interview).

According to the respondent: ‘we want to make sure people are keen on coming to work. Maybe it will sound a bit pompous, but we want our employees to feel like part of the family. We are a family business after all’ (quote from the interview). The other interviewees talked in a similar manner regarding the company. ‘We would not like to show off, as corporate responsibility is a mere duty, rather than PR. Therefore when media or contests’ organisers reach out to us we agree to engage, we rather do not seek for this ourselves, though’ (quote from the interview). ‘Sometimes we help people who would rather we did not talk about this for various reasons and we respect that’ (quote form the interview).

Representatives of local communities also praised the company: ‘One can’t complain. They don’t pollute, they help when needed. When there is a problem, they listen and help if they can (...), for example they built a road. On top of that they can give one a job’ (quote from the interview).

Summarising the results of the research in the condition of CSR in the Polish branch of the international corporation of Mars, the following conclusions can be drawn:

- Discrepancies between the global and local implementation of CSR do not affect the level of corporate responsibility of the Polish branch
- It is assumed on the local level (as well as on the global one) that the role of ethical code is fulfilled by five rules, which are the basis of everything the firm does
- The Polish branch has strong autonomy in creating and implementing CSR policy, which provides it with flexibility in responding to local needs
- Attentiveness to work conditions and atmosphere has an impact on employees' attitude towards the company
- Attentiveness to maintaining good relationships with local communities helps to avoid conflicts and leads to a friendly attitude towards the firm
- Polish branch of Mars corporation approaches social responsibility as a duty rather than an obvious obligation or a marketing effort

3.3 Avon Case

The story of Avon company resembles the American dream. It was established in 1886 by a door-to-door salesman in the U.S. who realised that his female clients preferred smelling well to reading books. Therefore the mission of the firm is to understand and fulfil women's needs by providing them with cosmetic products and giving opportunities for professional development (Avon 2018a).

The company produces cosmetic products and organises their direct sale via multilevel marketing method. Currently it operates in over 70 countries, engaging over 6 million consultants (Avon 2018b). The company employs around 25,000 employees, reaches the yearly revenue of 5.7 billion dollars, and the recognition of the brand on the major markets amounts to 90% (Avon 2017, 2018c). The official Avon Products, Inc. website informs in detail about the principles the company follows in implementing the CSR policy. Apart from the core values and principles that the firm follows, the website also informs about the areas where the declarations about CSR are being fulfilled, e.g. environment protection, battle against breast cancer, fight against domestic violence (Avon 2018d).

3.3.1 Analyse of CSR Communication

Unfortunately, there is no reference to all that on the Polish website of the company. Typing 'CSR' into their search engine brings 0 results (Avon Poland 2018a). 'Social responsibility' gives only one result, namely the 'cosmetician with a pink ribbon' (Avon Poland 2018b).

In the 'About us' section there is a brief note about pro-social actions, referred to as charity, like fight against breast cancer and domestic violence (Avon Poland 2018c). Slightly below there is a list of awards that the company has got in the category of 'best employer' and information about implemented social campaigns in 2011. The Polish site does not publish the code of conduct that is available in Polish language on the global site (Avon Poland 2018d). Considering the role that the Polish branch plays in the global structure it seems to be rather a modest display of CSR implementation, while Avon operates in Poland since 1992 in form of three independent partnerships dealing with cosmetics sales, production and distribution all over Europe, and also with financial services.

Avon Poland is quite weakly represented in the rankings of corporate social responsibility. Although the company is shown rather in a positive way in the media, it was unable to entirely avoid ethical flaws. In June this year there was much media coverage of a case of a female employee who claimed on facebook that she got fired after being diagnosed with breast cancer (Newsweek 2017). In the context of the declared fight against breast cancer and supporting women suffering from it that incident undermined the firm's reputation. Despite their immediate reaction, the principles they claim they follow were questioned, which can have an impact on the Avon's reputation among women who are the firm's main stakeholders.

3.3.2 Special Insights from the Interviews

Further research was undertaken in the form of interviews with higher-rank managers (two persons), external collaborators (two persons) and clients (one person), in order to verify the findings from available secondary sources.

Managers highlight the differences in implementation of CSR policy by individual partnerships within the Avon Poland Group. According to them Avon Cosmetics mainly focuses on cosmetics distribution and implements global CSR programs. Among them the following projects: 'Pink ribbon', whose objective is to fight against breast cancer and against domestic violence towards women. As one of the respondents said: 'Empowering women and helping them in difficult situations is not an empty slogan. Educational actions and sponsoring the research will surely contribute to an earlier detection of the disease' (quote from the interview).

Another interviewee stated: 'A woman who experiences domestic violence should not be left alone. Avon not only leads programs against violence, but also gives women the possibility to become financially independent' (quote from the interview). Most of the respondents confirm this position, which indicates a serious recognisability of key projects, i.e. fight against breast cancer, and fight against violence towards women and supporting entrepreneurship among women. One of the respondents told a story from her life, where Avon played a key role: 'I didn't have adequate education, I was a single mother. I worked very hard, but couldn't accomplish much. Only Avon gave me such a chance (. . .) Maybe not everybody is suitable for this kind of work, but that is the case with any job (. . .) Today I'm in a position I couldn't even dream about' (quote from interview).

Although it is related to the business model of the company, it can be recognised that the mission of helping women is being fulfilled. All the respondents agreed that Avon's prosocial actions are valuable. Interestingly, however, people see them as charity rather than CSR. None of the respondents associated them with CSR. Neither has any of them, including the high-rank managers, was able to say anything about the CSR policy, despite the fact that a globally published CSR report is available for the employees (Avon 2015). Also nobody could say anything about the ethical code or rather the lack of its Polish version.

3.3.3 Evaluation of Avon's CSR Engagement in Poland

Differences in implementing CSR programs were pointed out by respondents when talking about the aforementioned Avon Cosmetics and Avon Operations. The latter is responsible mainly for cosmetics production in the factory in Garwolin. There, as interviewees have claimed, the main emphasis is put on actions supporting local communities: 'We care about good relationships with our surrounding. That is why we support local initiative (. . .) mainly by sponsoring' (quote from the interview). The interviews with the employees show that many actions are replications of central programs, yet they are presented as local initiatives. Despite the lack of awareness of the company's global CSR policy, the respondents highlighted the importance the company attaches to transparency and the vigilance towards corruptive behaviours. According to one of the respondents: 'We cooperate with non-governmental organisations, but we avoid the governmental ones. We do not want to be seen as trying to influence the public administration' (quote from the interview).

Summarising the result of the research in CSR condition in the Polish branch of Avon multinational corporation one can conclude the following:

- There is a low awareness there of what CSR and its role is
- Mainly the global projects of the firm are recognised and implemented (fight against breast cancer, fight against violence towards women, supporting women's entrepreneurship) and seen as charity actions
- The Polish branch modifies and enriches global actions adjusting them to local needs
- The company aims at treating its employees well and at creating developmental opportunities for them
- Despite useful pro-social actions disproportions between emphasising business and socially responsible actions are visible

3.4 Conclusions from Own Research

Research results based on the analysis of the condition of CSR in two Polish branches of the company show a low awareness of the explicit meaning of CSR and its role from an employee perscpetive. Respondents often associate CSR with charity and sponsoring. Moreover, in their opinion social responsibility should result from responsibility for the nearest surrounding and its particular needs, and not from the need to fulfil some theoretical programs.

Review of global and local websites shows discrepancies in the approach to CSR policy. The websites of researched companies contain detailed CSR programs and ethical codes. The websites run by local branches mention that the company is socially responsible.

Some global programs are implemented locally. Polish branches are quite strongly autonomous in creating and implementing their CSR policies. The research shows that it is mainly used to create good relationships with the local community, which may mean that this is an important factor that motivates Polish branches of MNEs to pro-social behaviours. Research has not shown the Polish MBEs branches as particularly concerned with the likelihood of losing good reputation. Only Avon's quick reaction to the public opinion's protest about the sick employee dismissal can indicate otherwise.

The research has not indicated the awareness of the existence of a relationship between implementation of CSR policy and functioning of the firm. On the contrary, it has been shown that Polish market does not enforce efforts towards diligent CSR programs' implementation. That indicates lack of essential dialogue with stakeholders that would be helpful in recognising their needs.

4 Conclusions

The limitations of the research regarding its duration and applied methods do not allow one to draw general conclusions. They may only contribute to knowledge of the state of CSR in Polish divisions of international corporations and indicate direction of further exploration.

The research has heretofore given evidence for a steady growth of interest in implementation of CSR programs by Polish MNEs divisions, which can be seen in the growing number of published CSR reports (*Pięć lat raportowania...* 2012).

Despite the fact that there are many organizations, business-oriented as well as NGOs, that have long promoted running business responsibly, such as: Forum Odpowiedzialnego Biznesu, CSRinfo, RESPECT Index, the level of awareness and knowledge of CSR in Polish MNEs divisions is relatively low, which is shown by the research, and above all the still weak awareness of the importance of CSR in today's business (Kowalska 2015).

Among the corporate managerial staff still opinions are heard that CSR programs are costly and their implementation not fully justified (Tylec 2016), whereas the main points of focus for firms are environmental issues and charity.

"One conclusion seems inescapable: in light of the multinational's power, authority, and relative autonomy, the time-worn mandatory/voluntary dichotomy (of CSR and regulations) inhibits rather than advances our coming to grips with the challenges posed by corporate globalization (Ruggie 2017: 14)".

The discrepancy between knowledge, awareness and readiness to realize CSR programs between the headquarters and the Polish divisions can have multiple reasons. One of them may be the relatively low level of economic development as well as underdevelopment of civil society, but even more probably the Polish employees behave according to the international rules and regulations and as well do additional social and environmental friendly things locally, they just don't label it CSR, as it is for them natural. Similar behaviour can for example be observed in SMEs in Southern Germany.

Overall the engagement and especially the interviews showed that there is already a lot of engagement of multinational companies in Poland and that there only seems to be a lack of explicit communication that specific engagements are considered as CSR. But even if employees don't label actions as "CSR" it is even more important that there is local engagement according to those standards, which exists in Poland.

References

Batko-Tołuć, K., & Izdebski, K. (2012). *Organizacje strażnicze w Polsce. Stan obecny, wyzwania, perpektywy*. Warszawa: Instytut Spraw Publicznych.

Damali, J. (2008). A stakeholer approach to corporate social responsibility. A fresh perspective into theory and practice. *Journal of Business Ethics, 82*, 213–231.

Dancewicz, B. (2012). Ethics and social responsibility across cultures (chapter 7). In M. Rozkwitalska (Ed.), *Intercultural management issues*. Warszawa: Difin.

Dancewicz, B. (2016). *Zachowania etyczne w przedsiębiorstwach międzynarodowych a społeczna odpowiedzialność biznesu*. Łódź: SAN.

Davis, K. (1977). Business horizon 1975 – Five propositions for social responsibility. In A. B. Carroll (Ed.), *Managing corporate social responsibility*. Boston: Little, Brown and Company.

Domagała-Korona, B., & Herman, A. (Eds.). (2006). *Współczesne źródła wartości przedsiębiorstwa*. Warszawa: Difin.

Dyduch, J., & Krasodomska, J. (2017). Determinants of corporate social responsibility disclosure: An empirical study of polish listed companies. *Sustainability, 9*, 1934. https://doi.org/10.3390/su9111934.

Freeman, R. E., & Reed, D. (2010). Stockholders and stakeholders. A new perspective on corporate governance. In B. De Wit & R. Meyer (Eds.), *Strategy, process, content, context*. London: South-Western Cengage Learning.

Gallup Inc. (2016). *Global engagement report*. Washington. Accessed June 25, 2018, from http://www.shqiperiajone.org/sites/default/files/pdf/2016%20Global%20Civic%20Engagement%20Report.pdf

Global Finance Magazine. (2017). The world's richest and poorest countries. *Data collected from international monetary fund, world economic outlook database*. Accessed June 25, 2018, from https://www.gfmag.com/global-data/economic-data/worlds-richest-and-poorest-countries

Hąbek, P. (2017). CSR reporting practices in Visegrad group countries and the quality of disclosure. *Sustainability, 9*, 2322. https://doi.org/10.3390/su9122322.

Jędraszewski, M. (1984) Emmanuela Lévinasa filozofia Drugiego, Przegląd Powszechny, nr 12, s. 370.

Jones, I. W., Nyland, C. M., & Pollitt, M. G. (2004). *Multinationals in developing communities: How EU multinationals build social capital in Poland*. WP 285 March 2004, ESRC Centre for Business Research, University of Cambridge.

Kotler, P., & Armstrong, G. (1996). *Principles of marketing* (7th ed.). Upper Saddle River, NJ: Prentice Hall International Inc.
Krzysztofek, A. (2015). *Rozważania o pojęciu odpowiedzialności*. Prace Naukowe Uniwersytetu Ekonomicznego we Wrocławiu. ISSN 1899–3192 nr 401.
OECD (Organisation for Economic Co-operation and Development). (2011). OECD guidelines for multinational enterprises. OECD Publishing, Paris. Accessed June 25, 2018, from http://www.oecd.org/daf/inv/mne/48004323.pdf
Piasecki, R., & Gudowski, J. (2017). Corporate social responsibility: The challenges and constraints. *Comparative Economic Research, 20*(4). https://doi.org/10.1515/cer-2017-0032.
Rok, B. (2012). Społeczna odpowiedzialność biznesu. In W. Gasparski (Ed.), *Biznes, etyka, odpowiedzialność*. Warszawa: Wydawnictwa Profesjonalne PWN.
Roszyk-Kowalska, G. (2006). Reputacja jako kluczowa kompetencja przedsiębiorstwa. Prace Naukowe Akademii Ekonomicznej we Wrocławiu, nr 1141.
Ruggie, J. G. (2017). Multinationals as global institution: Power, authority and relative autonomy. In *Regulation and governance*. Milton, QLD: Wiley.
Smigielska, G., & Oczkowska, R. (2017). Retailers' responsibility towards consumers and key drivers of their development in Poland. *Administrative Sciences, 7*(3). https://doi.org/10.3390/admsci7010003.
Szczepankiewicz, E. I., & Mućko, P. (2016). CSR reporting practices of polish energy and mining companies. *Sustainability, 8*, 126. https://doi.org/10.3390/su8020126.
Szustek, A. (2008). *Polski sektor społeczny*. Warszawa: Uniwersytet Warszawski.
The Guardian. (2011). *Poland's steady economic progress matched by growth in social inequality*. Accessed June 26, 2018, from https://www.theguardian.com/world/2011/apr/06/poland-economic-progress-social-inequality

Electronic References

Avon. (2015). *Corporate responsibility report*. Accessed July 2, 2018, from http://www.avoncompany.com/corporate-responsibility/about-cr/reporting/reports/2015-avon-cr-report.pdf
Avon. (2017). *Form 10-K report*. Accessed July 2, 2018, from https://www.sec.gov/Archives/edgar/data/8868/000000886818000013/a2017form10-k.htm
Avon. (2018a). *Avon's history*. Accessed July 2, 2018, from http://rep.avon.de/PRSuite/history.page
Avon. (2018b). *About Avon*. Accessed July 2, 2018, from https://www.avon.de/uber-Avon/
Avon. (2018c). *Avon factsheet*. media.avoncompany.com/download/Avonfactsheet.pdf
Avon. (2018d). *Values*. http://www.avoncompany.com/aboutavon/history/values.html
Avon Poland. (2018a). https://www.avon.pl/szukaj/wyniki/?q=csr&cb=1429691627
Avon Poland. (2018b). https://www.avon.pl/szukaj/wyniki/?q=odpowiedzialno%C5%9B%C4%87%20spo%C5%82eczna&cb=1429691627
Avon Poland. (2018c). Accessed August 22, 2017, from http://www.kariera.avonpolska.pl/pl/about
Avon Poland. (2018d). *Code of conduct*. Accessed July 2, 2018, from http://www.avoncompany.com/documents/ethics_and_compliance/Code_of_Conduct_Polish.pdf
FOB. (2016). Raport „Odpowiedzialny biznes w Polsce 2016. Dobre praktyki". Accessed July 2, 2018, from http://odpowiedzialnybiznes.pl/publikacje/raport-2016/
Forbes. (2017). Accessed July 2, 2018, from https://www.forbes.com/companies/mars/ and https://www.statista.com/statistics/549091/largest-private-us-companies-by-revenue/
Forum Odpowiedzialnego Biznesu FOB. (2015). Początki FOB – grupa entuzjastów i trzy firmy. Accessed July 2, 2018, from http://odpowiedzialnybiznes.pl/aktualno%C5%9Bci/poczatki-fob-grupa-entuzjastow-i-trzy-firmy/

Mars. (2018a). *Quality*. Accessed July 2, 2018, from http://www.mars.com/global/about-us/five-principles/quality

Mars. (2018b). *Kim jesteśmy*. Accessed July 2, 2018, from http://www.mars.com/poland/pl/about.aspx

Mars. (2018c). *Principles in Action*. Accessed July 2, 2018, from https://www.mars.com/global/doing-our-part/principles-in-action

Newsweek. (2017). http://www.newsweek.pl/polska/spoleczenstwo/avon-zwolnil-chora-na-raka-piersi,artykuly,411646,1.html

Polska Agencja Inwestycji i Handlu. (2016). *List of major foreign investors in Poland – 2016/2017*. Accessed July 2, 2018, from https://www.paih.gov.pl/files/?id_plik=30926

Wyborcza. (2017). *Tiger drink advertisment*. Accessed July 2, 2018, from http://wyborcza.pl/7,155287,22211114,1-sierpnia-tiger-reklamowal-sie-srodkowym-palcem-i-haslem-chrzanic.html?disableRedirects=true

Relationship Between Sport and CSR in the Conditions of Functioning of Sport Organizations in Poland

Paweł Kuźbik

1 Introduction: Methodology of Research

The aim of this chapter is to show the relationship between sport (sports clubs) functioning under specific conditions, and the concept of social responsibility in Poland. The author poses the following research questions:

- Is there any relationship between sport, sports clubs, and the idea of CSR and what is its foundation?
- How the concept of CSR is understood in sports organizations?
- At what stage are the clubs in their implementing of the CSR concept? (CSR 1.0—CSR2.0—CSV)
- Can a model of sport-CSR relationship and its practical implementation be exposed based on the diagnosed club activities in the CSR area?

The presented study is the result of research conducted in 2015–2017 on 85 sports clubs representing four types of team games, i.e. football, volleyball, speedway and basketball. The research involved professional clubs, having the status of a joint-stock company and physical culture associations competing at the national level (Kurkowska-Budzan 2017).

The selection of this type of sports clubs was intentional, because their way of functioning allows for identification of specific management subsystems, and thus there is a supposition that the concept of social responsibility can and should apply to the working practice. Sports organizations have been the main goal of the author's research since 2007.

P. Kuźbik (✉)
Faculty of Management, University of Lodz, Łódź, Poland
e-mail: pawel.kuzbik@uni.lodz.pl

A. Długopolska-Mikonowicz et al. (eds.), *Corporate Social Responsibility in Poland*, CSR, Sustainability, Ethics & Governance,
https://doi.org/10.1007/978-3-030-00440-8_12

In 2016, the author assembled and sorted out the current knowledge on CSR and on the necessity to implement it into the conditions of the functioning of sports clubs on the example of football clubs (Kuźbik 2016). Therefore, in this paper, in the theoretical part, the author does not make a comprehensive review of the CSR concept, as it would be unfounded for the purpose of this study, but he focuses only on the understanding of the current definition (in accordance with the ISO 26000 standard) and on the current change trends from the concept CSR 1.0 to CSR 2.0 or CSV.

A survey questionnaire, analysis of documentation provided by clubs and an in-depth interview were used to collect the data. In addition, two forms of observation were used: participant and non-participant.

2 In Search of Responsibility

Questions about responsibility have been asked since the dawn of centuries. The theoretic and practical dimension of this issue was debated by politicians, leaders of local communities, clergymen, gurus, scientists or business representatives. Currently, in the twenty-first century, it is still not an easy task to clearly define responsibility. As H. Arendt writes, "nothing easier and nothing more difficult than to face the question of responsibility" (Arendt 1997), and J. Filek notes: "we do not know that we do not know what responsibility is. We ask about the subject of responsibility, we ask what he is responsible for, who he is responsible to, but what it would mean to be responsible, seems to be understandable to us by itself" (Filek 2003).

As a consequence, we fall into a dense web of the multidimensionality of the notion of responsibility, which becomes the main difficulty in its unequivocal interpretation. Nevertheless, what seems to be the most crucial is the fact that the responsibility should be connected with the human being and all his activities. Bearing responsibility is a factual state which, in a sense, automatically imposes itself on the one who performs a particular type of activity. We become responsible for the act since we undertook it and committed it (Ingarden 1987).

It seems, therefore, that the basis for understanding the idea of responsibility is to conceive it in the context of a positive experience, which allows for putting tasks in front of people and, moreover, it can be a bridge between human existence, human activity (including economic activity) and ethics (Filek 2002). In dictionary meaning, responsibility is "a moral or legal obligation to answer for own or somebody else's actions; it is taking on the duty of taking care of someone or something" (PWN 2017).

J. Tazbir, referring to the definition in Linde's Dictionary, goes further and writes that responsibility is "the obligation of explaining oneself to somebody about something. Explaining oneself means giving an account of the proper performance of duties resulting primarily from the office, and subsequently from the social status or religion." (Tazbir 1996). Put simply, it's about the organization's readiness to answer the community. E. Hope speaks in a similar tone, pointing out that: "Being responsible is recognizing human dignity in every human being, and at the same time it is respect for others expressed through professional integrity and the ability to perceive actions in a social perspective." (Hope 2000).

With the responsibility thus understood, it seems, therefore, that there should be no problem with the transition from the responsibility of an individual, or even of us as human beings, to the responsibility of an organization, which was created by the people and for the people. It is necessary to agree that the value and the need for responsibility are of a supra-historical and universal nature because we encounter it in all conditions of organized social life, and what who, for what and to who is held accountable, will depend on the social circle or industry as well as social needs. For this reason, responsibility leads to the complexities of interpretation and shows its multidimensionality.

It seems, therefore, risky to grasp the universal definition of the concept of responsibility, and not necessarily a needed, but it is worth emphasizing here the basic foundations which, regardless of the context, the responsibility will have in common. Every responsibility begins at the place where the act or omission of action begins, and it ends where the effects of such conduct appear or may appear. Responsibility is not only an element of measuring merit or blame for specific decisions but above all, the degree of moral maturity, social autonomy and self-awareness in the social space.

As a consequence, in relation to the organization, the responsibility should be attributed to the whole management system, and it itself becomes the bond of all actions taken to which the organization has committed (Fig. 1).

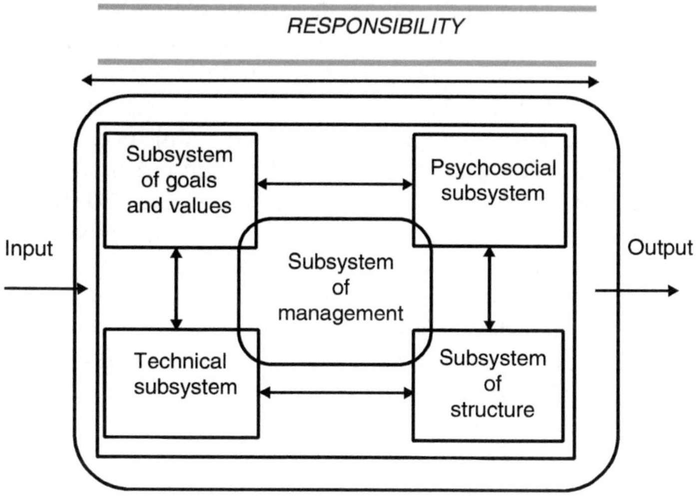

Fig. 1 Systemic approach to organization including the scope of responsibility. Source: Kuźbik (2017), based on Bielski (2005)

Since the organization does not work in a vacuum and is an open system, it should be connected with the environment and, hence, the community whose part it is also itself. Therefore, it will not be a mistake if we acknowledge the social responsibility of the organization as a dialogic relationship between man and man, that is, co-existence with others using specific means, tools and resources to create the overall socio-economic reality in a sustainable way. As a consequence, the definition of CSR (Corporate Social Responsibility) proposed by the ISO 26000 standard will be up-to-date, although it should be supplemented with a comment.

The ISO 26000 standard understands social responsibility as an organization's responsibility for the impact of its decisions and actions (products, services, processes) on society and the environment, through transparent and ethical behaviour, which (ISO 26000, 2010):

- contributes to sustainable development, including health and the well-being of society, (Pawlak et al. 2015)
- takes into account the expectations of stakeholders (people or groups that are interested in the decisions or actions of the organization),
- complies with applicable law and is consistent with international standards of conduct,
- is integrated with the activities of the organization and practised in its activities undertaken within its sphere of influence (PKN 2012)

The definition presented in this way makes social responsibility a tool that is applicable not only in large global companies but in any type of organization regardless of its size and scope. It determinates also other significant issues, namely the fact that it is impossible to create universal principles or tools that can be implemented in every organization and in all conditions.

The organizational practice has shown that what has been considered to be the foundations of CSR so far has not yielded the desired results, and the idea itself has earned the opinion of a cost centre rather than a profit centre (Socha 2017). In the minds of managers and, consequently, the surrounding environment, the concept of CSR and its programs have focused on the reputation of the organization and very often have little connection with its activities, which was usually difficult to justify and take into account in the strategic dimension of functioning, and thus take care for their functioning in the long run. As Visser writes, "Much more effective than wishful thinking about CSR as a fantastic tool to solve the problems of the world, it would be to see what it actually is: an outdated, obsolete artefact that was once useful but whose time has already passed. We must allow "old CSR" to die gracefully and provide it with a dignified burial. Let us give it what it deserves—praise for setting new limits of responsibility, it managed to achieve it in its glory days. Soon after, however, we shall look for a new generation of CSR—a newborn that will be able to continue the work." (Visser 2010).

This Visser's "newborn" is to be the concept of CSR 2.0 where according to the above-mentioned author the idea of social responsibility must cease to base its foundations on philanthropy directed towards PR, and instead become more interactive and open (Table 1) (Visser 2010).

Table 1 From CSR 1.0 to CSR 2.0

Traditional CSR (CSR 1.0)	New approach to CSR (CSR 2.0)
It is a way for companies to establish relationships with communities, to conduct philanthropic activities and to manage their image	Defined by the slogans: "global communities", "innovative partnerships", "stakeholder engagement"
It contains many pioneer start-ups, like Traidcraft, which later became a product for international corporations, such as shell	New tools include stakeholder panels, transparent reporting and a new wave of social entrepreneurship
Using the same standards, codes, norms and guidelines for all institutions	Recognizes the decentralization of power, as well as contextual changes: from a few large ones to many small ones and from exclusivity to community
CSR 1.0 (Principles)	*CSR 2.0 (Principles)*
Paternalistic	Community-based
Based on risk	Based on the prize
Based on the image	Based on results
Specialized	Integrated
Standardized	Differential
Marginal	Operating on a large scale
Western	Global
CSR 1.0 (Practices)	*CSR 2.0 (Practices)*
Premium markets	Markets from the "base of the pyramid"
Charity projects	Social entrepreneurship
CSR indices	CSR ratings
CSR sections	CSR incentives
Ethical consumption	Choice editing
Responsibility for the product	Contracts for the provision of services
CSR reporting cycles	Continuous flow of CSR data
Stakeholder groups	Social networks
Standards for processes	Standards for actions

Source: Visser (2010)

In order to make changes in the functioning of the organization in line with the new approach to CSR, Visser proposes five fundamental principles that should be complied with (Visser 2010):

- **Reassess**—measure the impact of your business to find out where you are. Do it honestly, not for PR purposes, but to know what can be improved throughout the production chain.
- **Realign**—check who you cooperate with; do you listen only to those people who agree with you; do you have contact with NGOs that call various things into question? We know that companies and governments themselves will not solve global problems. Therefore, COOPERATION is necessary.

- **Redefine**—think about whether your leaders are courageous (they admit that at the moment they do not solve existing problems) and ambitious in setting goals.
- **Redesign**—are you innovating in your products and services? Do you not only minimize the negative impact but create a positive one?
- **Restructure**—completely transform your business.

Such an approach is still a melody of the future on the Polish market, although noticeable is the trend associated with including CSR in the organizational strategy and with not treating it as an additional activity (Słowik 2017).

Interestingly enough, in 2007 M. Kramer and M. Porter wrote about this in an article entitled "Strategy and society: corporate social responsibility—a useful fashion or a new element of competitive strategy" where the authors have attempted to present an innovative approach to the business-society relationship.

They pointed out that corporate success and social well-being are not a zero-sum game. They then wrote: "The efforts of companies in terms of CSR were not as effective as expected, for two reasons. First of all, business and society are treated as two hostile forces; while actually, they are interdependent. Secondly, companies are expected to use the CSR concept in relation to general social issues instead of forcing them to think in these categories with reference to their own strategies." (Porter and Kramer 2007). The CSR concept has become a good PR tool; it was simply convenient for managers to create the company's image in this way. However, it was known that the opportunities generated by the strategic approach to social responsibility will eventually have to be noticed, and social needs and expectations will force a change in the way it functions.

The answer to these allegations is certainly to some extent the CSR 2.0 concept proposed by W. Visser in 2010, but M. Kramer and M. Porter a year later presented and described in detail the concept of creating shared value (CSV) (Porter and Kramer 2011). In the article, they explain that organizations belong to ecosystems, whose various conditions may affect the absorption of markets and thus the efficiency of these organizations. Then they write: "Enterprises must initiate collective impact projects involving all players from their ecosystems. For this, it is important to meet five conditions: ensuring a joint program, a common measurement system, mutually supporting activities, constant communication and targeted, basic support from one or several independent organizations" (Socha 2017). Finally, these authors define the shared value (Table 2) as a strategy and operational practices that improve the competitiveness of the enterprise, while elevating the living conditions of local communities in the environment where it operates (Porter and Kramer 2011).

It is therefore obvious that the future of social responsibility must be created in its practical, strategic dimension. Whether it will be a CSR 2.0 or CSV concept, it does not matter as long as the benefits of the organization's functioning will be felt in its entire ecosystem, and the goals and needs of all stakeholders are balanced in an appropriate way.

Table 2 Differences between CSR and CSV

Concept of CSR	Concept of CSV
Value—doing good	Value—economic and social benefits in relation to costs
Civic attitude, philanthropy, sustainable development	Value creation by the company together with the local community
At the discretion or in response to external pressure	Inextricably linked to competition
In isolation from maximizing profit	Inextricably linked to profit maximization
Action plan defined by outside information and personal preferences	The action plan is specific to the company and flows from its internal needs
The effects limited by the negative impact of the company on the environment and the community as well as the budget of the CSR program	It entails the reconstruction of the entire company's budget
In both cases, it is assumed that the law and ethical norms are respected and the harms caused by the company's activities are mitigated	

Source: Porter and Kramer (2011)

3 Sports Club in the Structures of Polish Sport

Sport changes both in terms of the technology used, infrastructure, number and type of sports disciplines, as well as forms and organization in individual countries and regions. What is unchangeable is the fact that sport must be responsible (Kuźbik 2014). Its social role, reach and influence on society is huge, as the father of the modern Olympic games Pierre De Coubertin mentioned, stressing that this is our heritage and nothing can replace it.

This fact is also underlined by the European Commission, which in 2007 already recognized that: "Sport is a dynamic sector and still gaining importance (...) It can serve as a tool for local and regional development, economic regeneration of cities or rural development [...]. Sport is not only important for improving the health of European citizens but also has an educational dimension and plays a social, cultural and recreational role. [...] Sport contributes significantly to economic and social cohesion and to greater integration of societies. All citizens should have access to sport. The special needs and situations of groups that are not sufficiently represented should be addressed; account should be taken of the specific role that sport can play for young people, people with disabilities and people from disadvantaged backgrounds. Sport can also favour the integration of immigrants and people of foreign origin in society and support intercultural dialogue (BK 2007).

The European Commission defines sport as "all forms of physical activity which, through ad hoc or organized participation, aim to develop or improve a physical and mental condition, develop social relations or achieve sports results at all levels" (BK 2007). The White Paper on Sport was supposed to unify the rules on sport in the Member States, which is why an identical definition of sport can be found in the Polish Sport Act 2010. In addition, the indicated law considers as sport also competition based on intellectual activity, the aim of which is to achieve a sports

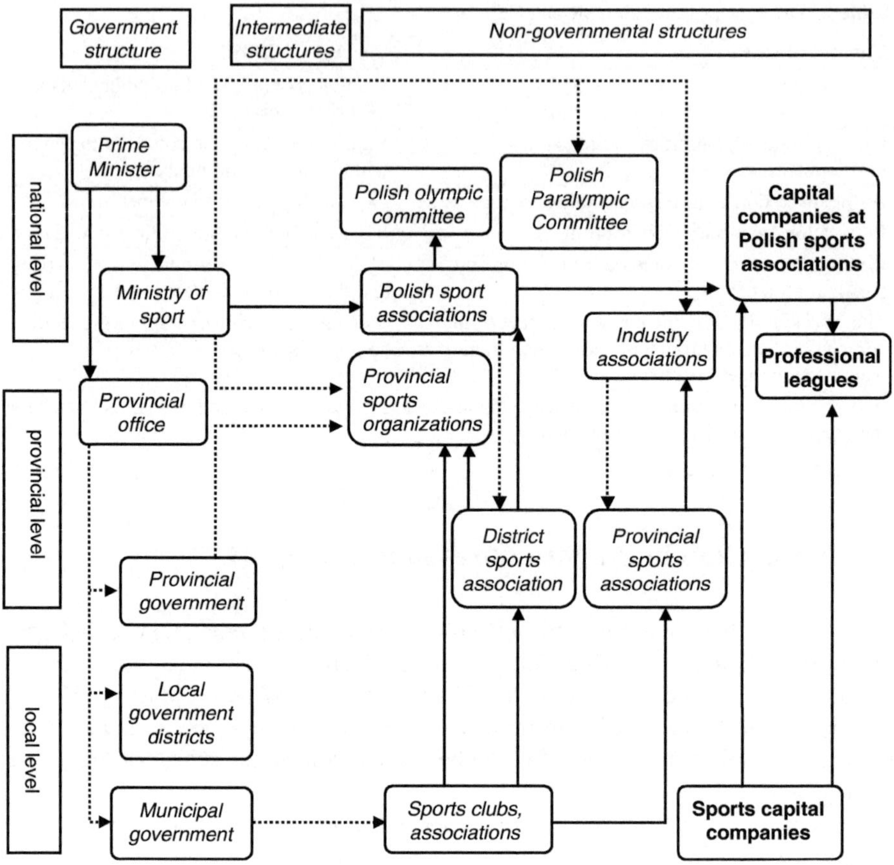

Fig. 2 Sports clubs in the structure of Polish sport. Source: Żyśko (2008)

result. Sport together with physical education and physical rehabilitation makes up the physical culture in Poland [Journal of Laws of 2010 No. 127 item 857].

The sports club is considered to be the basic organizational unit implementing the principles of physical culture (Breuer et al. 2017). In Poland, the activity of sports clubs is strictly conditioned by the aforementioned Sport Act. This Act does not specify what legal form a sports club should have, does not state its definition, it only indicates that it must have legal personality. The creators of a sports club can decide for themselves on its legal form depending on their needs (assumed goals and game level) and possibilities.

Sports clubs can operate in the form (Fig. 2):

- Associations or foundations registered in the National Court Register (formed and registered like any other associations/foundations)
- Student sports clubs
- Associations entered into the register of the head of the county competent for the seat of the club (on condition that they do not run a business, and what is more,

conducting business cannot be provided for in the statute of such an association, because then they would have to register in the National Court Register)
- Limited liability companies, i.e. limited liability companies registered in the National Court Register (KRS), (these entities are enterprises conducting a business activity; many football clubs operate in this legal form).

According to the latest report of the Central Statistical Office (GUS) from 2017, in Poland in 2016 there were 14,858 sports clubs and it was 6.1% more than in 2014. These clubs had more than 1019.6 thousand members, with over 1030.2 thousand people practising in them.[1] Compared to 2014, the structure of sports clubs did not change significantly. Students' sports clubs still have the largest share—40% (in 2014—42.3%). The popularity of clubs belonging to the "Ludowe Zespoły Sportowe" Association also remained at a similar level. The number of LZSs increased slightly (by 65), and their share accounted for 17.6% of all sports clubs. Other sports divisions, such as the Academic Sports Association, School Sports Association and religious clubs, accounted for 1.5% of all sports clubs. Invariably, in Poland, the most popular are team sports (58.4% of people exercising), among them football, which was practised by 39.5% of all those who train (40.8% with beach soccer and indoor football). The number of people practising this sport increased in relation to 2014 by 15.5%. Volleyball took second place—7.5% of all those who train (7.8% with beach volleyball), but as the only one of the most popular disciplines in Poland, it noted a decrease in the number of players compared to 2014 by 3.8% (Malinowska 2017). Over the past 20 years, the number of sports clubs in Poland has increased fivefold.

However, it should be clearly stated that of the number of sports clubs in Poland, only about 6000 participate in the national competitions, and only a small number of them has the status of professional clubs which at the same time set the standards for functioning in the areas of represented sports.[2]

4 CSR and Sports Clubs

Analysing the use of the CSR concept in sports organizations, one should first of all answer the question regarding the conditions in which they function today, what characterizes the market on which they move and why their relationship with the concept of social responsibility should be sought.

Sports clubs in Poland, over the last years, have gone through a long and difficult path of transformation. In the 1960s and 1970s, the sports administration phase

---

[1]The number of sports club members is not the same as the number of people exercising in clubs, as non-club members may also participate in the exercises.

[2]About 60 sports companies are registered in the National Court Register. According to the requirement of the Sport Act, a club wishing to compete in the professional league (the highest level of competition) must have the status of a capital company (most often a joint-stock company).

prevailed, where sports clubs were built at large production plants or appropriate government structures (such as the army or the police). It was crucial that the infrastructure and equipment were properly used. The next period of 1970s–1980s is the transition to the sports sponsorship phase, where expectations for the sports clubs from the sponsor or fans began to appear, and in the competition itself began to be more than just a sports spectacle. The natural successor was the phase of sports marketing, which began to dominate in the 1980s and 1990s, and the key assumption was to organize sports events to bring benefits to sponsors. The current trend already assumes a phase of marketing and responsible sports management, where the following phenomena come to the fore:

- professionalization of sport—a process involving the professionalization of athletes and sports clubs in competitive sport,
- commercialization of sport—a process in which sport turns into a business generating huge income, sports spectacle becomes a well-paid product, players become goods and faces of advertisements for products and services, and sports facilities and broadcasts from the competition are used to promote companies or product brands,
- medialisation of sport—the process of wide sharing of participation in sports events through broadcasts from the competition, access to sports people on TV, the Internet, radio and press,
- internationalization of sport—the process of international transfer of players and trainers, as a result of which the national character of clubs, organizations and sports competitions disappears,
- globalization of sport—the process of broadly understood internationalization of sport, mainly competitive, initiated by the IOC and global and continental federations in individual sports disciplines, regulating the rules of running competitions and organizing international competitions, as well as the development of sports tourism or the international transfer of players and trainers (Smoleń and Pawlak 2015)

In 2005, Z. Waśkowski presented the results of own studies conducted on approximately 300 clubs, determining their degree of marketing orientation. It turned out that vast majority ranks between the sports sponsoring phase and the sports marketing phase (Waśkowski 2005).

In 2018, not much has changed in this area, but more and more sports organizations and clubs, especially professional clubs from the studied sports disciplines, are already in the marketing sport management phase. This is of course related to the global development of sport in the world, where the most popular sports disciplines, such as football, have become huge businesses.

This type of trend entailed, apart from positive aspects, also negative ones, and sport like never before has experienced many pathologies. Such phenomena as doping, racism, setting sports results or corruption have become commonplace. Along with the increase in social expectations of different types of stakeholders and the impact and scope of sport and sports clubs, some standards, transparency and responsibility began to be demanded. Sports organizations have noticed that in order to survive and develop further, in order to be competitive, they should look for new solutions that would allow them to achieve the assumed sporting, economic and social goals in a sustainable manner.

It turned out that the sports club should be thought of in a comprehensive way in the areas of the mentioned goals and, returning to the roots, should be built on a social foundation (Fig. 3). Sports clubs from the very beginning were created as those organizations that in the first place, through sport, were supposed to solve various social problems (Comporek 2015).

It is obvious that sports organizations such as sports clubs have the overriding objective of functioning, that is creating conditions for sports and organizing sport events (Szwarc (ed.) 2000), but in the era of the marketing management of sport and the need to be responsible, such understanding of a sports club is insufficient. Professional clubs participating in sports competition at the highest gameplay/competitory levels are subjected to similar rules of the market economy, as organizations in the traditional sense (Waśkowski 2011: p. 253), but additionally, they have to face the specificity of the market on which they mainly move. In addition to the fact that the most important clients of sports clubs are supporters whose emotional attachment to this type of organization puts high expectations as to the products and services offered or the directions of operation, the following elements,

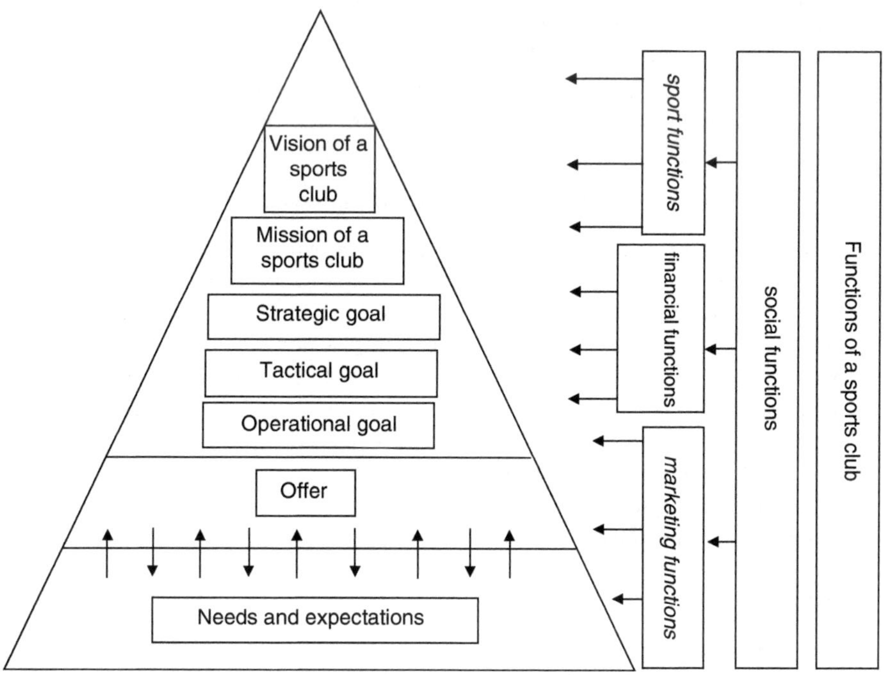

Fig. 3 Functions of a sports club. Source: Own elaboration based on Perechuda (2000)

defining the specificity of the market on which sports clubs move (Sznajder 2008) should also be mentioned:

- Sports organizations operate on the sports market on the supply side and are strongly diversified (in terms of objectives, legal form, organizational structure, scope of activity, relevance to the specific sport, decision-making methods, etc.)
- Goals of a sports organization (Foster et al. 2006: p. 8) are more diverse than the goals of enterprises operating in other industries (sports result, financial profit, satisfaction with trophies, local community integration, social mission, etc.)
- Decisions of the governing bodies of sports organizations are quickly commented on by the media and fans
- Professional clubs, on the one hand, compete with each other and on the other hand depend on each other (player transfers, exchange of experience and technology, sports result, infrastructure, finance)
- The sports market is regulated. This means that clubs with the character of commercial law companies must respect not only generally applicable regulations, but also specific rules regarding sport (Sports Act in Poland, regulations of Polish Sports Associations, Regulations of national companies managing professional leagues, international regulations established by UEFA and FIFA, IOC, etc.)
- The sports market is doubly dual. Clubs operate simultaneously on the sports market as well as on the advertising and sponsoring market. Sales take place on the primary market (sports spectacle) and on the secondary market (sale of a sports product through the media, other business activity)
- The products offered by the clubs vary depending on whether they are offered to consumers (sports spectacle, club souvenirs, products resulting from additional business activities, etc.) or institutional buyers (television rights, image rights, etc.)

In addition, it is also worth adding the following aspects (Waśkowski 2015):

- Sport is by nature an activity based on physical activity, so from a marketing point of view it is an intangible and subjectively assessed product, and consumers (viewers or participants) buy impressions.
- Sport evokes a strong emotional experience among consumers, strengthens personal ties because it is mostly consumed in public. In addition, it is a universal medium of communication and refers to all sections of life (geographical, demographic, cultural, economic, religious, political, etc.).
- The result of sporting competition is unpredictable and the rivalry cannot be repeated. This means that the sports organization manager cannot provide consumers with a specific effect, both in the form of the outcome of the game and its level. This has important consequences in the process of quality assurance of services provided.

Taking into account all these conditions, it seems a necessity to balance sporting and economic goals with social ones, and in addition to using the right methods and management techniques, also to implement the idea of social responsibility in the strategy of operation. Thus, the relationship of sport—sports clubs with the concept of social responsibility is a natural consequence of the specificity of the market, ways of operating, characteristic groups of stakeholders and their needs/expectations and objectives set by clubs and international organizations.

5 CSR in the Practical Dimension of Sports Clubs: Research Results

As it has already been mentioned, the research was conducted on sports clubs of the professional nature, competing at the national level and representing the most popular sports in Poland, i.e. football, volleyball, speedway and basketball. 85 clubs participated in the survey; among which the majority were joint-stock companies, the rest were physical culture associations.

It should be clearly emphasized that in the sport-CSR relation there are two ways to be considered: CSR through sport, where various types of companies, in order to pursue their own missions in the areas of social responsibility, support sport in various forms (most often financial), and CSR within sport, where we are talking about socially responsible activities of sports organizations themselves. From the point of view of the purpose of this article, the research will deal with the latter formula.

Answering the questions posed in the questionnaire, 92% of the respondents (decision-makers, club representatives) emphasize that acting in accordance with generally functioning social norms (moral and ethical norms) plays a very important role in managing a sports club and in their opinion the clubs they represent are socially responsible. However, attempting to determine activities that would indicate that the club could be said to work in accordance with the concept of social responsibility is no longer so obvious (Graph 1).

According to the respondents, the socially responsible club is, first of all, a club that cares for the youngest adepts, providing them with the right conditions for development, honest, taking into account the needs of all stakeholders and regularly fulfilling its obligations. On the other hand, issues related to environmental protection are of less importance.

In the assessment of the club's activities in the areas of social responsibility, respondents declare that these are usually charitable activities, socially engaged marketing, social campaigns, and that the club respects human rights and is tolerant (Graph 2).

At the same time, only 37% believe that they have a coherent CSR concept and 44% state that it is not part of the club's strategy as it is merely an additional activity (Graph 3).

In the further section, respondents were asked to indicate the reasons why they engage in socially responsible activities (Graph 4).

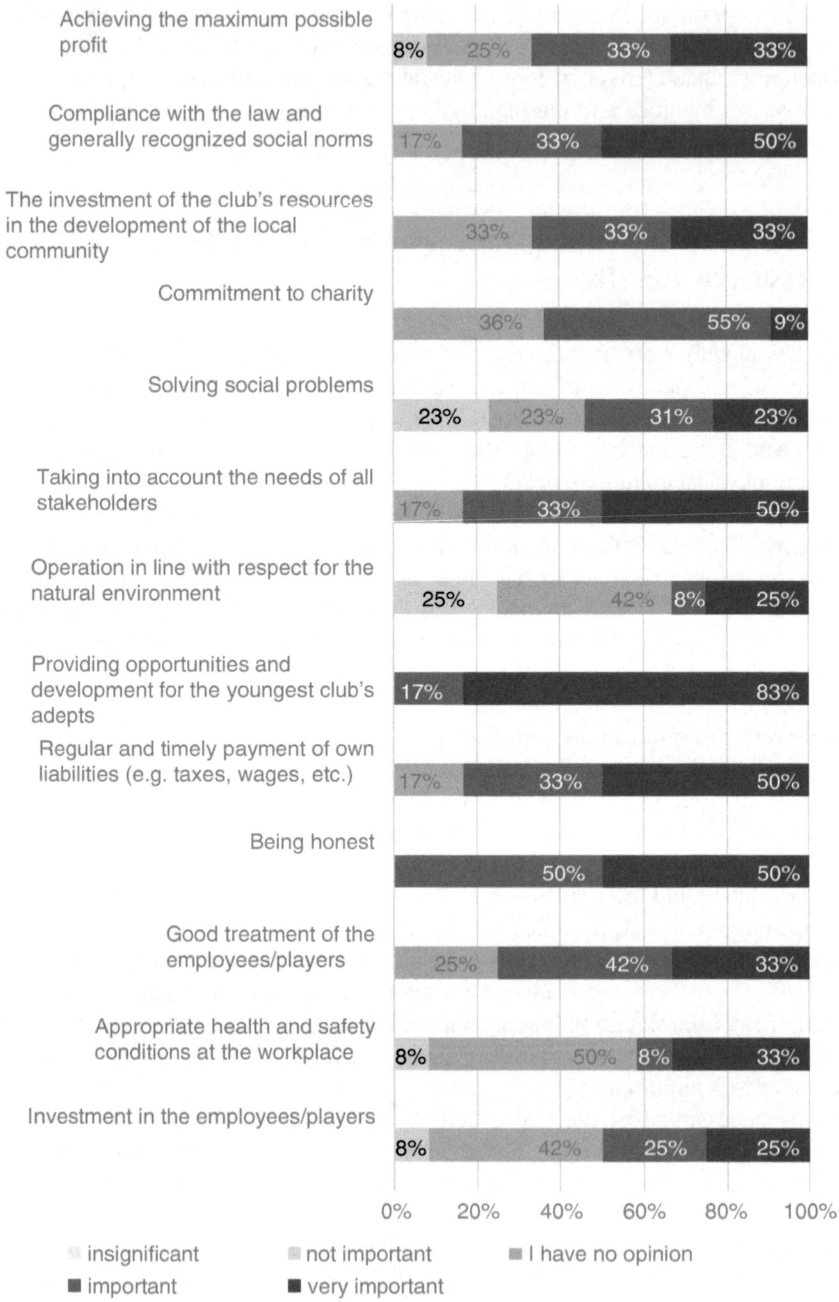

Graph 1 Actions qualifying the club as socially responsible. Source: Own research

Graph 2 Declared actions within a social responsibility of clubs. Source: Own research

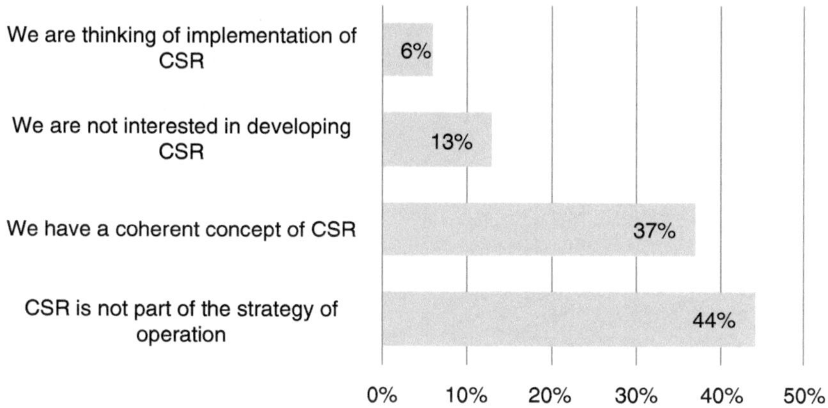

Graph 3 The role of responsibility in managing a sports club. Source: Own research

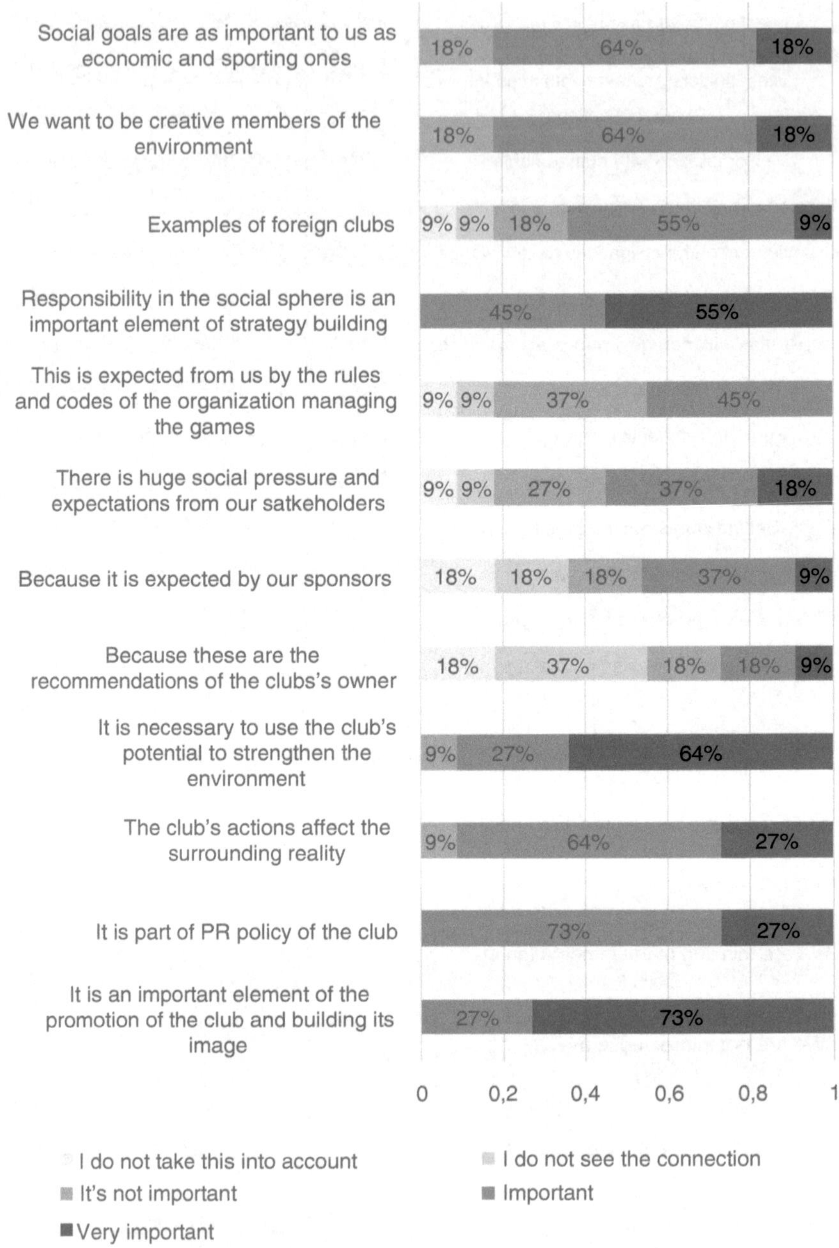

Graph 4 Reasons for engaging in socially responsible activities. Source: Own research

Sports clubs are aware of functioning in a specific environment and of its impact, which is why strengthening this environment and the issues of PR and image play the key roles here (it is the basic inspiration for the start of activities under CSR for 46% of clubs). Therefore, decisions about undertaking socially responsible activities come in 36% from the club management and in 28% from the marketing department.

Regarding the question to whom the CSR activities of the club are addressed, who is the beneficiary, the decisive dominant form is external responsibility, and the main beneficiaries are the local community and customers (fans). Less than 10% indicated that the main beneficiaries are young players of the club (Graph 5).

It is also worth emphasizing the fact that respondents' answers are declarative in nature and often the willingness to perceive themselves as socially responsible, affect the answers given and, as a consequence, falsify the image of real actions. In the hitherto achievements of the CSR idea, documents and standards which it has developed, it is difficult to seek well-established knowledge among decision-makers.

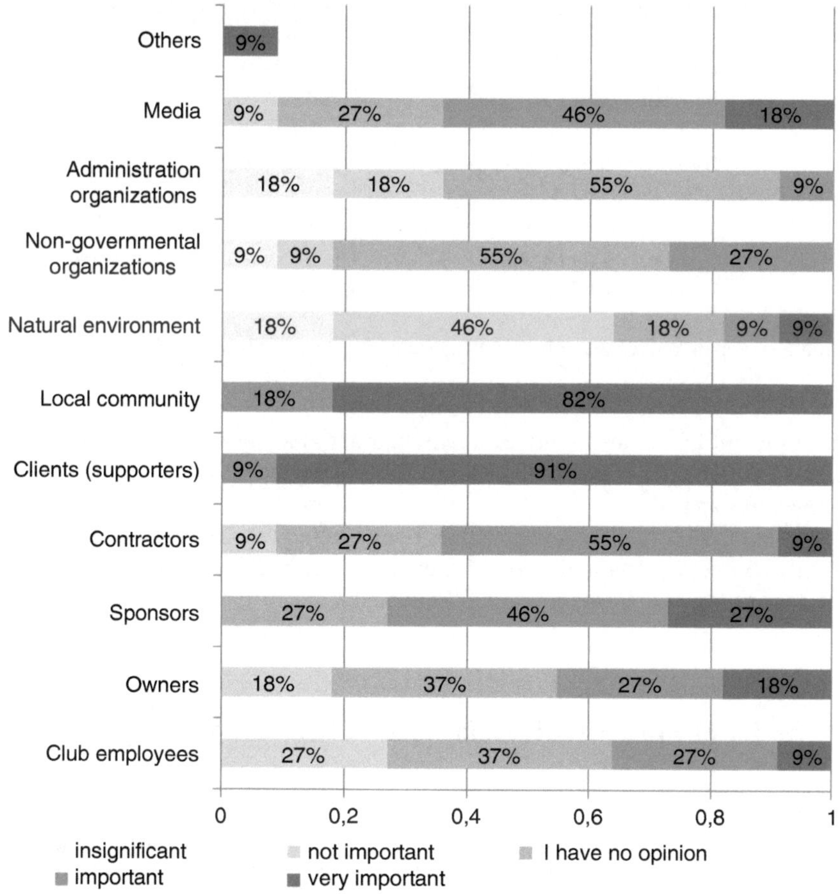

Graph 5 Beneficiaries of CSR activities. Source: Own research

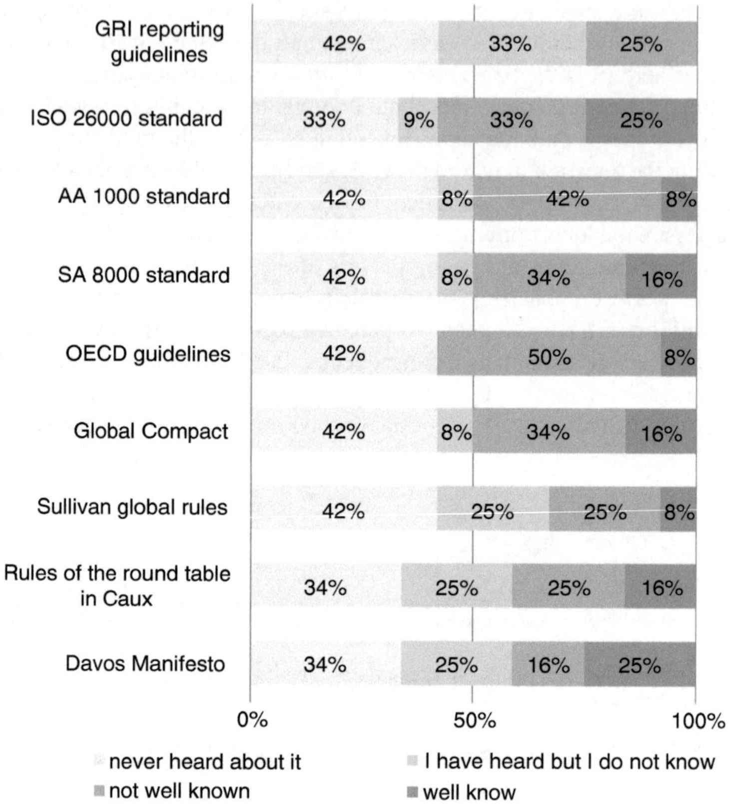

Graph 6 Level of knowledge of basic CSR documents. Source: Own research

It turns out that clubs highly evaluate their activities in the context of CSR, but the basic documents or recommendations offered by the CSR concept are virtually unknown (Graph 6).

Moreover, 94% of clubs indicate their own official website as the main and often the only source of communication of their activities in CSR areas, and this already says a lot of how the concept is actually treated.

6 Model of Sport: CSR Relations—Conclusions

The concept of social responsibility is undoubtedly at the stage of transformations. What has been considered fundamental so far has not worked in practical terms. Therefore, the world of the organization has begun searching for new solutions that would make CSR a real strategic support.

Given the fact that sport in the professional and occupational dimension has become a business participating in a normal market game, it was obvious that the CSR concept would also go to this industry. The relationship of sport and CSR is obvious and results from the specifics of this type of organization, its goals, a characteristic group of stakeholders or the specificity of the market.

Analyzing the level of using CSR principles by Polish clubs, it should be stated that in this matter, just like traditional organizations, they chose only what was convenient for the club and what contributed to building its image.

It seems, therefore, that the construction of sports club goals, based on balancing goals from sports areas with economic and social ones, brings to the fore the need to build proper rules and relationships with stakeholder groups, and this means that existing activities in the field of social responsibility have become insufficient (Kuźbik 2015).

If we agree with Visser (2010), that the previous development of CSR can be divided into five periods[3]:

The Era of Greed Focusing on profits and economic growth.

The Era of Philanthropy Beneficence and charity. As Visser noted, the very term "giving back" something to the community is problematic, because it means that the company took something earlier. According to Visser, companies should be more interested in the impact of their activities on society and the environment when making profits, and not sharing them after making them.

The Era of Marketing Focusing on the PR and marketing effects of CSR.

The Era of Management The inclusion of the company's impact on society and the environment to management strategies.

The Era of CSR 2.0 CSR 2.0 is about creating and implementing a sustainable and responsible business model, which, supported by a reformed financial and economic system, will make creating a better world easier, more natural and profitable [...] CSR 2.0 boils down to one thing: explanation and reorienting the goals that organizations face [...] Creating a positive contribution to society is the essence of CSR 2.0—it is not about additional activities, but about the way of doing business.

Then it should be recognized that sports clubs in Poland are stuck somewhere between the era of philanthropy and marketing, in principle not reaching even the era of management. This conclusion, of course, applies to those clubs that represent the most popular sports disciplines in Poland in the top classes. However, the lower the gameplay/sports level, the more the sports clubs tend to become beneficiaries of socially responsible actions of other organizations trying to implement the CSR strategy, rather than themselves including activities of this type in the practice of functioning (Kuźbik 2017a).

[3]The first four eras are connected with the concept of CSR 1.0.

Conducted research has shown that at a declarative level, clubs see themselves as socially responsible and even actions are indicated that are supposedly performed in these areas. Very soon, however, it turns out that CSR is not a strategic one, and the most common cell implementing CSR activities is the marketing department. What's more, it is difficult to talk about CSR in the conditions of functioning of Polish clubs, because by analyzing real actions, we can only identify fragments of this idea; the observation of everyday practice shows that these are mainly charitable and PR actions. However, Polish sport still struggles with much pathology that have nothing to do with social responsibility, which again shows the idea of CSR as a façade only.

Analysis from the observer's perspective, review of the websites of the examined clubs, which are indicated as the main source of communication with stakeholders, clearly places Polish clubs at the CSR 1.0 stage, and so only in its fragmented assumptions (Kuźbik 2014a). So if we can talk about any model, it is only in the area of current activities and the right directions of change (Fig. 4).

The assumptions of this model in the simplest way show the current relationship between sport and social responsibility in Poland. At the first stage, there appears the so-called CSR initiator who may want to pursue his own concept of social responsibility, seeing the sense of supporting sport in a broader context. Most often it is CSR through sport in the relation Business—sports club. Very often, however, for the club, this kind of relationship is only an element of sponsorship and contributes more to achieving the goal of survival than further development (most often it concerns associations of physical culture or clubs at lower levels which do not have ambitions of sports development).

On the other hand, there may also appear a partner coming from the local community, organizations managing the particular competition, a national sports association or a partner from the NGO or local government, who will see in a sports club or an individual sportsperson a full-fledged partner in the implementation of tasks that will support both parties involved and the environment in which these partnerships coexist. Then a network of initiators and CSR beneficiaries is created, and this is the creation of a shared value through dialogue and properly built partnerships (CSV).

Grouping of key criteria such as economic criterion, legal criterion, ethical criterion, ecological criterion, social criterion, sports criterion, becomes the starting point, to prepare the conditions allowing for proper dialogue with the wider environment (Kuźbik 2016).

The key activities to build the dialogue are (Gasiński 2011):

Defining stakeholders in terms of strategic goals and projects, followed by processes within individual subsystems; Analysis of the impact of individual partners on the sub-objectives and the effects of the implemented project (including the organization's goals); Choosing from the stakeholder group the most important ones, having the greatest impact on the implemented or expected goals; Conducting a preliminary analysis of the expectations, concerns, needs and objectives of particular groups of key stakeholders before meetings with their representatives;

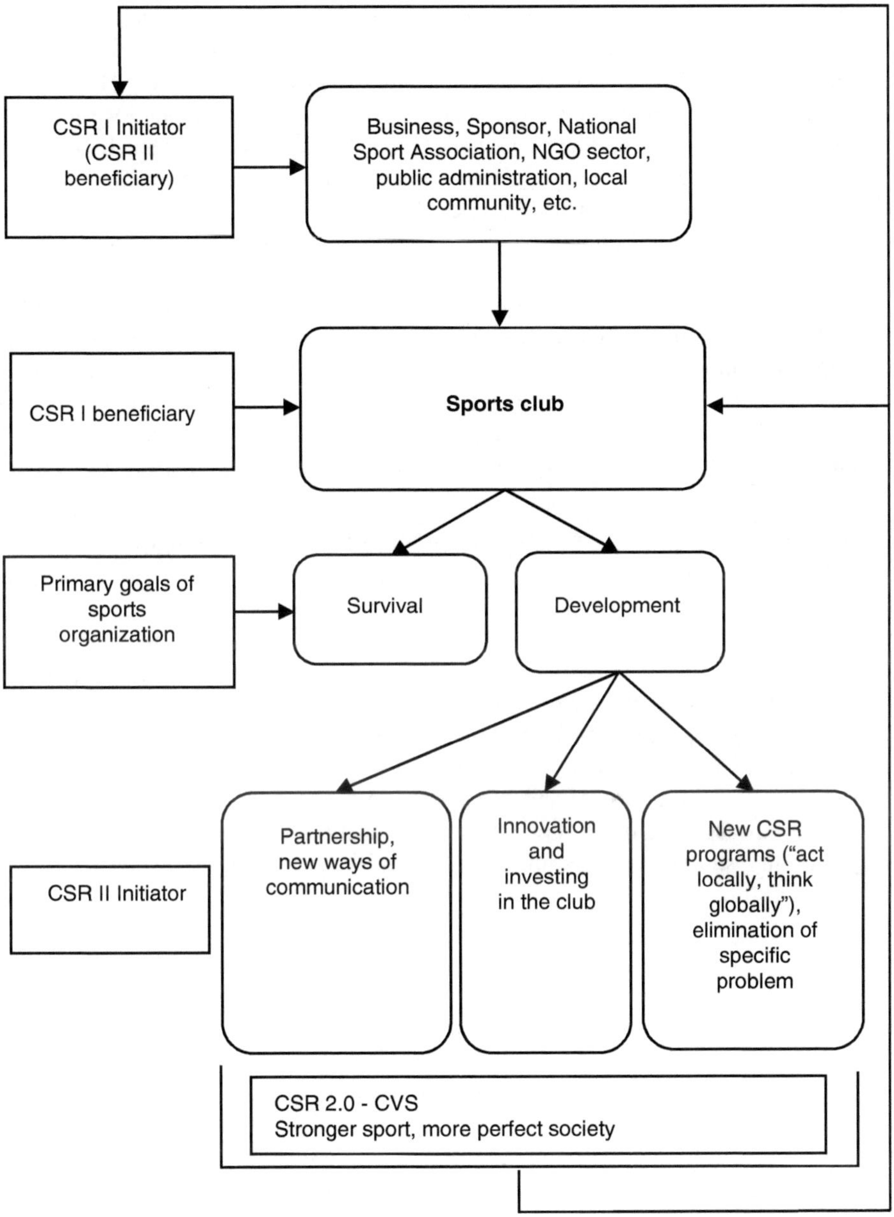

Fig. 4 Model of sport—CSR relations. Source: Own elaboration

Designing and planning this dialogue through (identifying opinion leaders among selected stakeholders, identifying education needs and informing the public about the progress and results of the dialogue, preliminary determination of the ways and frequency of conducting a dialogue); Developing and agreeing, together with the most important dialogue partners, rules for its conduct; Developing and agreeing on areas of common goals—organization and environment—and uniform identification of risk factors; Developing different ways, proposals, declarations in the field of avoiding threats, resolving disputes between the company and stakeholders;
Implementation of jointly developed solutions and informing on the results of the undertaken commitments and actions, as well as monitoring the process of implementation of specific common goals of the dialogue, including the others, belonging to a wide range of stakeholders; Conducting a periodic analysis of stakeholders' perceptions regarding the assessment of the effectiveness of the communication method adopted and the effects of the dialogue in order to improve the relationship.

From the point of view of the sports club itself, apart from being a beneficiary, regardless of previous actions undertaken with other partners, for the sake of its strategy, if it does not want only to survive, but also to develop, it should become the initiator of CSR and build its own network of initiators and beneficiaries. The inclusion of the concept of social responsibility in accordance with the assumptions of CSR 2.0 or CSV to the strategy of operation becomes an absolute necessity. The basic direction of action is then thinking from outside to inside, so that in the first place to be able to identify the right areas of activity and, above all, to know the specificity of the environment and its expectations (Fig. 5).

Due to the specificity of the sports club functioning, when building a social responsibility strategy, one should take into account the criteria thanks to which

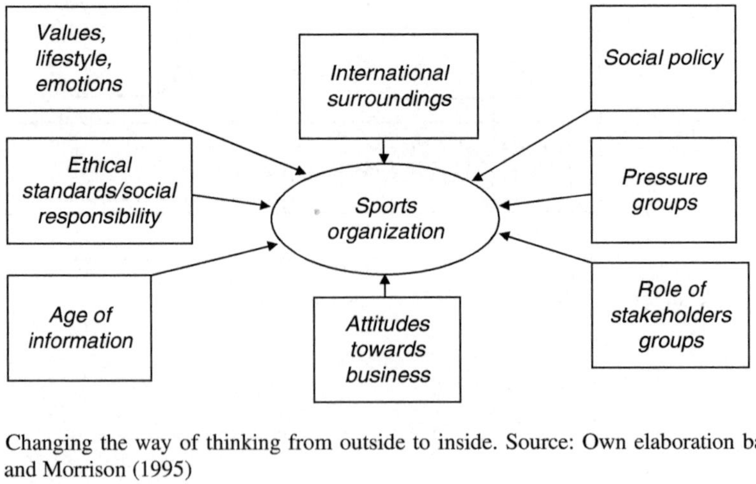

Fig. 5 Changing the way of thinking from outside to inside. Source: Own elaboration based on Ashley and Morrison (1995)

the club will determine the importance of particular areas and within them will be able to indicate priorities for the policy of social responsibility.

The conditions for the implementation of social responsibility in managing a sports club should be flexible, transparent and understandable, and most importantly, tailored to the specificity of the sports industry.

Therefore, it seems that the best way to identify key areas of social responsibility within selected criteria will be first of all to identify real problems affecting sports clubs in Poland so that the expectations of the broadly understood environment could be met, and then to undertake a dialogue with all stakeholders through diversified communication channels.

In the article the author put the following research questions

- Is there any relationship between sport, sports clubs, and the idea of CSR and what is its foundation?
- How the concept of CSR is understood in sports organizations?
- At what stage are the clubs in their implementing of the CSR concept? (CSR 1.0—CSR2.0—CSV)
- Can a model of sport-CSR relationship and its practical implementation be exposed based on the diagnosed club activities in the CSR area?

As a result, as a result of the conducted research, it should be stated that sport, and in particular sports clubs as organizations, must be responsible. Depending on the stage of development on which they are located (Beech and Chadwick 2004), their responsibility will manifest themselves differently, however, it will result strictly from the fundamental social, economic and sporting objectives, creating the proper relationship between the club and the environment. In Poland, however, the use of CSR has no strategic dimension, and its understanding is mainly focused on charitable activities, but it is possible to point out recurring procedures that allow to create an up-to-date model of activities in the area: sports clubs—CSR (Fig. 5).

It seems, therefore, that such planned activities will enable the creation of strong sports organizations that, at the same time, will strengthen the environment and entire communities, and the idea of social responsibility will not only be an additional activity but a key element of the strategy in line with the expectations of all stakeholders.

References

Arendt, H. (1997). To nie ja, to system, Gazeta Wyborcza, 15–16 lutego.

Ashley, W. C., & Morrison, J. L. (1995). *Anticipatory management*. Leesburg: Issue Action Publications Inc.

Beech, J., & Chadwick, S. (2004). *The business of sport management*. Harlow: Pearson Education.

Bielski, M. (2005). *Podstawy teorii organizacji i zarządzania*, Wyd. C.H. Beck, Warszawa.

Breuer, C., Feiler, S., Llopis-Goig, R., & Elmose-Østerlund, K. (2017). *Characteristic of European sports clubs*. Odense: Centre for Sports, Health and Civil Society University of Southern Denmark.

Comporek, M. (2015). Sport corporate responsibility jako cel funkcjonowania klubów piłkarskich, Zarządzanie i Finanse. *Journal of Management and Finance, 13*(1), 99–116.
Filek, J. (2002). *O wolności i odpowiedzialności podmiotu gospodarującego*, Wyd. AE w Krakowie, Kraków.
Filek, J. (2003). *Filozofia odpowiedzialności XX wieku*, Wyd. Znak, Kraków.
Foster, G., Greyser, S. A., & Walsh, B. (2006). *The business of sports*. Mason: Thomson Higher Education.
Gasiński, T. (2011). *Praktyczne narzędzia w budowaniu efektywnego dialogu*, Wyd. HBR, Odpowiedzialny biznes 2011 – CSR 2.0, Warszawa.
Hope, E. (2000). Etyka biznesu etyką odpowiedzialności, Filozofia No 5/2000, Zeszyty Naukowe Politechniki Gdańskiej.
Ingarden, R. (1987). *Książeczka o człowieku*, Wyd. Literackie, Kraków.
ISO 26000. (2010). www.iso.org
BK. (2007). Komisja Europejska, Biała księga na temat sportu (BK), Bruksela, 2007, www.msit.gov.pl [dostęp 2018].
Kurkowska-Budzan, M. (2017). Writing the history of sport in Poland. The underestimated potential of the cultural history perspective. *The International Journal of the History of Sport, 34*, 992–1002.
Kuźbik, P. (2014). Odpowiedzialność społeczna w sporcie, „Prace Naukowe Uniwersytetu Ekonomicznego we Wrocławiu", No 338.
Kuźbik, P. (2014a). Social media and the development of corporate social responsibility in sports organizations. In J. Reichel (Ed.), *CSR trends. Beyond business as usual*. Łódź: Centrum Strategii i Rozwoju Impact.
Kuźbik, P. (2015). Zarządzanie klubem sportowym w warunkach kryzysu, „Marketing i Rynek", 2015, nr. 5.
Kuźbik, P. (2016). *Społeczna odpowiedzialność w przestrzeni organizacyjnej klubu piłkarskiego*, Wyd. SIZ, Łódź.
Kuźbik, P. (2017). Role and impact of stakeholders on the environment of a sports club. In: Social responsibility of organizations. Old – New Stakeholders?, Research Papers of Wrocław University of Economics No 464/2017.
Kuźbik, P. (2017a). CSR 2.0 jako narzędzie zmian w organizacji sportowej, „Zeszyty Naukowe Uniwersytetu Szczecińskiego Ekonomiczne Problemy Usług", 2017, No 1 (126).
Malinowska, E. (2017). *Kultura fizyczna w Polsce w 2016 r.* Rzeszów: GUS.
Pawlak, G., Łasiński, G., & Głowicki, P. (2015). Intersectoral cooperation of business organizations and sports club. In A. Rudnicka (Ed.), *CSR trends. Making a difierence*. Łódź: CSR Impact.
Perechuda, K. (2000). Metodologia restrukturyzacji klubu sportowego, Człowiek i ruch No 2/2000.
PWN. (2017). www.sjp.pwn.pl
PKN. (2012). Polska Norma PN-ISO 26000:2012 *Wytyczne dotyczące społecznej odpowiedzialności* została opracowana w KT 305 ds. *Społecznej odpowiedzialności* i opublikowana 5 listopada 2012 r. https://www.pkn.pl/informacje/2013/09/iso-26000
Porter, M., & Kramer, M. (2007). Strategia a społeczeństwo: społeczna odpowiedzialność biznesu pożyteczna moda czy nowy element strategii konkurencyjnej?, HBRP nr 52, czerwiec.
Porter, M., & Kramer, M. (2011, January–February). „Creating Shared Value", Harvard Business Review.
Słowik, J. (2017). Biznes mądrze odpowiedzialny, Ed. HBR No 170, April 2017.
Smoleń, A., & Pawlak, B. (2015). *Strategia zawodowych klubów sportowych w Polsce*, Ed. OW SGH, Warszawa.
Socha, J. (2017). *Jak rozumieć tworzenie wartości wspólnej*, Wyd. Harvard Business Review, Wyd. specjalne, Nowy wymiar społecznej odpowiedzialności.
Sznajder, A. (2008). *Marketing sportu*. Wyd. PWE, Warszawa.
Szwarc, A. (Ed.), (2000). *Status prawny polskich klubów i związków sportowych*, Wyd. Poznańskie, Poznań.
Tazbir, J. (1996). Odpowiedzialność przed Rzecząpospolitą, Res Publica No. 11/96.
Visser, W. (2010). CSR 2.0. In M. Pohl & N. Tolhurst (Eds.), *Responsible business: How to manage a CSR strategy successfully*. www.odpowiedzialnybiznes.pl.

Waśkowski, Z. (2005). Rola i możliwości marketingowego zorientowania klubów sportowych w Polsce, Nowoczesny klub sportowy – Grać by wygrać, Grupa Podium, Wrocław.
Waśkowski, Z. (2011). Kompetencje menedżera sportu i determinanty ich kształtowania'. In B. Sojkin & Z. Waśkowski (Eds.), *Zarządzanie polskim sportem w gospodarce rynkowej*, Wyd. UE w Poznaniu, Poznań.
Waśkowski, Z. (2015). Wykorzystanie teorii interesariuszy w procesie kształtowania strategii marketingowej organizacji sportowych, Polityki Europejskie. Finanse i Marketing No 13 (62).
Żyśko, J. (2008). *Zmiany we współczesnych systemach zarządzania sportem wyczynowym w wybranych krajach europejskich*, Wyd. AWF, Warszawa.

Part V
CSR Reporting

CSR Reporting Practices in Poland

Justyna Fijałkowska and Małgorzata Macuda

1 Introduction

CSR is an idea of constructing a long-term strategy by a company that voluntarily takes into account the social, ethical and ecological aspects in its business operations, and subsequently discloses it. Such a company takes responsibility for decisions made and activities undertaken, which have an impact on the local community and the environment, and also leads a dialogue with the internal and external stakeholders (Macuda et al. 2015). In the context of implementation of CSR, Jaglińska (1984) stated that accounting should measure the company's responsibility and allow to reveal how its' activities influenced the level of environmental pollution, usage of non-renewable resources, public security, health and education, civil rights and many other aspects of social life—namely to report about activities in the social, environmental and economic areas related to the functioning of the company. Therefore, according to her, it was necessary to create a system of internal and external reporting (clear, easy to understand and rewarding for users) in order to inform all stakeholders about all aspects and results of corporate activity. Studies on CSR disclosures in Poland are thought to have been pioneered by Jarugowa (1984), according to whom, financial data had to be supported by non-financial information.

In today's global business environment it is becoming increasingly clear that traditional reporting is not meeting the requirements of its users. Uncertainty, high risk and frequent financial scandals make stakeholders more demanding for high

J. Fijałkowska (✉)
Katedra Finansów i Rachunkowości, Społeczna Akademia Nauk, Łódź, Poland
e-mail: jfijalkowska@san.edu.pl

M. Macuda
Katedra Rachunkowości, Uniwersytet Ekonomiczny w Poznaniu, Poznań, Poland
e-mail: Malgorzata.Macuda@ue.poznan.pl

A. Długopolska-Mikonowicz et al. (eds.), *Corporate Social Responsibility in Poland*, CSR, Sustainability, Ethics & Governance,
https://doi.org/10.1007/978-3-030-00440-8_13

quality, credibility and transparency of the financial and non-financial information disclosed by companies. Companies should not only focus on the multiplication of the financial capital value, but ought to concentrate on the creation and development of intellectual and social capital. As stated by Central Europe (CE) CSR managers, responders to Top 500 survey conducted by Deloitte (2016), the impact of the business community would determine the social and economic progress of their respective countries—not just in terms of competitiveness, but also the effect of enterprise on the local labor market, the knowledge-based economy and intellectual capital growth. The knowledge derived from measuring this impact has allowed CE companies to manage their activities so as to drive positive societal change. 84% of CSR managers believed that the businesses had facilitated to find positive solutions for social problems within their countries. Sustainability and social responsibility are certainly gaining ground within the CE business community.

According to the "Annual CSR Outlook in Poland 2017", less than half (46%) of large and medium-sized companies in Poland currently carry out CSR activities. Another 15% declare that they are considering CSR engagements, thus in the coming years the development of CSR concept and its implementation in Polish companies can be expected. The share of enterprises involved in CSR is higher among these with foreign capital, assuming the highest value (61%) for companies with only foreign capital. CSR activities are undertaken more and more often by large and very large companies—two thirds of these that employ over 1000 people operate in accordance with the CSR concept. In the group of smaller companies in terms of employment (up to 249 people), only a quarter carry out CSR activities.

Reporting issues are tightly connected with CSR involvement. The number of companies in Poland that make endeavors to communicate on social, environmental and economic activities in the form of CSR reports is increasing in recent years. According to the results of Central Europe Top 500 survey (Deloitte 2016), Polish companies are preparing relatively most CSR reports compared to other countries from the region—26% of all reports. The number of potential readers of CSR reports has widened, covering external and internal stakeholders, including shareholders (Kolk 2008). External stakeholder engagement has been considered so far as the primary motivation behind voluntary CSR reporting (Yim et al. 2018).

Due to the Directive 2014/95/EU on the disclosure of non-financial and diversity information, which applies to fiscal years starting on first of January 2017, the largest companies will be required to report their sustainability policies development and a responsible approach to running a business (Fijałkowska 2016a). For many of them it is going to be a great challenge. According to the latest research related to the reporting practices in Poland of Matuszak and Różańska (2017a), in the majority of cases, CSR disclosure of companies were not compliant with the new requirements, hence the new reporting obligation should increase the extend and quality of non-financial disclosure among Polish companies.

2 Reporting of CSR Involvement in Poland: Empirical Analysis

CSRInfo database comprises 357 CSR reports. Some of the Polish companies just started their experience with the disclosure of CSR information in the form of separated reports. For others it is already a routine practice (e.g. Bank Millennium—11 reports, PKN Orlen—10 reports, Coca Cola HBC Polska—9 reports, Grupa Lotos—9 reports, Kompania Piwowarska—9 reports). The empirical analysis shows an increased interest in publishing CSR reports over the last 10 years. The first and only non-financial report related to social, environmental and economic activities was published in 2005. In subsequent years a growing number of CSR reports issued by Polish companies may be observed, until the year 2017, when their number amounted to 47 (Fig. 1). CSR is no longer perceived as part of company's PR but rather as an inherent element of its long-term strategy (Krasodomska and Cho 2017). The number of reports based on a strategic approach to CSR is growing—from 18% in 2007 to 84% in 2016 (Forum Odpowiedzialnego Biznesu 2017a).

The majority of companies publishing CSR reports in 2017 are from the fuel industry (12.6%), banking sector (12%) and food industry (11.6%). Other leading sectors are the following: power engineering (9.5%), transport and logistics (7%), other services (6.3%) and healthcare (6%). All the other 29 sectors are marginal in reporting related to social and environmental issues and constitute all together 35% of all CSR reports (Fig. 2). The tendency to report more on CSR activities by the companies operating in an environmentally sensitive sector, which was observed by Bonsón and Bednárová (2015), is confirmed as well in the case of Polish companies. Macuda (2016a) noticed the expansion of the concept of CSR and its reporting from the private sector (private companies) to public organizations, also specific entities, such as hospitals.

Since 2005, about 100 companies, 10 NGOs, one university and one city have published reports concerning CSR involvement. Among these reports, 50 followed postulate stating that the content of CSR reports should be substantively integral

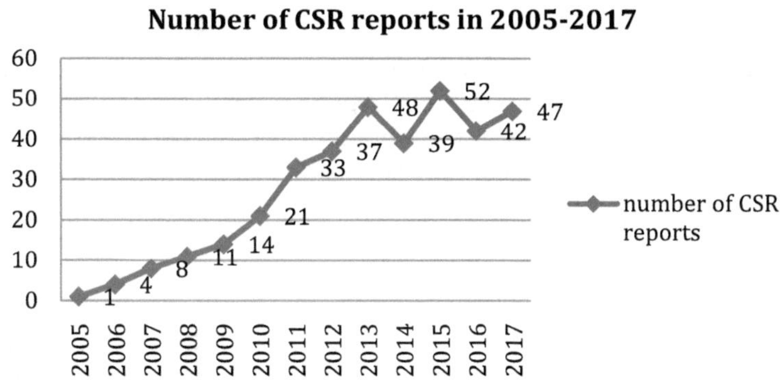

Fig. 1 Number of CSR reports in 2005–2017 in Poland. Source: Own elaboration based on CSRinfo.org

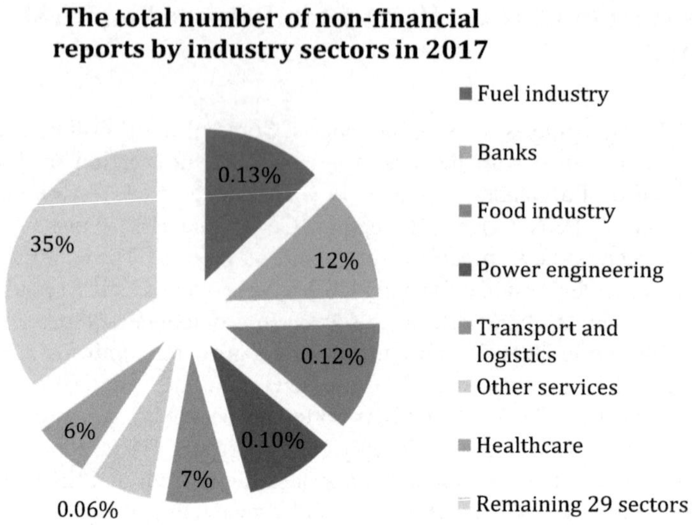

Fig. 2 The total number of non-financial reports by industry sector in 2017 in Poland. Source: CSRinfo.org (20 of March 2017)

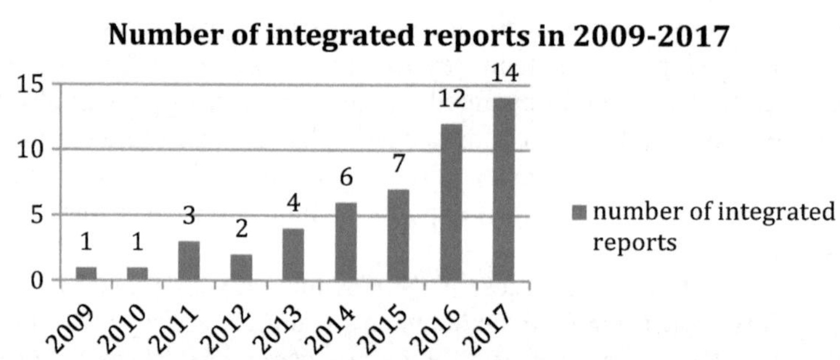

Fig. 3 Number of integrated reports in 2009–2017 in Poland. Source: Own elaboration based on CSRinfo.org

with financial reports—financial and non-financial data in a form of integrated report. In Poland, the idea of integrated reporting has been applied in practice since 2009, initially by one company, while in 2017, 14 integrated reports were published (Fig. 3). Some companies prepared just one integrated report. However, there are examples in Polish CSR reporting practices like Grupa Lotos that already issued seven integrated reports, uninterruptedly since 2010, or ANG, KGHM and Pelion—each of them published four integrated reports.

Samelak (2013) was the first in Poland who attempted to establish a conceptual framework for integrated reporting by a socially responsible company. Such integration is based on the annual report concept, in which the operational report is the key part of the integrated report, or on the integrated report concept, in which a single document integrates all non-financial information (including the CSR report) with financial information by indicating relations and interrelations between said information types. It is worth mentioning that with the increasing number of Polish companies that prepare integrated reports, more and more researchers investigate their practices in this regard. According to the literature review of Krasodomska (2015a) and Waniak-Michalak et al. (2016), informative content of the following companies' integrated reports have been analyzed in Polish literature: LOTOS, GAZ SYSTEM and Scanmed Multimedis (Krasodomska 2013), LOTOS, GAZSYSTEM, Scanmed Multimedis, TAURON Polska Energia and Grupa RWE SA (Szczepankiewicz 2014), LOTOS (Tatarska 2013), Zakłady Azotowe SA (Michalczuk and Mikulska 2014). The integrated reports of above mentioned companies present the information in a diverse manner and follow the Integrated Reporting Framework guidelines in a varied way—more or less strictly.

In 2010, the concept of integrated reporting emerged mainly owing to the GRI guidelines and the development of sustainability reporting over the years (Macuda 2016b). The idea of CSR has drawn attention to the importance of non-financial information for a comprehensive assessment of business activities. However, there are some important differences between the CSR and integrated reporting. CSR reporting targets a wider stakeholder audience than does integrated reporting and it focuses on impacts on the environment, society and the economy rather than on the value creation over time (Waniak-Michalak et al. 2016).

Although CSR disclosures have been so far voluntary, the number of reporters in Poland is still growing (Wójcik-Jurkiewicz 2017). The companies involved in CSR disclosure chose to follow international reporting standards—mainly Global Reporting Initiative (GRI) guidelines or International Integrated Reporting Council (IIRC) guidelines. In the period of 2015–2017, 76% of all reports were prepared in accordance with GRI standards, while in the period of 2012–2016, 78% of them, and

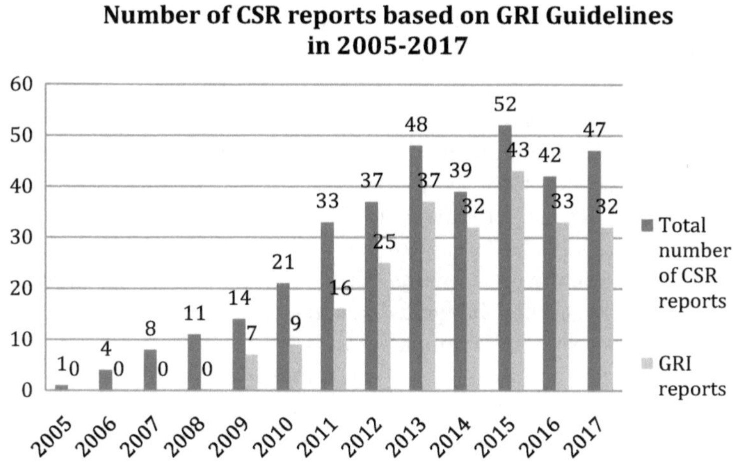

Fig. 4 Number of CSR reports based on GRI Guidelines in 2005–2017 in Poland. Source: Own elaboration based on CSRinfo.org

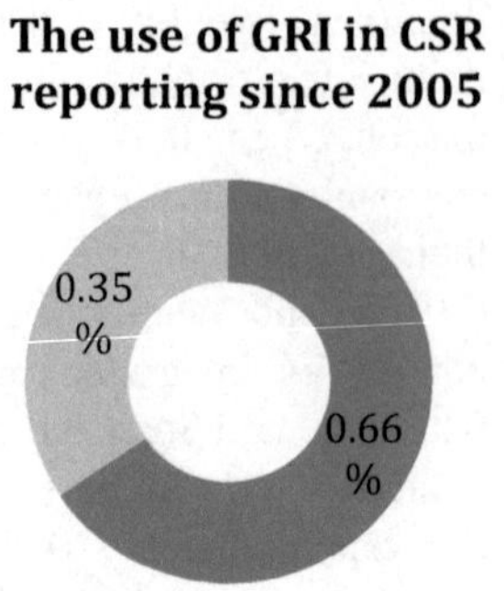

Fig. 5 The % use of GRI in CSR reporting in Poland, since 2005. Source: Own elaboration based on CSRinfo.org

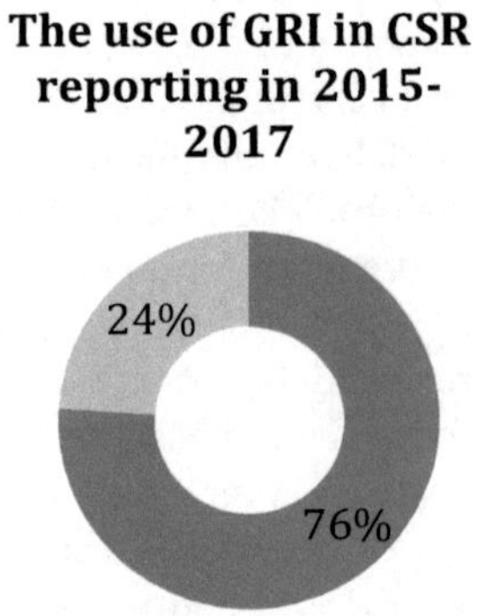

Fig. 6 The % use of GRI in CSR reporting in Poland. Source: Own elaboration based on CSRinfo.org

in the whole sample analyzed (358 CSR reports) more than 65% (235 reports) (Figs. 4, 5 and 6). Thus the position of Poland is above the world average, as according to GRI data, 60% of CSR reports in the world are based on their guidelines.

Other standards used by Polish companies to prepare CSR reports were the following: UN Global Compact, PN-ISO 26000:2012, OECD guidelines for multinational enterprises, Guidelines of the Carbon Disclosure Project and Integrated Reporting Standard.

Among 46 of Polish companies, which published their CSR reports in CSRinfo database in 2017, 16 of them prepared their report for the first time, 7 for the second time, 3 for the third time, 5 for the fourth time, 2 for the fifth time, 4 for the sixth time, 4 for the seventh time, 3 for the ninth time, 1 for the eleventh time and 1 for the thirteenth time. 9 reports have only online version, while 35 were prepared and published in a traditional way. 7 reports (15%) were examined by external auditors. 14 reports (30%) constituted integrated reports. The review of characteristics of 46 CSR reports issued by Polish companies in 2017 are presented in the Table 1. The table presents the name and the sector of the company disclosing CSR activities, the number of pages of CSR report, the guidelines according to which these reports were prepared, the information on the number of reports until 2017, the statement if the latest report is the integrated one, if it was verified by the external auditor, and if it contains the contact person data.

Table 1 Characteristics of CSR reports issued in 2017

Lp.	Name	Sector	No of pages	Guidelines	No of CSR reports	Integrated reporting	External audit	Contact person
1	DB Schenker	Transport &Logistics	Online	GRI-Core	4	–	√	√
2	PKN Orlen	Fuel	521	GRI-G4-Core	13	√	√	√
3	PGE	Energetic	174	GRI-G4-Core	6	–	–	–
4	Nestle Poland	Food	74	GRI-G4-Core	3	–	–	√
5	Bank Millennium	Banks	126	GRI-Core	11	√	–	√
6	Grupa Kęty	Construction materials	Online	GRI-G4-Core	2	√	–	√
7	Grupa Enea	Energetic	82	GRI-G4-Core	7	–	–	√
8	Carrefour Polska	Retail trade	60	–	2	√	–	√
9	ING Bank Śląski	Banks	Online	GRI-G4-Core	7	√	–	√
10	BZ WBK	Banks	Online	GRI-G4-Core	4	–	√	√
11	Bank BGŻ Paribas	Banks	Online	–	6	–	–	–
12	Pelion	Healthcare	Online	GRI-G4-Core	5	√	–	–
13	Grupa Allegro	Retail trade	137	GRI-G4-Core	1	–	–	√
14	Energa	Energetic	90	GRI-G4-Core	6	–	–	–
15	SuperDrob	Food	25	–	1	–	–	√
16	Budimex	Constructions	122	GRI-G4-Core	9	√	√	√
17	Fundacja Medicover	NGO	34	–	2	–	–	–
18	Grupa GPEC	Energetic	Online	–	1	–	–	–
19	Urząd Transportu Kolejowego	Transport &Logistics	34	–	1	√	–	√
20	Margostal- Rewolińscy	Metal	9	GRI-G4-Core	1	–	–	√

(continued)

Table 1 (continued)

Lp.	Name	Sector	No of pages	Guidelines	No of CSR reports	Integrated reporting	External audit	Contact person
21	Zarząd Morskiego Portu Gdańsk	Transport &Logistics	30	GRI-G4-Core	1	–	–	√
22	Fundacja Podaj Dalej	NGO	50	GRI-G4-Core	1	–	–	√
23	Porta KMI	Wood and paper	35	–	1	–	–	–
24	Tesco	Retail trade	11	–	4	–	–	–
25	Lafarge	Construction materials	68	GRI-G4-Core	1	–	–	√
26	Rossmann	Retail trade	29	GRI-G4-Core	2	–	–	√
27	Bogdanka	Raw materials	210	GRI-G4-Core	4	√	–	√
28	Fundacja Zaradni	NGO	49	–	1	–	–	√
29	mBank	Banks	Online	GRI-G4-Core	2	√	–	√
30	Signal Group	Informatics	60	–	1	–	√	–
31	PCC Exol	Chemical	Online	GRI-G4-Core	2	–	–	√
32	ArcelorMittal Poland	Metal	71	GRI-G4-Core	7	–	–	√
33	Interkadra	Other services	15	–	2	–	–	–
34	Cemex Polska	Construction materials	112	–	4	–	√	√
35	ANG Spółdzielnie	Other finance	70	GRI-G4-Core	3	√	–	√
36	Orange Polska	Telecommunication	225	GRI-G4-Core	7	√	√	√
37	Tauron Polska Energia	Energetic	Online	GRI-G4-Core	5	√	–	–
38	Kompania Piwowarska	Food	76	GRI-G4-Core	9	–	–	√
39	Orbis	Other services	54	GRI-G4-Core	1	–	–	√
40	Roche Polska	Healtcare	120	–	3	–	–	–

41	Grupa Żywiec	Food	110	GRI-G4-Core	9	–	–	–
42	Bibus Menos	Other services	19	–	1	–	–	–
43	Grupa Unibed	Constructions	90	GRI-G4-Core	1	–	–	–
44	KGHM Polska Miedź	Raw materials	214	GRI-G4-Core	6	√	–	–
45	Polenergia	Energetic	71	GRI-G4-Core	1	–	–	–
46	Nutricia Medyczna	Healthcare	40	GRI-G4-Core	1	–	–	√

Source: Own elaboration based on reports from CSRinfo.org

3 Content of CSR Reports

A content analysis of CSR reports published in CSRinfo database in 2017 by Polish companies (names and sectors of these companies were presented in Table 1) have been applied in order to examine the area of disclosures (Table 2). The main area of disclosures are the following: local society (98%—45 reports), employees (96%—44 reports), environment & ecology (89%—41 reports), clients (82%—38 reports), code of ethics (80%—37 reports), stakeholder (76%—35 reports), values (76%—35 reports), corporate governance (65%—30 reports), anticorruption and bribery (59%—27 reports), risk (52%—24 reports), human rights (52%—24 reports), negative impact (44%—20 reports), business model (35%—16 reports), financial data on CSR (22%—10 reports), and value created EC-1 (11%—5 reports).

According to the study analysing CSR reporting practices in selected EU member states performed by (Hąbek and Wolniak 2016), Polish reports achieved higher level of quality than the British and Danish reports and were top rated in several assessment criteria: key stakeholders, integration with business processes, basic reporting principles and stakeholders dialogue outcomes.

As it may be noticed, the content information of CSR reports prepared by Polish companies differs. All that makes any comparison difficult. This observation fall in line with the previous studies concerning the content of CSR reports (Fijałkowska 2013; Hąbek 2014; Wolniak and Hąbek 2016). Similar results, indicating the lack of comparability between particular reports and diversity in the content of information disclosed, were also observed in banks (Krasodomska 2015b; Fijałkowska et al. 2017; Matuszak and Różańska 2017b) in energy and mining industry (Szczepankiewicz and Mućko 2016), and in tanning enterprises (Śmiechowski and Lament 2017).

It is however important to underline that considerable diversity of reporting patterns results from the nature of CSR reporting, because it concerns highly entity-specific issues (Szczepankiewicz and Mućko 2016). CSR covers a huge number of different activities, which impedes the comparability of the data disclosed in the reports. The difficulty in establishing a clear connection between CSR involvement, on the one hand, and growth, productivity, the company's value creation, and competitive advantage on the other, has to do with the complexity of the CSR concept. Nevertheless, all these impediments should not in any way restrict efforts relating to the attempt to measure CSR. The effectiveness of CSR activities and the value that is created within the CSR initiatives are important for a company—its internal and external purposes (Fijałkowska and Macuda 2017). The three most important barriers in the presentation of CSR data, pointed by Chojnacka and Wiśniewska (2017), are the following: too high costs, lack of the system of data collection and lack of competent personel.

Popowska (2015), while analysing the CSR disclosure practices among top 30 non-financial companies listed on the WSE at the end of 2013, concludes that CSR reporting activity is very poor, almost absent in the enterprises with entirely Polish capital. Dyduch and Krasodomska (2017) stated, while analysing the CSR disclosure

Table 2 Area of disclosures in CSR reports issued by Polish companies in 2017

Lp.	Stakeholders	Clients	Employees	Local society	Environment & ecloogy	Values	Cororate governance	Code of ethics	Risk	Human rights	Business model	Anticorruption and bribery	Financial data on CSR	Negative impact	Measurement of CSR	Value created EC-1
1	√	√	√	√	√	√	–	√	√	–	–	√	–	–	–	–
2	√	√	√	√	√	√	√	√	√	√	√	√	√	√	–	√
3	√	√	√	√	√	√	√	√	√	√	√	√	–	√	–	–
4	√	√	√	√	√	√	√	√	–	√	–	–	√	√	–	–
5	√	√	√	√	√	√	√	√	√	√	–	√	–	–	–	–
6	√	√	√	√	√	√	√	√	√	√	–	√	√	√	–	–
7	√	√	√	√	√	√	√	√	–	–	–	√	–	√	–	–
8	–	√	√	√	√	–	√	√	–	–	–	√	–	–	–	–
9	–	√	√	√	√	√	√	√	√	–	√	√	–	–	–	–
10	√	√	√	√	√	√	√	√	–	√	–	√	–	–	–	√
11	–	√	√	√	√	√	–	√	√	–	–	√	√	–	–	–
12	√	√	√	√	√	√	√	–	√	–	√	–	–	–	–	–
13	√	√	√	√	√	–	√	√	√	√	–	√	–	–	–	–
14	√	√	√	√	√	√	√	√	√	√	√	√	√	√	–	√
15	–	–	√	√	√	–	–	√	–	–	–	–	–	–	–	–
16	√	√	√	√	√	√	√	√	√	√	√	√	–	√	–	√
17	–	–	–	√	–	–	–	–	–	–	–	–	–	–	–	–
18	–	–	√	√	–	√	–	–	–	–	–	–	–	–	–	–
19	√	√	√	√	√	√	√	√	√	–	–	√	–	–	–	–
20	√	√	√	√	√	–	–	–	–	–	–	–	–	–	–	–
21	√	√	√	√	√	–	√	√	–	–	–	–	–	–	–	–
22	–	–	√	√	–	√	–	–	–	√	–	–	–	–	–	–
23	√	√	√	√	√	–	√	–	–	–	–	–	–	–	–	–
24	–	√	√	√	√	√	–	√	–	√	–	–	–	–	–	–
25	√	√	√	√	√	√	√	√	–	√	–	√	√	–	–	–
26	√	√	√	√	√	√	√	√	–	√	–	–	–	–	–	–

(continued)

Table 2 (continued)

Lp.	Stakeholders	Clients	Employees	Local society	Environment & ecloogy	Values	Cororate governance	Code of ethics	Risk	Human rights	Business model	Anticorruption and bribery	Financial data on CSR	Negative impact	Measurement of CSR	Value created EC-1
27	√	√	√	√	√	–	√	√	√	√	√	√	√	√	–	√
28	–	√	–	√	–	√	–	–	–	–	–	–	–	–	–	–
29	√	√	√	√	√	√	√	–	√	–	√	√	√	√	–	–
30	√	–	√	√	√	–	–	√	–	–	–	–	–	–	–	–
31	√	–	√	√	√	√	√	√	–	–	–	√	–	–	–	–
32	√	√	√	√	√	√	√	√	–	–	–	√	√	√	–	–
33	–	–	√	√	√	–	–	√	–	√	–	–	–	–	–	–
34	√	√	√	√	√	√	√	√	√	√	√	√	–	√	–	–
35	√	√	√	–	–	–	–	√	√	–	√	–	–	–	–	–
36	√	√	√	√	√	√	√	√	√	√	√	√	–	√	–	–
37	√	√	√	√	√	√	√	√	√	–	√	–	√	√	–	–
38	√	–	√	√	√	√	√	√	√	√	–	√	–	√	–	–
39	√	√	√	√	√	√	–	√	√	√	√	√	–	√	–	–
40	–	√	√	√	√	√	–	√	–	–	–	–	–	√	–	–
41	√	–	√	√	√	√	√	√	√	√	√	√	√	√	–	–
42	√	√	√	√	√	√	–	–	–	√	–	–	–	–	–	–
43	√	√	√	√	√	√	√	√	–	√	√	√	–	√	–	–
44	√	√	√	√	√	√	√	√	√	√	√	√	–	√	–	–
45	√	√	√	√	√	√	√	√	√	√	–	–	–	√	–	–
46	√	√	√	√	–	√	–	√	√	–	–	√	–	–	–	–

Source: Own elaboration based on reports from CSRinfo.org

practices from 2014 among 60 companies (top 30 and 40 medium size non-financial companies listed on the WSE, excluding entities with the average number of employees lower than 500 on their balance sheet date—31 December 2014), that the significant factors influencing CSR disclosure level are: turnover, industry environmental sensitivity, duration of the stock exchange listing, inclusion in the Respect Index portfolio and foreign capital share.

After 4 years we may add that even though some positive changes are observed (e.g. the doubled number of integrated reports) we may still see the differences in approach to responsibility activities between companies from Western Europe and from Central Eastern Europe (Fijałkowska et al. 2018). The "old" EU member states seem to share a rather homogenous view of the future of CSR—they expect its role to increase. In contrast, the new EU member states (CEE countries) remain more cautious about the future growth of CSR. These differences can be explained by the fact that the CEE countries have a relatively short history of corporate responsibility (Kudłak et al. 2018). Moreover, due to the socialist heritage, there is a general perception that social responsibility and social caring is the primary role of government, thus the existing "corporate social responsibility gap" between the two European regions (Steurer and Konrad 2009). Moreover, there are fears about reliability of CSR reports. Greenwashing, conscious attempts at communicating positive social, environmental and employee information in order to improve the reputation and social image of companies, is quite a widespread practice (Śmiechowski and Lament 2017). Polish companies, if they want to have a say in an increasing competitive market, will definitely have to join the general trend of CSR engagement (Popowska 2015) and a transparent and intensified CSR reporting.

4 Respect Index and CSR Reporting

Warsaw Stock Exchange (WSE) is one of the active promotors of CSR disclosures. The current concept of the RESPECT index (More about the Respect index, see: Dyczkowska et al. 2016) project is a follow-up to the activities of the WSE which took place in 2009, resulting in the establishment of the first index of responsible companies in the Central and Eastern Europe. RESPECT index is one of the most important projects in the area of CSR promotion, embracing listed companies which adopted CSR standards. It is designed to select companies that are managed responsibly and sustainably, but also strongly emphasizes the attractiveness of a company, quality of reporting, level of investor relations and information disclosure. The RESPECT index was published for the first time in November 2009 and included 16 companies. The process of qualifying for this index is quite complex. First of all, WSE selects companies that are part of the indexes WIG20, mWIG40, and sWIG80, then analyzes the quality of information disclosure and verifies the extent to which individual companies are engaging in stakeholder involvement. This process is carried out by monitoring company reports and reviewing their web sites. The most important step is an audit of the survey completed by a company. The companies that obtain the highest number of points are accepted into the index. It is revised every 6 months and the number of companies included varies. Currently it is 25 companies. The companies presently accepted into the RESPECT Index are the following: KGHM (Raw materials, 11.66% of share in the index portfolio), PKN Orlen (Fuel, 10.04%), PZU (Finance, 9.80%), Pekao (Finance, 9.52%), BZ WBK (Finance, 9.25%), PGNIG (Fuel, 8.64%), PGE (Power engineering, 8.34%), ING BSK (Finance, 4.94%), Lotos (Fuel, 3.76%), Millennium (Finance, 3.63%), Tauron (Power engineering, 2.87%), Orange Polska (Telecommunication, 2.60%), Grupa Azoty (Chemistry, 2.50%), Budimex (Constructions, 2.28%), Handlowy (Finance, 2.28%), Energa (Power engineering, 1.93%), Forte (Industry, 1.17%), GPW (Finance, 1.09%), Bogdanka (Raw materials, 0.79%), Apator (Industry, 0.78%), Kogeneracja (Power engineering, 0.58%), Trakcja (Industry, 0.52%), Elektrobudowa (Constructions, 0.47%), Pelion (Trade, 0.36%), and PCC Rokita (Chemistry, 0.19%).

5 New Directive: Outlooks

As an institutional answer to the requirements of higher transparency and accountability of companies concerning their responsibility and diversity, the European Union decided to issue the new regulation (Fijałkowska 2016a, b). 15 November 2014 the Directive 2014/95/EU of the European Parliament and of the Council of 22 October 2014 amending Directive 2013/34/EU as regards disclosure of non-financial and diversity information by certain large undertakings and groups has been published in the Official Journal of the European Union (Parliament 2014). In Poland the new regulation, related to the disclosure requirements in terms of non-financial reporting, is applicable to about 300 entities since 2017. The provisions of the UE Directive have been incorporated into the Polish law, however Polish regulations focus on promoting the topic, and they do not force the CSR reporting. Preparing a separate social report is just one of the possible options. It is worth underlining that nowadays the majority of Polish companies not only do not publish social reports but they usually do not have sufficient information related to social and employee issues to meet the requirements of the new European regulation (Forum Odpowiedzialnego Biznesu 2017b). The introduction of minimum level of mandatory CSR disclosures into Polish accounting standards seems to be a necessity to move toward the right direction (Krasodomska and Cho 2017). In Poland the provisions of the Directive are implemented in the framework of:

- Amendment of the Accounting Act of 15 December 2016 (Journal of Laws of 2017, item 61),
- Regulation of the Minister of Finance of 25 May 2016 amending the regulation on current information and periodic disclosures by issuers of securities and the conditions for recognizing as equivalent information required by the laws of a

non-member state (Journal of Laws of 2016, item 860). Ordinance entered into force on June 29, 2016.

The implementation of the Directive 2014/95/EU has drawn attention of the accounting environment to the non-financial disclosures and calls for research in this area (Albu et al. 2013).

6 Summary

The CSR report is a part of a broader business process. It is important to engage employees in CSR reporting and use the process of CSR reporting to increase the knowledge of a company as well as to define its business opportunities and risks. It's important to root CSR reporting in the enterprise and to indicate to managers how they can use the reporting process in their work area. A decade ago many managers in Poland did not understand the meaning of CSR reporting. Such reporting was new and the business was accepting them with reservations and doubts. Today, corporate awareness concerning CSR is much greater, many companies got involved in CSR activities, several dozens of organizations every year publish a CSR report and understand its usefulness for business. The belief that a social report is a valuable source of information about the impact of business on people, the environment and the society, as well as of the fact that it discloses a comprehensive knowledge on the company to its stakeholders has increased.

There are strong institutional activities aimed at supporting non-financial reporting in Poland and helping to meet the requirements of the new Directive. Since 2016, the GRI standard is available free of charge in Polish language. The decision to choose a standard may be supported by a free publication, "Non-Financial Reporting. Guidance for Reporters," available at website of the Polish Ministry of Development. Also, the Association of Stock Exchange Issuers is actively engaged in the process of promoting CSR reporting—it organizes conferences and congresses related to this subject. In addition, this association, in order to facilitate the implementation of non-financial reporting, has initiated the development of a non-financial information standard that will be a very useful reporting tool for companies that have not previously published CRS reports.

The organizations that promote and support CSR involvement and disclosure in Poland are e.g. the following: Responsible Business Forum, Academy for the Development of Philanthropy in Poland, Center for Business Ethics, Corporate Social Responsibility Institute and the Foundation Center CSR.pl (Hąbek 2014; Wołczek 2015). The Warsaw Stock Exchange, Polish Accounting Standards Committee and the Big 4 audit firms are also actively involved in fostering the CSR concept in Polish (Dyczkowska et al. 2016). The Polish Chamber of Statutory Auditors (PIBR) encourage CSR reporting by organizing and co-organizing many conferences devoted to this subject, such as the "Business and Social, Environmental and Economic Challenges of the Country, Role and Responsibility" Conference, together with the Ministry of Economy, the National Fund for Environmental Protection and Water Management, Deloitte and the Responsible Business Forum. Moreover, PIBR dedicated the third conference in the series "Audit for Business Security and Development: Responsible Reporting" to the theme of CSR reporting. It organized a series of nationwide seminars for auditors related to CSR reporting and its verification. Furthermore, the Polish Chamber of Statutory Auditors became involved financially in the translation of the latest financial reporting standards developed by the Global Reporting Initiative. The Polish Government is also a key player in the promotion of CSR reporting by mandating, facilitating, partnering, endorsing, and demonstrating CSR disclosure (Hąbek and Wolniak 2013).

Altogether, there are around 90 companies in Poland that have engaged in CRS reporting. In 2017 almost half of them issued CSR report. Their practice, knowledge and experience gained in reporting of non-financial information, are now the benchmark and source of inspiration for about 300 of the largest companies in Poland, that will have to face the new requirements concerning extended non-financial reporting as of 1 January 2017.

References

Albu, N., Albu, C. N., Dumitru, M., & Dumitru, V. F. (2013). Plurality or convergence in sustainability reporting standards? *Amfiteatru Economic, XV*, 729–742.

Bonsón, E., & Bednárová, M. (2015). CSR reporting practices of Eurozone companies. *Revista de Contabilidad, 18*, 182–193. https://doi.org/10.1016/J.RCSAR.2014.06.002.

Chojnacka, E., & Wiśniewska, J. (2017). Reporting CSR data – The benefits and barriers to involvement in the idea of corporate social responsibility. *Journal of Positive Management, 8*, 3–16. https://doi.org/10.12775/JPM.2017.001.

CSRInfo CSRInfo. Accessed July 24, 2017, from http://www.rejestrraportow.pl/

Deloitte. (2016). *Central Europe top 500 an era of digital transformation*. Accessed July 14, 2017, from https://www2.deloitte.com/content/dam/Deloitte/global/Documents/About-Deloitte/central-europe/ce-top-500-2016.pdf

Dyczkowska, J., Krasodomska, J., & Michalak, J. (2016). CSR in Poland: Institutional context, legal framework and voluntary initiatives. *Accounting and Management Information Systems, 15*, 206–254.

Dyduch, J., & Krasodomska, J. (2017). Determinants of corporate social responsibility disclosure: An empirical study of Polish listed companies. *Sustainability, 9*, 1934. https://doi.org/10.3390/su9111934.

Fijałkowska, J. (2013). Zakres ujawnień w raportach społecznej odpowiedzialności i zrównoważonego rozwoju przedsiębiorstw. *Pr Nauk Uniw Ekon we Wrocławiu, 311*, 85–94.

Fijałkowska, J. (2016a). New regulation of non-financial disclosure in business reporting. *Entrepreneurship and Management, 17*, 27–39.

Fijałkowska, J. (2016b). Raportowanie informacji niefinansowych zgodnie z nową dyrektywą UE 2014/95/EU jako wyzwanie dla przedsiębiorstw. *Pr Nauk Uniw Ekon we Wrocławiu, 436*, 115–122. https://doi.org/10.15611/pn.2016.436.12.

Fijałkowska, J., & Macuda, M. (2017). Value creation disclosure in CSR reports: Evidence from Poland. *Zesz Teoretyczne Rachun, 93*(149), 61–78. https://doi.org/10.5604/01.3001.0010.3189.

Fijałkowska, J., Zyznarska-Dworczak, B., & Garsztka, P. (2017). The relation between the CSR and the accounting information system data in Central and Eastern European (CEE) countries – The evidence of the Polish financial institutions. *Journal of Accounting and Management Information Systems, 16*, 490–521. https://doi.org/10.24818/jamis.2017.04004.

Fijałkowska, J., Zyznarska-Dworczak, B., & Garsztka, P. (2018). Corporate social-environmental performance versus financial performance of banks in Central and Eastern European countries. *Sustainability, 10*, 772. https://doi.org/10.3390/su10030772.

Forum Odpowiedzialnego Biznesu. (2017a). 10 years of CSR reporting in Poland.

Forum Odpowiedzialnego Biznesu. (2017b). Analiza tematyczna Nr 1/2017. Nowe obowiązki prawne związane z ujawnianiem informacji niefinansowych i polityki dotyczącej różnorodności. Accessed July 14, 2017, from https://www.mf.gov.pl/c/document_library/get_file?uuid=2d9f133c-75a9-4541-bbe8-a30ebe3360d8&groupId=764034

Hąbek, P. (2014). Evaluation of sustainability reporting practices in Poland. *Quality and Quantity, 48*, 1739–1752. https://doi.org/10.1007/s11135-013-9871-z.

Hąbek, P., & Wolniak, R. (2013). Analysis of approaches to CSR reporting in selected European Union countries. *International Journal of Economic Research, 4*, 79–95.

Hąbek, P., & Wolniak, R. (2016). Assessing the quality of corporate social responsibility reports: The case of reporting practices in selected European Union member states. *Quality and Quantity, 50*, 399–420. https://doi.org/10.1007/s11135-014-0155-z.

Jaglińska, Z. (1984). Rachunkowość odpowiedzialności społecznej przedsiębiorstw kapitalistycznych. *Zesz Teoretyczne Rachun, 7*, 15–21.

Jarugowa, A. (1984). Rachunkowość społeczna na tle tendencji rozwoju rachunkowości. *Zesz Teoretyczne Rachun, 7*, 5–14.

Kolk, A. (2008). Sustainability, accountability and corporate governance: Exploring multinationals' reporting practices. *Business Strategy and the Environment, 17*, 1–15. https://doi.org/10.1002/bse.511.

Krasodomska, J. (2013). Sprawozdawczość zintegrowana – aspekty teoretyczne i praktyczne. In M. Remlein (Ed.), *Rachunkowość odpowiedzialności społecznej : teoria i praktyka* (pp. 65–87). Poznań: Zakład Poligraficzno-Wydawniczy M-DRUK.

Krasodomska, J. (2015a). Sprawozdawczość zintegrowana jako nowy obszar badań naukowych w rachunkowości. *Zesz Teoretyczne Rachun, 82*(138), 77–92. https://doi.org/10.5604/16414381.1155811.

Krasodomska, J. (2015b). CSR disclosures in the banking industry. Empirical evidence from Poland. *Social Responsibility Journal, 11*, 406–423. https://doi.org/10.1108/SRJ-02-2013-0019.

Krasodomska, J., & Cho, C. H. (2017). Corporate social responsibility disclosure. *Sustain Accounting, Management Policy Journal, 8*, 2–19. https://doi.org/10.1108/SAMPJ-02-2016-0006.

Kudłak, R., Szőcs, I., Krumay, B., & Martinuzzi, A. (2018). The future of CSR – Selected findings from a Europe-wide Delphi study. *Journal of Cleaner Production, 183*, 282–291. https://doi.org/10.1016/j.jclepro.2018.02.119.

Macuda, M. (2016a). Corporate social responsibility – A new trend in hospital accounting? *Stud Ekon Zeszty Nauk Uniw Ekon w Katowicach, 284*, 76–86.

Macuda, M. (2016b). Le nouveau défi de la comptabilité: la divulgation d'informations RSE – vers le rapport intégré. *Rev Int des Econ Lang Française, 1*, 227–244.

Macuda, M., Matuszak, Ł., & Różańska, E. (2015). The concept of CSR in accounting theory and practice in Poland: An empirical study. *Zesz Teoretyczne Rachun, 84*(140), 115–138. https://doi.org/10.5604/16414381.1173808.

Matuszak, Ł., & Różańska, E. (2017a). CSR disclosure in Polish-listed companies in the light of directive 2014/95/EU requirements: Empirical evidence. *Sustainability, 9*, 2304. https://doi.org/10.3390/su9122304.

Matuszak, Ł., & Różańska, E. (2017b). An examination of the relationship between CSR disclosure and financial performance: The case of Polish banks. *Journal of Accounting and Management Information Systems, 16*, 522–533.

Michalczuk, G., & Mikulska, T. (2014). Sprawozdanie zintegrowane jako produkt finalny rachunkowości społecznej. *Zesz Nauk Uniw Szczecińskiego Finans Rynk Finans Ubezpieczenia, 69*, 197–207.

Parliament E. (2014). Directive 2014/95/EU of the European Parliament and of the Council of 22 October 2014 amending Directive 2013/34/EU as regards disclosure of non-financial and diversity information by certain large undertakings and groups. In Official Journal of the European Union L330/1. Accessed October 22, 2017, from https://publications.europa.eu/en/publication-detail/-/publication/406ad3b8-6c9f-11e4-9680-01aa75ed71a1/language-en

Popowska, M. (2015). Europe, between mandatory and voluntary approach to CSR: Some new evidence from the Polish listed companies. *Journal of Yaşar University, 10*, 38–45. https://doi.org/10.19168/jyu.13091.

Samelak, J. (2013). *Zintegrowane sprawozdanie przedsiębiorstwa społecznie odpowiedzialnego*. Poznań: Wydawnictwo Uniwersytetu Ekonomicznego.

Śmiechowski, K., & Lament, M. (2017). Impact of Corporate Social Responsibility (CSR) reporting on pro-ecological actions of tanneries. *Journal of Cleaner Production, 161*, 991–999. https://doi.org/10.1016/j.jclepro.2017.05.104.

Steurer, R., & Konrad, A. (2009). Business–society relations in Central-Eastern and Western Europe: How those who lead in sustainability reporting bridge the gap in corporate (social) responsibility. *Scandinavian Journal of Management, 25*, 23–36. https://doi.org/10.1016/j.scaman.2008.11.001.

Szczepankiewicz, E. I. (2014). Zintegrowane sprawozdanie przedsiębiorstwa jako narzędzie komunikacji z interesariuszami. *Pr Nauk Uniw Ekon we Wrocławiu, 329*, 271–281. https://doi.org/10.15611/pn.2014.329.29.

Szczepankiewicz, E., & Mućko, P. (2016). CSR Reporting Practices of Polish Energy and Mining Companies. *Sustainability, 8*, 126. https://doi.org/10.3390/su8020126.

Tatarska, M. (2013). Sprawozdawczość zintegrowana na przykładzie Grupy Lotos. *S.A. Pr Nauk Uniw Ekon we Wrocławiu, 314*, 187–199.

Waniak-Michalak, H., Macuda, M., & Krasodomska, J. (2016). Corporate Social Responsibility and Accounting in Poland: A Literature Review. *Accounting and Management Information Systems, 15*, 255–303.

Wójcik-Jurkiewicz, M. (2017). Role of CSR Reporting. Evidence from Poland. *Zesz Teoretyczne Rachun, 94*(150), 173–188. https://doi.org/10.5604/01.3001.0010.5001.

Wołczek, P. (2015). Development of the CSR concept in Poland – Progress or stagnation? *Pr Nauk Uniw Ekon we Wrocławiu, 387*, 200–214.

Wolniak, R., & Hąbek, P. (2016). Quality Assessment of CSR Reports – Factor Analysis. *Procedia – Social and Behavioral Sciences, 220*, 541–547. https://doi.org/10.1016/J.SBSPRO.2016.05.530.

Yim, D., Liu, X., & Khuntia, J. (2018). Following the CSR footprints in sustainability reporting. In Proceedings of SIG GREEN Workshop.

Non-financial Reporting. Conceptual Framework, Regulation and Practice

Maria Aluchna and Maria Roszkowska-Menkes

1 Introduction

Non-financial reporting has developed significantly in the past four decades although in emerging and post-transition countries (such as Poland) it remains at the very early stage of development. It originated from CSR and environmental disclosure to provide a complex communication to various stakeholders. In its current version non-financial reporting is evolving towards integrated reporting that delivers a complete picture of company multidimensional performance and reveals interdependencies between business, society and environment. It also exemplifies the advanced corporate communication to numerous groups of stakeholders and constitutes a strategic response to social, cultural, institutional and regulatory pressures. Non-financial disclosure is perceived as a systemic progress of assessing company performance beyond traditional financial reporting emphasizing its long-term impact and strategic development. With a set of measures and guidelines non-financial reporting adds to the operationalization of the CSR and sustainability concepts and assures standards for a comparative companies' assessment.

The development of non-financial disclosure increases transparency, improves implementation of sustainability principles and enhances company accountability and legitimacy in the relations with various constituencies (Eccles et al. 2011; Fernandez-Feijoo et al. 2014). Yet, as studies reveal the quality, content and assurance of non-financial reporting is to large extent determined by institutional environment and significantly embedded in business practice.

In recent years non-financial reporting has developed from CSR and environmental communication to sustainability disclosure towards integrated reporting

M. Aluchna (✉) · M. Roszkowska-Menkes
Warsaw School of Economics, Warszawa, Poland
e-mail: maria.aluchna@sgh.waw.pl

A. Długopolska-Mikonowicz et al. (eds.), *Corporate Social Responsibility in Poland*, CSR, Sustainability, Ethics & Governance,
https://doi.org/10.1007/978-3-030-00440-8_14

(Higgins and Coffey 2016). From a narrative description of social issues such as employee rights, community issues and environmental matters of resource use and waste management, non-financial reporting has evolved into strategic communication. Reports integrate company capitals, cover social and environmental concerns, provide assessment of company impact and accountability of organizational performance and examine multidimensional performance within the strategic development.

With growing interest in non-financial reporting there is still a gap in the understanding of the patterns of its adoption, effectiveness and efficiency (de Villiers et al. 2014b). Existing studies fail also to explain how non-financial reporting contributes to the implementation of CSR. We would like to add to the literature indicating the role of non-financial reporting in the implementation of CSR principles addressing the existing literature to the case of Poland. Specifically, we draw upon the development of non-financial disclosure pointing at its link with CSR and sustainability concepts. In addition, we discuss the practice of non-financial reporting in Poland revealing results of a research on the popularity of non-financial disclosure on the sample of listed companies.

The remainder of this chapter is as follows. In the first section we present the concept of non-financial reporting discussing its origins, motivations for and benefits of its adoption. The second section addresses standards of non-financial reporting referring to its quality and assurance. In the third section we discuss the practice of non-financial reporting in Poland using the existing studies as well as presenting the result of empirical analysis conducted on the sample of companies listed on the Warsaw Stock Exchange between years 2010–2014. Final remarks are presented in the discussion and conclusion section.

2 The Concept of Non-financial Reporting

2.1 Definition and Origins

The term of non-financial reporting refers to the voluntary, solicited or mandatory disclosure of social, economic and governance information of a company. It is defined as "the process of communicating the social and environmental effects of organizations' economic actions to particular interest groups within society and to society at large" (Gray et al. 1987, ix as quoted in Kotonen 2009). It is viewed as the important innovation in the disclosure of company operation and performance providing a multidimensional picture of its social and environmental impact. The existing literature and corporate reports adopt a number of different phrases for non-financial reporting (Kotonen 2009) such as CSR reporting (Tschopp and Huefner 2015), social and environmental reporting (Moneva and Cuellar 2009), sustainability reporting (Steurer and Konrad 2009; Hahn and Kühnen 2013; KPMG 2016), ESG standards, social accounting (Owen and Swift 2001), social and environmental disclosure (Deegan 2002), sustainability disclosure (Joseph and Taplin

2011; Deloitte 2016), social auditing (Gray 2000), social review (O'Dwyer and Owen 2005). Non-financial reporting is placed in the theoretical framework of corporate social responsibility, stakeholder management, triple bottom line and sustainability (Skouloudis et al. 2009; Thorne et al. 2014). In recent years scholars adopted also signaling theory (Deegan 2002) and neo-institutional and legitimacy perspectives (Thorne et al. 2014) to study the practice of non-financial reporting in the organizational and environmental contexts. Conceptually non-financial reporting is perceived as

1. a corporate response to stakeholder pressure resulting in the changing paradigm for business role in economy and society (Manetti and Toccafondi 2012; Fernandez-Feijoo et al. 2014),
2. the operationalization of sustainable business/sustainable development (Mio and Venturelli 2013) and
3. the reaction of accounting practice to develop formal ways in order to capture the value of intangible assets (Gray 2010), representing "enlightened" version of shareholder theory that assumes there is growing number of social and environmental factors that influence firm's ability to create value in the long term, and they should be addressed by the managers in their strategies and reporting.

First, the emergence of non-financial reporting mirrors the fundamental change in the strategic goal of business shifting from the shareholder value focus towards the concept of creating shared value (Crane and Matten 2007). With the dominance of the principal-agent theory and value based management framework (Vermalen 2009) companies were expected to increase its share price to assure return on investment for shareholders (Lazonick and O'Sullivan 1997; Vogel 2005). The growing awareness of limitations of financial performance measures as well as the increased stakeholder activism and regulatory pressure gave rise to the development of concepts such as corporate social responsibility and sustainable development/business (Aluchna and Mikołajczyk 2016). Non-financial reporting represents a practical case of active engagement by stakeholders to improve the content and assurance of corporate disclosure (Manetti 2011; Manetti and Toccafondi 2012; Fernandez-Feijoo et al. 2014). A wider framework to assess performance was meant to capture the complete company impact (Strumińska-Kutra and Woźniczko 2010; Kemper and Martin 2011) and to develop the criteria and indicators to measure company success with respect to the social and environmental performance (Mayer 2002).

Second, while CSR and sustainable development promote environmental protection, lowering ecological footprint, implementation of resource-efficient and low emission technologies, adoption of standards for customer safety and employee rights (Geels 2011), the concepts failed to develop a clear framework to measure company progress in these areas. The gap was partially filled by the CSR/sustainability indexes and ratings, yet it is the frameworks of non-financial reporting to offer standards for assessing the social and environmental performance. In this sense non-financial reporting constitutes a transition from financially focused short-term thinking to long-term sustainable value-based business philosophy (Ballou et al. 2012;

Beattie and Smith 2013). It provides a set of measures to operationalize and assess sustainable development (Mio and Venturelli 2013).

Third, non-financial reporting, especially in the form of integrated reporting offers an operationalization of measuring value, which is omitted in the traditional financial reports. It uses the Global Reporting Initiative (GRI) framework to categorize the nonfinancial information by reference to the six categories of non-financial capital including economic, environment, labor, human rights, product responsibility, and society indicators (Milne and Gray 2013) and aims at showing interdependencies between them and firms value. Thus, non-financial reporting is viewed as the extension of corporate disclosure beyond traditional financial information (Gray and Bebbington 2000; Tregidga and Milne 2006; Gray 2010) capable to capture the dynamics of intangible value.

2.2 *Motivations and Benefits*

The idea of voluntary non-financial disclosure is driven by a number of systemic reasons, which are to address the shortcomings of conventional corporate communication with stakeholders. As noted by Kotonen (2009) and Graham et al. (2005) the adoption of non-financial reporting (1) promotes transparent and complete disclosure of company operation, (2) reduces information risk of the company valuation and (3) mitigates deficiencies of mandatory reporting.

Existing studies examine the contextual factors, organizational determinants (Adams and McNicholas 2007; Fifka 2013; de Villiers et al. 2014a; Waris et al. 2017) and effects (Fifka 2013; Ioannou and Serafeim 2014) of the adoption of non-financial reporting. They look at the variable of the adoption, extent and quality of non-financial reporting and indicate numerous quantitative and qualitative benefits for the company.

While the first companies started to publish sustainability reports 30 years ago (Higgins and Coffey 2016), the number of reports rose significantly in the twenty-first century. The review of existing studies by Hahn and Kühnen (2013) indicate that in general companies adopting non-financial tend to be larger. While the measures of social and environmental performance reveal mixed results, the variables of financial performance and ownership structure are found to be statistically insignificant for the adoption, extent and quality of sustainability reporting (Hahn and Kühnen 2013). Referring to external determinants media exposure indicates positive links with the adoption, extent and quality of non-financial reporting. Also sector affiliation plays a positive role for extent and quality of non-financial disclosure, yet for its adoption studies reveal indifferent results. Specifically, since firms with higher impacts on the environment respond to higher stakeholder pressures for transparency, companies operating in the environmentally sensitive industries such mining, oil and gas extraction, paper manufacturing, chemical manufacturing belong to those most frequently issuing non-financial reports (Fernandez-Feijoo et al. 2014; Lock and Seele 2016; Stacchezzini et al. 2016).

The adoption of non-financial reporting is believed to be positively viewed by stakeholders and shareholders and have a positive impact on company performance and value. The existing literature draws upon investor reaction and variation of financial results for companies, which introduced sustainability/CSR disclosure. As suggested by Boiral (2013: 1040) "voluntary disclosure (. . .) makes it possible for a firm to inform stakeholders of the organization's sustainability performance, to distinguish itself from poor-performing competitors, and to increase its reputation by shedding light on hard-to-imitate sustainability strategies". Additionally some authors (Eccles et al. 2014; Serafeim 2014) argue that investors are increasingly interested in environmental, social and governance performance metrics and policies when making investment decisions, but they consider sustainable information as more reliable and relevant if its disclosed in an integrated report.

Yet, the existing studies still deliver mixed evidence regarding the benefits of non-financial disclosure (Berthelot et al. 2012). While Healy and Palepu (2001) indicate the positive impact of non-financial reporting for corporate profitability, a negative effect of non-financial reporting on performance was noted by McWilliams and Siegel (2000). In their recent study Xu and Liu (2017) observed on the sample of Chinese firms that share price volatility after CSR disclosure is lower than before CSR disclosure and that stock liquidity significantly improves after CSR disclosure. Both trends are non-linear. Moreover, however, share price volatility increases first and then decreases. Additionally, by dividing CSR disclosure into economic (hard) disclosure and generic (soft) disclosure, they find that the reduction in information asymmetry is higher for economic disclosure than CSR disclosure. The research shows different market reaction to the issuance of the reports revealing positive (Schadewitz and Niskala 2010) or no significant effect (Guidry and Patten 2010). Some companies, including those operating in U.S., Canada, China, India and Korea, "remain skeptical about the importance of sustainability in their strategies, to some extent exacerbated by the short-term nature of their capital markets" (Eccles and Serafeim 2011: 80–81). Studies also note that non-financial reporting appears to be not important to investors (Solomon and Solomon 2006; Berthelot et al. 2012) and does not meet their expectations (Murray et al. 2006; Cormier and Magnan 2007; Moneva and Cuellar 2009).

Companies adopting non-financial reporting may also experience the indirect effect of benefits, which are qualitative in nature. In general introducing non-financial reporting is seen as a progress in the way to communicate company information to stakeholder (Moravcikova et al. 2015) and as the mean to increase transparency (Deloitte 2016). It can also represent a strategic reaction to competitive environment (McWilliams and Siegel 2000) addressing changing stakeholder expectations and the need to provide a complete assessment of company operation. In addition, non-financial reporting serves as the evidence to document company engagement and performance in social and environmental aspects of its core business operation and improves company image and reputation (Bebbington et al. 2008; Boiral 2013). While providing non-financial information often results from pressure and facilitates dialog with various entities (Fernandez-Feijoo et al. 2014), it can also be an effective tool to manage powerful stakeholders and control the national environmental agenda (Kotonen 2009).

Increasing transparency with non-financial reporting allows to increase accountability to investors signaling the company openness to communicate its performance. It attains legitimacy amongst various constituencies with whom the company cooperates and in this way it allows to reduce external costs (Caron and Turcotte 2009; Ballou et al. 2006a, b; Thorne et al. 2014). In the perspective of neo-institutional theory (Meyer and Rowan 1977; DiMaggio and Powell 1983; Mizruchi and Fein 1999) it is defined as an element of organizational isomorphism. The companies' ability to survive and to grow is determined to large extent by how well they conform to rules, norms and belief systems prevalent in their operating environment (Wild and van Staden 2013). Hence, even if the company has no marginal return from sustainability practices and integrated reporting, it might still decide to engage in these activities as a result of regulation or social pressure, creating differences among countries produced by the demand for sustainability (Fernandez-Feijoo et al. 2015; Nazari et al. 2015). Finally, it can integrate business community (Moon 2002), add to the emergence of new standards and improve their adoption.

3 Reporting Practice

3.1 Standards

Companies adopt the framework of non-financial reporting addressing stakeholder expectations (Matten and Moon 2008; Milne and Gray 2013; Eccles et al. 2014; Fernandez-Feijoo et al. 2014). For years CSR and sustainability disclosure was based on voluntary basis and the adoption of different reporting standards (Searcy and Buslovich 2014). The most often used guidelines include:

- The standards published by the Global Reporting Initiative (GRI) (Searcy and Buslovich 2014; Zimara and Eidam 2015; Lock and Seele 2016; GRI 2017)
- The standards by International Integrated Reporting Council (IIRC 2017)
- The standards of International Petroleum Industry Environmental Conservation Association (IPIECA), which supports reporting in entities operating in oil and gas industry (KPMG 2016; IPIECA 2016)

Global Reporting Initiative (GRI) as independent organization helps understand the consequences of their business on sustainability issues location (GRI 2017) and offers internationally the most prominent and most widely used principles and standards for non-financial data disclosure (Clayton et al. 2015). More precisely, the GRI provides standards on sustainability reporting and disclosure, which helps businesses, governments, and other entities to make better decisions regarding sustainability issues (GRI 2017). The core guidelines of GRI are also supported by sector- and country-specific additions (Searcy and Buslovich 2014; GRI 2017). It has been designed to encourage organizations to take into consideration the whole impact of their operations.

GRI introduces ten Reporting Principles, which are fundamental to achieving high quality sustainability reporting. An organization is required to apply the Reporting Principles if it wants to claim that its sustainability report has been prepared in accordance with the GRI Standards. The principles are divided into two groups (GRI 2016):

1. Principles for defining report content including stakeholder inclusiveness, sustainability context, materiality and completeness
2. Principles for defining report quality including accuracy, balance, clarity, comparability, reliability and timeliness

Regarding the external assurance for sustainability report, it advised, but not required in order to make a claim that a report has been prepared in accordance with the GRI Standards (GRI 2016).

GRI's sustainability accounting approach is based on the Elkington's (1997) triple bottom line (TBL) concept (or three pillars theory—People, Planet, Profit) (Robins 2006). TBL adds to the traditional, economic bottom line two other balance sheets focusing on the conservation of social, natural and economic capitals, giving them equal importance. The framework, however, by using three separate bottom lines, fails to track interconnections between the various types of capitals and is more focused on disclosing decreases in these capitals rather than on value creation (Adams 2015). Additionally, as the GRI standards became more complex and started to cover a broad range of social, environmental and governance issues, sustainability reports compiled in accordance with them also became more complex, lengthy and detailed, hindering identification of linkages between different policies and practices (de Villiers et al. 2014b).

The most recent significant global development in the area of non-financial reporting is the formation of the International Integrated Reporting Council (IIRC), coalition of regulators, investors, companies, standard setters, accountants and NGOs, that calls calls for a reorientation of the focus of corporate reporting from short-term, backward-looking financial information to forward-looking, connected and strategic information that discusses an organization's ability to generate value over time (Adams and Simnett 2011). The goal of integrated reporting movement is to overcome one of the major weaknesses of financial and sustainability reporting (Lodhia 2015) that fail to provide stakeholders with information on interdependencies between various areas of company's operations (Clayton et al. 2015). In its framework (International IR Framework) published in 2013 IIRC defines integrated report as "a concise communication about how an organization's strategy, governance, performance and prospects, in the context of its external environment, lead to the creation of value over the short, medium and long term" (IIRC 2013: 7). It is built around the notion of value that is generated from six categories of capital, not necessarily owned by the company: financial, manufactured, intellectual, human, social and relationship, and natural (IIRC 2013). "In essence, integrated reporting is a hybrid practice that spans between the different worlds of financial reporting and sustainability reporting" (van Bommel 2014: 1158).

The IIRC's framework (IIRC 2013: 16–23) provides seven guiding principles underpinning preparation and defining the content and form of an integrated report:

1. Strategic focus and future orientation—an integrated report should provide insight into the organization's strategy, how it creates value and how it effects particular capitals
2. Connectivity of information—an integrated report should present the holistic picture of the company's value creation process
3. Stakeholder relationships—an integrated report should provide information on how and to what extent the organization understands, takes into account and responds to their legitimate needs and interests
4. Materiality—an integrated report should disclose information about matters that substantively affect the organization's ability to create value over the short, medium and long term
5. Conciseness—an integrated report should be concise
6. Reliability and completeness—an integrated report should include all material matters, both positive and negative
7. Consistency and comparability—the information in an integrated report should be presented in a consistent way, enabling comparison with other organizations

The content of the integrated report, according to IIRC's template (IIRC 2013: 24–32), should include eight elements:

1. Organizational overview and external environment
2. Governance structure and how it supports organization's ability to create value in different time perspectives
3. Organization's business model
4. Risks and opportunities affecting the organization's ability to create value and how the organization is dealing with them
5. Future-oriented information regarding strategy and resource allocation
6. Performance related to strategic objectives and impact on the capitals;
7. Outlook, i.e. challenges and uncertainties that the organization is likely to face and their potential implications for its business model and future performance
8. Basis of preparation and presentation of the report's content

However, as being principle-based, the IIRC's framework does not provide companies with any specific tools for non-financial and financial data integrated disclosure. Similarly to GRI framework, non-financial data third-party assurance is viewed by IIRC as a fundamental mechanism enhancing credibility and reliability of integrated reports, however it is not required (IIRC 2014). High level of integrated reports external assurance is difficult to obtain because of the lack of audit regulations and KPIs, and possible high costs of such audits (Oprisor 2015). Other challenges relate to the time frame of integrated reports, namely assuring future-oriented information (Huggins et al. 2015). To address these challenges IIRC (2014) encourages discussion concerning development of specific assurance standards.

3.2 Assurance

The concept of non-financial disclosure is shared by the domains of sustainability/CSR and stakeholder management as well as accounting and reporting. According to Kotonen (2009) non-financial reporting practice needs to be based on the following assumptions: accounts are to be formal and need to be prepared by an organization. In addition, reports are centered around certain areas of activities or ethical issues which may have a significant impact on the natural environment, employees, consumers and products as well as local and international communicates. Finally, it is assumed that reports are published and communicated to internal and external constituencies of the organization.

The voluntary character of non-financial disclosure provides a lot of flexibility and discrepancy to reporting companies. Thus, with the growing number of reports, their quality emerged as an issue to be studied and analyzed. While the "early voluntary reporting of supplementary information tended to emphasize narrative discussion on selected environmental, community and employee matters within the conventional annual report to shareholders" (Milne and Gray 2013: 17), a significant progress was achieved with the formulation and adoption of the non-financial disclosure as mentioned in the previous section. The concept of credibility understood "as a multilayer construct long CSR and communication theories" (Lock and Seele 2016: 186) addresses the issues of reports content and quality. It is based on the central argument of validity of communication and requires four main characteristics of the reports (Lock and Seele 2016; Moravcikova et al. 2015):

- Truth of the statements made
- Sincerity understood as truthfulness
- Appropriateness viewed as the rightness of the message in its context where sender and recipient agree on the communication
- Understandability defined as intelligibility or comprehensibility of the message

In addition, the report needs to be complete with respect to the segments and dimensions of business operations and the geographical scope. It should provide complete information about the major areas that of the organization's impact on society and the environment (Moravcikova et al. 2015). The report is also expected to be of certain significant which is translated in the use of quantitative and qualitative indicators to assess its social responsibility.

As it was discussed above with respect to the adoption, extent and quality of non-financial disclosure, the assurance of reporting remains contextual and is determined by a number of factors such as the use of formal standards, the number of adopted guidelines, the number of application levels (materiality matters), the report length, the company size, the experience of reporting, the country regulation and the sector of operation (Lock and Seele 2016). The assurance of non-financial reporting can be naturally increased by the adoption of interdependent external auditor which is found more frequent in the case of companies which document stronger stakeholder orientation (Simnett et al. 2009).

3.3 Limitations and Criticism

Existing studies indicate that despite growing business interests in the adoption of CSR and sustainability and the increase of number of non-financial disclosure, the quality of the reports and the motivation of reporting rise significant concerns (Moneva et al. 2006; Gray 2010). Analyses reveal "stressed the opacity of sustainability reports, their questionable connection with the firm's real situation, and their often superficial nature" (Boiral 2013: 1040). Specifically, Boiral (2013) enumerates three main pleas to the practice of non-financial reporting which include: the growing disconnectivity between reality and its representations, the control and manipulation of information and the influence exerted by the proliferation of misleading images.

Companies appear to be driven by instrumental and legitimacy motivations, which represents "business interests rather than a genuine concern for transparency and accountability" (Boiral 2013: 1040). "Externally available disclosures reflect a pragmatic approach to legitimacy, as the disclosures produced satisfy the perceived needs of specific stakeholder groups" (Dumay et al. 2015: 4). Companies tend to manipulate data following the strategy of cherry picking when they report progress or positive information, follow optimistic rhetoric, while omitting data of their negative impact on society and environment (Dumay et al. 2015). The overload of information in large reports hinder the its quality, communication and transparency (Fernandez-Feijoo et al. 2014). Instead the reports are viewed as camouflage for the unsustainable nature of some of the activities (Moneva et al. 2006; Gray 2010; Boiral 2013).

In result, non-financial reports are viewed by many as the glossy product of self-admiration (Porter and Kramer 2006) isolated from the organization as a whole (Brown-Liburd and Zamora 2015) and "merely exacerbating the already overwhelming amount of disclosure provided without adding any further insight" (Adams and Simnett 2011: 294). They lack credibility (Dando and Swift 2003), remain pseudo transparent (Coombs and Holladay 2013) and pursue company greenwashing (Seele and Lock 2015).

Some authors (Adams and Simnett 2011; Eccles and Krzus 2010) argue that the shortcomings of the non-financial disclosure can be tackled with the introduction of integrated reporting that represents an opportunity for improving transparency, governance and decision making for organizations of all types. Burritt (2012: 391) states that "if integrated reporting is both required and successfully adopted throughout the world (. . .) environmental performance accountability (. . .) would no longer be a subservient supplement to the main financial accounts and reports in the way that environmental and sustainability reporting have emerged until now". Others, however, acknowledge that integrated reporting concept, at least the one developed by IIRC, does not satisfy the needs of broad stakeholder groups, as it suffers from lack of integration between financial and non-financial metrics (Atkins et al. 2015). The IIRC's framework is constructed around notion of value to investors and not value to society (Thomson 2015) and guides companies to address only those social and environmental effects of business operations that have material impact on their ability to create value (Flower 2015). It has been criticized for business case framing, one-sided approach to assessing and reporting on sustainability issues and serving the interests of finance capital far more than wider public (Brown and Dillard 2014).

4 Non-financial Reporting in Poland

4.1 Practice in the 2005–2016

Poland with its characteristics of post-socialist economy lags in term of CSR/sustainability implementation behind the EU leaders (Steurer and Konrad 2009). While in line with the economy transition the environmental and social standards have significantly improved over the last 20 years, the dominance of coal in the energy production, insufficient legal enforcement and lower social awareness and understanding for CSR and sustainability places Poland amongst the emerging economies.

The concept of non-financial reporting has been evolving in Poland since 2005 when the first CSR report was published. The studies of non-financial disclosure carried out by different NGOs indicate that:

- of 69 companies which participated in the Ranking of Responsible Firms by Gazeta Prawna, 27 (39.1%) published a report, including 20 firms reporting in accordance with GRI standards, 7 adopting an external independent audit (Mikulska and Michalczuk 2014)
- of 500 companies ranked by Polityka magazine 136 firms provided information on their CSR engagement, including 24, which revealed their impact on social and economic and natural environment in the form of publicly available report prepared in accordance with internationally accepted standards (e.g. GRI or others) (Mikulska and Michalczuk 2014)
- over 10 years in sum 317 reports were published. The reports in vast majority followed GRI standards—67% reports in the 10-year period jumping to 78% reports for the last 5 years (2011–2016) (CSRInfo 2017)

In addition to GRI Polish companies adopt the guidelines of UN Global Compact, PN-ISO 26000:2012, Guidelines for multinational corporations and Carbon Disclosure Project (CSRInfo 2017). The leaders of non-financial disclosure include Bank Millennium, PKN ORLEN, Coca Cola HBC Polska, Grupa LOTOS, Orange Polska and Kompania Piwowarska. The breakdown of reporting companies by sector and by organization type is presented in Table 1.

As shown in Table 1 companies dominate as the issuers of non-financial reports recruiting mostly from industries of significant environmental impact (oil, energy) or highly regulated sectors (banking). Referring to the form of the report Polish companies publish CSR/environmental reports, integrated reports (e.g. LOTOS), interactive

Table 1 The breakdown of companies reporting non-financial information (number of reports, 2005–2016)

Reporting by sector	Reporting by organization type
Oil industry—40	Companies—110
Banking sector—38	Business organizations—2
Food industry—37	Foundations—4
Energy sector—29	City—1
Transportation and logistics—22	University—1
Service sector—20	Associations—1
Health care sector—19	
Retail—14	
Construction—12	
Telecommunication—11	
Natural resources extraction—10	

Source: CSRInfo (2017)

reports (e.g. PwC, Schenker) or reports in the form of an audiobook (Danone) (Mikulska and Michalczuk 2014).

Currently we observe a significant literature gap on non-financial reporting in Poland. The literature review indicates there is a limited number of analysis on the role and impact of CSR but the area of non-financial reporting remains significantly neglected. The gap is most likely caused by relatively small corporate activity in this field. Thus there is no evidence to discuss determinants of non-financial reporting or its impact on financial performance.

In coming years a significant improvement in the scope and quality of non-financial and diversity reporting by Polish companies is expected in the reaction to the regulation by the European Parliament and Council—Directive 2014/95/EU as the amendment of directive 2013/34/EU (EC; European Parliament 2014). The regulation applies to companies of significant public relevance that hire more than 500 employees, local companies based in the EU and foreign companies traded on the EU stock markets, which have a balance sheet total of at least €20 million or a net turnover of at least €40 million. Non-financial directive addresses listed companies, credit institutions, insurance companies and all other organizations designated by an EU member state as such due to their size or the nature of their business (EC 2014). Companies need to disclose the information on environmental impact, social and employee aspects, anti-corruption and bribery matters, as well as governance issues (EC 2014; Aluchna and Mikołajczyk 2016). This information should either be integrated in the annual management report or be prepared as a separate sustainability report.

Since the Non-Financial Reporting (NFR) requires affected companies in the European Union to disclose an annual non-financial statement in 2018, first reports will cover fiscal year of 2017 and will be published around June–August 2018 making it a historical date for non-financial reporting.

4.2 The Empirical Analysis. Goals, Sample and Methodology

The goal of the research was to identify the popularity of non-financial reporting amongst Polish listed companies. Specifically, we analyze how many companies listed on the Warsaw Stock Exchange published a CSR/sustainability or environmental report and how many of them were integrated reports and were covered by an external independent audit. The sample covers the whole population of listed firms. Since the concept of CSR was introduced at the Polish stock market in 2009 with the formulation of RESPECT, the CSR Index, we included data for 2010–2014. The data on general information (sector of operation, ownership structure, financial performance) was retrieved from IQ Capital base. The information on issuance of non-financial reports was collected by hand from companies' websites.

For the purpose of the study we formulated the following research questions:

1. What is the popularity of non-financial in Poland by companies?
2. What is the dynamics of non-financial reporting in Poland by listed companies in recent years?
3. What is the popularity/frequency of non-financial reports in Poland by listed companies with respect to:

   CSR/environmental reports.
   Inclusion in the RESPECT index.
   External audit of non-financial reports.
   Integrated reports.
   Other forms of non-financial disclosure—e.g. CSR/environmental communication on corporate/investor relations website

4. What is the characteristics of companies, which issue non-financial reports with respect to:

   Capitalization and size.
   Sector of operation (financial versus non-financial sector).
   Financial performance (Q, net income, ROA, dividend yield).
   Ownership structure (ownership concentration, free float, ownership by industry investors, financial investors and state).

The research was conducted between May and August 2017. The descriptive analysis was carried out with the use of standard MS Office Excel software.

4.3 Results and Discussion

First, we examined the popularity of non-financial reporting. Of the whole population of companies listed on the Warsaw Stock Exchange estimated at between 385 firms in 2010 and 472 firms in 2014 non-financial reports were issued by 17, 19,

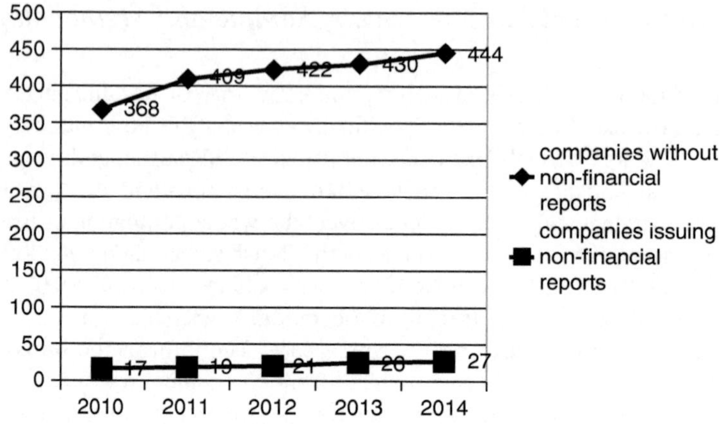

Fig. 1 Non-financial reporting in Poland (number of listed companies, 2010–2014)

Table 2 Types of non-financial disclosure (number of listed companies, 2010–2014)

Type of report (no. of companies)	2010	2011	2012	2013	2014
CSR/environmental report	17	19	21	26	27
CSR report audited	2	5	7	11	10
Integrated report	1	1	1	5	6
Other CSR communication	80	88	92	93	97
Respect index[a]	17	21	20	20	23

[a]Number of companies included in Respect index

21, 26 and 27 in 2010, 2011, 2012, 2013 and 2014 respectively. The data is presented in Fig. 1.

The data indicates a slight increase in number, from 17 companies reporting in 2010 to 27 reporting in 2014. This means an increase from 4.6% of companies in 2010 to 6% in 2014.Yet, the non-financial reporting has still remained marginal over the analyzed period.

We also examined the frequency of non-financial reports in Poland by listed companies with respect to different types of disclosure. The data is revealed in Table 2.

As shown on Table 2 non-financial disclosure includes the issuance of CSR/environmental report, issuance of integrated report and CSR communication on the corporate website. While the publication of integrated reports remains marginal, the communication on the corporate website appear to be the most popular type of non-financial disclosure. In addition, the external and independent audit was carried out only in the case of 10–37% non-financial reports. Table 2 reveals also the number of companies included in Respect Index which itself represents certain engagement in sustainability, however not all RESPECT companies publish a non-financial report.

Table 3 Companies reporting non-financial information versus companies not reporting non-financial information (selected characteristics, 2010–2014)

	Companies without non-financial reports	Companies issuing non-financial reports
General information		
Market cap (average)	248.33	3064.41
Assets (average)	749.08	8168.03
Financial sector (no/yes)	185	30
Financial performance		
Q (average)	1.15	0.58
Net income (average)	21.72	299.28
ROA (average)	−0.07	0.04
Dividend yield (average)	0.19	1.09
Ownership structure		
First largest shareholder (average stake)	39.56	49,061
Free float (average)	22.96	15,65,927,273
Industry investors (average stake)	54.16	54.62
Financial institutions (average stake)	22.87	29.72
State (average stake)	0.82	21.59

Finally, we compared the characteristics of companies reporting non-financial information versus companies not reporting non-financial information with respect to general information, financial performance and ownership structure as presented in Table 3.

The analysis indicates that companies reporting non-financial information versus companies not reporting non-financial information are:

- Larger, both in terms of market cap and assets
- Are likely to operate in the financial sector
- Perform better as measured by net income, ROA and dividend payout and reveal lower Q
- Characterized with stronger ownership concentration and lower free float
- Characterized with larger stake by financial institutions
- Characterized with larger stake by the state

Numbers leave no doubt—non-financial reporting has become a mainstream practice among world's largest companies. Among 100 largest companies in Americas, Europe, Africa and Middle East, and Asia Pacific 77%, 74%, 53%, 79% respectively were publishing sustainability report in 2015 (KPMG 2016). In many regions non-financial disclosure is becoming mandatory. Among 71 countries studied by KPMG (2016) 80% have introduced some form of regulatory sustainability reporting instrument, one third of which apply exclusively to large listed companies. Still, however, there are territories, like Poland, where non-financial reporting remains marginal practice of few largest and most successful corporations. It is hoped that regulation introducing mandatory disclosure of non-financial data will mainstream sustainability and its reporting in Polish companies and capital market. However Brown and Dillard (2014) suggest that overcoming shortcomings of the dominating accounting model is not just the matter of introduction of regulations. With no fundamental rethink of accounting theory, policy and practice mandatory non-financial disclosure will only institutionalize neo-liberal logic in social and environmental reporting. Despite the increasing regulatory initiatives for sustainability reporting around the world, the practice is still largely criticized either for being merely a box-ticking exercise providing investors and other stakeholders with hard to read documents overloaded with disconnected information or for being simply a lie or window dressing (Atkins et al. 2015; van Bommel 2014).

In the language of neo-institutional theory we would say that sustainability reporting is strongly institutionalized in many markets. In response to institutional pressures (new laws, but also stakeholder expectations) organization implement new practice. DiMaggio and Powell (1983) argue that in search for legitimacy (social fitness) organizations structure themselves according to the characteristics of the environment they operate in. This facilitates institutional isomorphism among organizations—the growing similarity of organizations in a given field resulting from "competition for political and institutional legitimacy as well as market position" (Mizruchi and Fein 1999: 657). However, as argued by Meyer and Rowan (1977) in their study on organization myth and ceremony, companies, for the purpose of attaining legitimacy within their environments, are prone to create some formal structures, implement new practices and construct stories about their actions that on the one hand correspond to socially prescribed dictates about what organization should do (Mizruchi and Fein 1999: 656), but on the other, are decoupled from actual business operation (Meyer and Rowan 1977). Empirical evidence on the poor quality of the majority of sustainability reports published worldwide suggests that this mechanism was also present in the institutionalization process of non-financial disclosure. Neo-institutional theory provides further explanation for this state of affairs. There are three mechanisms of isomorphic organizational change: coercive, mimetic and normative isomorphism. Coercive isomorphism constrained by the owners of firm resources results from formal and informal pressures of other organizations upon which the firm is dependent. Mimetic isomorphism is a result of uncertainty that encourages organization to model themselves intentionally or unintentionally (through influence of consultants and employees hired from other companies) on more legitimate or successful peers. Finally, the normative isomorphism is driven by the similar education (graduate and postgraduate) of professionals and strengthened by their interactions within growing professional networks, across which new models diffuse rapidly. DiMaggio and Powell (1983) note that while the three types of isomorphic pressures often intermingle, they may lead to different outcomes.

Some authors (Scott 1987; Tolbert and Zucker 1983) argue that coercive institutional power foster rather superficial changes in the organizational structures and practices, encouraging only formal introduction of changes (Meyer and Rowan 1977).

Therefore, we argue that regulatory coercion will drive diffusion of non-financial disclosure among Polish companies, but only on a minimum and shallow level. Similar outcome is to be produced by diffusion through mimetic isomorphism mechanisms. Under the conditions of uncertainty, which is increased by growing expectations and power of different stakeholder groups, as well as accelerating globalization processes and resource scarcity, companies imitate structures and practices that have proven to be successful in other organizations (Mizruchi and Fein 1999). Indeed, it has been argued that while adopting sustainability reporting many companies desire to make sure either that they did not fall behind the competition on CSR grounds or to take a leadership role (Boiral 2013). Another source of mimetic isomorphism is the growth of CSR consulting industry promoting CSR standards, codes of best practice and management systems. In other words diffusion driven by mimetic isomorphism means that companies implement sustainability reporting only because "everybody else does it" without taking into consideration actual benefits that this practice might have on company's effectiveness if its incorporated into the core business model. Although the lack of reporting standards, especially those for integrated disclosure, is one of the major obstacles for further development in this area, if the diffusion is based solely on this mechanism, it will lead to situation where sustainability reporting is limited to blind compliance with standards.

Our analysis reveals that the practice of non-financial reporting in Poland remains at the very early stage of development. Amongst listed companies which naturally are in the public spotlight and often represent large firms only 6% issue non-financial report. As long as we observes a slight increase in the analyzed period, the numbers are still very low. The overall communication is implemented by approximately 20% of analyzed companies which suggests a potential for an improvement. Yet, however such communication does not follow standards of non-financial disclosure and may be used by companies to report only positive information or data they wish to disclose, not necessarily required variables on environment and social impact.

We argue that virtual change in this area and true integration of sustainability reporting with business practice in Polish companies (and in general) is possible only from within organizations through normative isomorphism mechanism—the process that corresponds with value infusion (Selznik 1957). Creation of ethical and sustainable organizational cultures is driven by values of managers and owners, who believe that the engagement in social and environmental activity and taking responsibility by the company for its overall impact is simply the right thing to do. CSR and sustainability need to be "taken-for-granted" and set the most obvious and natural way to conduct business operations (Berger and Luckmann 1967). For this to happen it is not enough to introduce regulations and create accounting standards (which of course will be needed for operationalization of the new business philosophy). What is needed is the change in mindset of accounting and management practitioners that can be initiated only through proper sustainable education.

In response to the growing awareness on the role of business in society and the criticism of the neoliberal doctrine dominance, institutions of higher education have expanded sustainability in their curricula (Sherif 2015). Poland is no exception here. However, many universities still face challenge related to the petrification of the so-called hidden curriculum in management educational programs (Blasco 2012). Hidden curriculum is a subtext message to the students about what actually is important and matters, what is serious business problem and what is merely marginal topic. Thus, adding CSR or sustainability courses to the program without reframing of already existing courses and changing the way university itself is managed, will be simply inefficient or even detrimental. "Hence, the major challenge regarding sustainability education in management schools results from the necessity of redesigning educational approach as a whole, redefining the existing, petrified concepts of business' goals and practice" (Dembek and Roszkowska-Menkes 2018). Overcoming this challenge and initiating virtual change in corporate reporting practices in Poland requires multi-stakeholder co-operation between regulators, professional associations, business, NGOs and broad academic community representing various economic disciplines. A potentially promising movement is the declaration signed by 23 Polish universities to promote the concept of CSR (Ministry of Science and Higher Education 2017).

5 Conclusion

Conceptually non-financial disclosure has been based on stakeholder theory and sustainable development, and has offered a framework for their operationalization. The most recent development in this area represented by integrated reporting incorporates also shareholder perspective and calls for the reconciliation of shareholder and stakeholder theory (Eccles et al. 2014). The motivations to implement non-financial reporting vary and include compliance with new regulations, addressing shareholders and other stakeholders expectations for transparency, pursuit of legitimacy, competition with other companies, but also support for strategic management. However, empirical studies on benefits of sustainability reporting provide inconsistent results. Yet, non-financial reporting is slowly becoming a mainstream practice with growing number of companies disclosing ESG data and increasing regulatory effort worldwide. Poland, where only 27 out of 444 companies listed on the Warsaw Stock Exchange main market published a CSR or environmental report in 2014, lags far behind the rest of Europe. While it is hoped that new mandatory reporting rules in EU (Directive 2014/95/EU) will force more companies to disclose non-financial information, the questions on how to encourage them to move from minimum compliance to strategic approach to sustainability reporting and how to ensure high quality of such disclosure still remain open.

We believe that our paper contributes to the understanding of non-financial reporting in countries which lag behind in the field of sustainability disclosure and partially fills the literature gap on Poland. Our analysis reveals some limitations—corporate disclosure is a long process. Companies disclose their performance of the given year and issue the reports by July or August the following year. The collection and analysis of data require additional time what hinders the fast capturing of trends and dynamic of non-financial disclosure. The future direction for research should encompass a longer period of time and focus on the identification of stimulators and inhibitors of non-financial reporting in the form of regression models.

References

Adams, C. A. (2015). The International Integrated Reporting Council: A call to action. *Critical Perspectives on Accounting, 27*, 23–28.

Adams, C. A., & McNicholas, P. (2007). Making a difference. Sustainability reporting, accountability and organisational change. *Accounting, Auditing & Accountability Journal, 20*(3), 382–402.

Adams, S., & Simnett, R. (2011). Integrated reporting: An opportunity for Australia's not-for-profit sector. *Australian Accounting Review, 21*(3), 292–301.

Aluchna, M., & Mikołajczyk, O. (2016). Akcjonariusze vs. interesariusze. Przypadek raportowania zintegrowanego. *Prace Naukowe Uniwersytetu Ekonomicznego we Wrocławiu, 436*, 11–20.

Atkins, J., Atkins, B. C., Thomson, I., & Maroung, W. (2015). 'Good' news from nowhere: Imagining utopian sustainable accounting. *Accounting, Auditing & Accountability Journal, 28*(5), 651–670.

Ballou, B., Heitger, D., & Landes, C. (2006a). The future of corporate sustainability reporting. *Journal of Accountancy, 202*, 65–74.

Ballou, B., Heitger, D., & Landes, C. (2006b). The future of corporate sustainability reporting: A rapidly growing assurance opportunity. *Journal of Accountancy, 22*(3), 5–74.

Ballou, B., Casey, R., Grenier, J. H., & Heitger, D. (2012). Exploring the strategic integration of sustainability initiatives: Opportunities for accounting research. *Accounting Horizons, 26*(2), 265–288.

Beattie, V., & Smith, S. J. (2013). Value creation and business models: Refocusing the intellectual capital debate. *The British Accounting Review, 45*(4), 243–254.

Bebbington, J., Larrinaga, C., & Moneva, J. (2008). Corporate social reporting and reputation risk management. *Accounting Auditing and Accountability Journal, 21*(3), 337–361.

Berger, P. L., & Luckmann, T. (1967). *The social construction of reality*. New York: Doubleday.

Berthelot, S., Coulmont, M., & Serret, V. (2012). Do investors value sustainability reports? A Canadian study. *Corporate Social Responsibility and Environmental Management, 19*, 355–363.

Blasco, M. (2012). Aligning the hidden curriculum of management education with PRME: An inquiry-based framework. *Journal of Management Education, 36*(3), 364–388.

Boiral, O. (2013). Sustainability reports as simulacra? A counter-account of A and A+ GRI reports. *Accounting Auditing & Accountability Journal, 26*(7), 1036–1071.

Brown, J., & Dillard, J. (2014). Integrated reporting: On the need for broadening out and opening up. *Accounting Auditing & Accountability Journal, 27*(7), 1120–1156.

Brown-Liburd, H., & Zamora, V. L. (2015). The role of corporate social responsibility (CSR) assurance in investors' judgments when managerial pay is explicitly tied to CSR performance. *Auditing: A Journal of Practice & Theory, 34*(1), 75–96.

Burritt, R. L. (2012). Environmental performance accountability: Planet, people, profits. *Accounting Auditing & Accountability Journal, 25*(2), 370–405.

Caron, M., & Turcotte, M. (2009). Path dependence and path creation: Framing the non-financial information market for a sustainable trajectory. *Accounting Auditing and Accountability Journal, 22*(2), 272–297.

Clayton, A. F., Rogerson, J. M., & Rampedi, I. (2015). Integrated reporting vs. sustainability reporting for corporate responsibility in South Africa. *Bulletin of Geography Socio–Economic Series, 29*, 7–17.

Coombs, W. T., & Holladay, S. J. (2013). The pseudo-panopticon: The illusion created by CSR-related transparency and the internet. *Corporate Communication An International Journal, 18*(2), 212–227.

Cormier, D., & Magnan, M. (2007). The revisited contribution of environmental reporting to investors' valuation of a firm's earnings: An international perspective. *Ecological Economics, 62*, 613–626.

Crane, A., & Matten, D. (2007). *Business ethics*. Oxford: Oxford University Press.

CSRInfo. (2017). *Raportowanie niefinansowe w Polsce 2016* [Non-financial reporting in Poland 2016]. www.csrinfo.org

Dando, N., & Swift, T. (2003). Transparency and assurance minding the credibility gap. *Journal of Business Ethics, 44*(2–3), 195–200.

de Villiers, C., Low, M., & Samkin, G. (2014a). The institutionalisation of mining company sustainability disclosures. *Journal of Cleaner Production, 84*, 51–58.

de Villiers, C., Rinaldi, L., & Unerman, J. (2014b). Integrated reporting: Insights, gaps and an agenda for future research. *Accounting Auditing & Accountability Journal, 27*(7), 1042–1067.

Deegan, C. (2002). Introduction – The legitimising effect of social and environmental disclosures – A theoretical foundation. *Accounting Auditing & Accountability Journal, 15*(3), 282–311.

Deloitte. (2016). *Sustainability disclosure getting ahead of the curve.* Accessed August 25, 2017, from https://www2.deloitte.com/content/dam/Deloitte/us/Documents/risk/us-risk-sustainability-disclosure.pdf

Dembek, A., & Roszkowska-Menkes, M. (2018). Business in society. Why is teaching sustainability in business schools still difficult. In W. Amann (Ed.), *Phronesis in business schools: Reflections on teaching and learning*. Charlotte, NC: Information Age Publishing.

Dimaggio, P., & Powell, W. (1983). The iron cage revisited: Institutional isomorphism and collective rationality in organizational fields. *American Sociological Review, 48*, 147–160.

Dumay, J., Frost, G., & Beck, C. (2015). Material legitimacy: Blending organizational and stakeholder concerns through non-financial information disclosure. *Journal of Accounting and Organizational Change, 11*(1), 2–23.

EC. (2014). *Directive 2014/95/EU of the European Parliament and of the Council of 22 October 2014 amending Directive 2013/34/EU as regards disclosure of non-financial and diversity information by certain large undertakings and groups*. Accessed August 25, 2017, from http://755eur-lex.europa.eu/legal-content/EN/TXT/PDF/?uri=CELEX:32014L0095&from=EN

Eccles, R., & Krzus, M. (2010). *One report: Integrated reporting for a sustainable strategy*. Hoboken, NJ: John Wiley & Sons.

Eccles, R. G., & Serafeim, G. (2011). Accelerating the adoption of integrated reporting. In F. de Leo & M. Vollbracht (Eds.), *CSR index*. Zurich: Innovation Publishing Ltd.

Eccles, R.G., Ioannou, I., & Serafeim, G. (2011). *The impact of corporate sustainability on organizational processes and performance* (Working paper, Harvard University). Accessed June 4, 2016, from http://www.hbs.edu/faculty/Publication%20Files/SSRN-id1964011_6791edac-7daa-4603-a220-4a0c6c7a3f7a.pdf

Eccles, R., Ioannou, I., & Serafeim, G. (2014). The impact of corporate sustainability on organizational processes and performance. *Management Science, 60*(11), 2835–2857.

Elkington, J. (1997). *Cannibals with Forks: The triple bottom line of 21st century business.* Stoney Creek, CT: New Society Publishers.

Fernandez-Feijoo, B., Romero, S., & Ruiz, S. (2014). Effect of stakeholders' pressure on transparency of sustainability reports within the GRI framework. *Journal of Business Ethics, 122*, 53–63.

Fernandez-Feijoo, B., Romero, S., & Ruiz, S. (2015). Multilevel approach to sustainability report assurance decisions. *Australian Accounting Review, 75*(4), 346–358.

Fifka, M. S. (2013). Corporate responsibility reporting and its determinants in comparative perspective – A review of the empirical literature and a meta analysis. *Business Strategy and the Environment Journal, 22*(1), 1–35.

Flower, J. (2015). The international integrated reporting council: A story of failure. *Critical Perspectives on Accounting, 27*, 1–17.

Geels, F. (2011). The multi-level perspective on sustainability transitions: Responses to seven criticisms. *Environmental Innovation and Societal Transitions, 1*, 24–40.
Graham, J., Harvey, C., & Rajgopal, S. (2005). The economic implications of corporate financial reporting. *Journal of Accounting and Economics, 40*(1–3), 3–73.
Gray, R. (2000). Current developments and trends in social and environmental auditing, reporting and attestation: A review and comment. *International Journal of Auditing, 4*(3), 247–268.
Gray, R. (2010). Is accounting for sustainability actually accounting for sustainability and how would we know? An exploration of narratives of organisations and the planet. *Accounting Organizations and Society, 35*(1), 47–62.
Gray, R., & Bebbington, J. (2000). Environmental accounting managerialism and sustainability. *Advances in Environmental Accounting and Management, 1*, 1–44.
Gray, R., Owen, D., & Maunders, K. (1987). *Corporate social reporting – Accounting and accountability*. Hemel Hempstead: Prentice-Hall.
GRI. (2016). *Consolidated set of GRI sustainability reporting standards*. Amsterdam: Global Reporting Initiative.
GRI. (2017). *Global reporting initiative*. Accessed August 25, 2017, from https://www.globalreporting.org/standards
Guidry, R. P., & Patten, D. M. (2010). Market reactions to the first-time issuance of corporate sustainability reports: Evidence that quality matters. *Sustainability Accounting Management and Policy Journal, 1*(1), 33–50.
Hahn, R., & Kühnen, M. (2013). Determinants of sustainability reporting: A review of results, trends, theory, and opportunities in an expanding field of research. *Journal of Cleaner Production, 59*, 5–21.
Healy, P., & Palepu, K. (2001). Information asymmetry, corporate disclosure and the capital markets: A review of the empirical disclosure literature. *Journal of Accounting and Economics, 31*(1–3), 75–102.
Higgins, C., & Coffey, B. (2016). Improving how sustainability reports drive change: A critical discourse analysis. *Journal of Cleaner Production, 136*, 18–29.
Huggins, A., Simnett, R., & Hargovan, A. (2015). Integrated reporting and directors' concerns about personal liability exposure: Law reform options. *Company and Securities Law Journal, 33*, 176–195.
IIRC. (2013). The International <IR> Framework. International Integrated Reporting Council.
IIRC. (2014). Assurance on <IR>. An exploration of issues. International Integrated Reporting Council.
IIRC. (2017). *Integrated reporting*. Accessed August 25, 2017, from https://integratedreporting.org/
Ioannou, I., & Serafeim, G. (2014). The impact of corporate social responsibility on investment recommendations: Analysts' perceptions and shifting institutional logics. *Strategic Management Journal, 36*(7), 1053–1081.
IPIECA. (2016). *About us, IPIECA*. Accessed August 25, 2017, from http://www.ipieca.org/about-us/
Joseph, C., & Taplin, R. (2011). The measurement of sustainability disclosure: Abundance versus occurrence. *Accounting Forum, 35*, 19–31.
Kemper, A., & Martin, R. (2011). After the fall: The global financial crisis as a test of corporate social responsibility theories. *European Management Review, 7*, 229–239.
Kotonen, U. (2009). Formal corporate social responsibility reporting in Finnish listed companies. *Journal of Applied Accounting Research, 10*(3), 176–207.
KPMG. (2016). *Carrots and sticks. Global trends in sustainability reporting regulation and policy*. Accessed August 25, 2017, from https://assets.kpmg.com/content/dam/kpmg/pdf/2016/05/carrots-and-sticks-may-2016.pdf
Lazonick, W., & O'Sullivan, M. (1997). Maximizing shareholder value: A new ideology for corporate governance. *Economy and Society, 29*(1), 13–35.

Lock, I., & Seele, P. (2016). The credibility of CSR (corporate social responsibility) reports in Europe. Evidence from a quantitative content analysis in 11 countries. *Journal of Cleaner Production, 122*, 186–200.

Lodhia, S. (2015). Exploring the transition to integrated reporting through a practice lens: An Australian customer owned bank perspective. *Journal of Business Ethics, 129*(3), 585–598.

Manetti, G. (2011). The quality of stakeholder engagement in sustainability reporting: Empirical evidence and critical points. *Corporate Social Responsibility and Environmental Management, 18*, 110–122.

Manetti, G., & Toccafondi, S. (2012). The role of stakeholders in sustainability reporting assurance. *Journal of Business Ethics, 107*, 363–377.

Matten, D., & Moon, J. (2008). Implicit and explicit CSR: A conceptual framework for a comparative understanding of corporate social responsibility. *Academy of Management Review, 33*(2), 404–424.

Mayer, C. (2002). *Firm commitment: Why the corporation is failing us and how to restore trust in it*. Oxford: Oxford University Press.

McWilliams, A., & Siegel, D. (2000). Corporate social responsibility and financial performance: Correlation or misspecification? *Strategic Management Journal, 21*(5), 603–609.

Meyer, J. W., & Rowan, B. (1977). Institutionalized organizations: Formal structure as Myth and Ceremony. *American Journal of Sociology, 83*(2), 340–363.

Mikulska, T., & Michalczuk, G. (2014). Raportowanie społecznej odpowiedzialności przedsiębiorstwa [Reporting of corporate social responsibility], Zeszyty Naukowe Uniwersytetu Szczecińskiego, 827 – Finanse, Rynki Finansowe, Ubezpieczenia, 69, 209–219.

Milne, M., & Gray, R. (2013). W(h)ither ecology? The triple bottom line, the global reporting initiative, and corporate sustainability reporting. *Journal of Business Ethics, 118*, 13–29.

Ministry of Science and Higher Education. (2017). *23 szkoły wyższe podpisały Deklarację Społecznej Odpowiedzialności Uczelni* [23 universities signed the declaration on social responsibilities of university]. Accessed March 21, 2018, from http://www.nauka.gov.pl/aktualnosci-ministerstwo/23-szkoly-wyzsze-podpisaly-deklaracje-spolecznej-odpowiedzialnosci-uczelni.html

Mio, C., & Venturelli, A. (2013). Non-financial information about sustainable development and environmental policy in the annual reports of listed companies: Evidence from Italy and the UK. *Corporate Social Responsibility and Environmental Management, 20*, 340–358.

Mizruchi, M., & Fein, L. (1999). The social construct of organizational knowledge: A study of the uses of coercive, mimetic and normative isomorphism. *Administrative Science Quarterly, 44*, 653–683.

Moneva, J., & Cuellar, B. (2009). The value relevance of financial and non-financial environmental reporting. *Environment Resource Economics, 44*, 441–456.

Moneva, J., Archel, P., & Correa, C. (2006). GRI and the camouflaging of corporate unsustainability. *Accounting Forum, 30*(2), 121–137.

Moon, J. (2002). Business social responsibility and new governance. *Government and Opposition, 37*(3), 385–408.

Moravcikova, K., Stefanikova, L., & Rypakova, M. (2015). CSR reporting as an important tool of CSR communication. *Procedia Economics and Finance, 26*, 332–338.

Murray, A., Sinclair, D., Power, D., & Gray, R. (2006). Do financial markets care about social and environmental disclosure? Further evidence and exploration from the UK. *Accounting Auditing & Accountability Journal, 19*(2), 228–255.

Nazari, J. A., Herremans, I. M., & Warsame, H. A. (2015). Sustainability reporting: External motivators and internal facilitators. *Corporate Governance, 15*(3), 375–390.

O'Dwyer, B., & Owen, D. L. (2005). Assurance statement practice in environmental, social and sustainability reporting: A critical evaluation. *The British Accounting Review, 37*(2), 205–229.

Oprisor, T. (2015). Auditing integrated reports: Are there solutions to this Puzzle? *Procedia Economics and Finance, 25*, 87–95.

Owen, D., & Swift, T. (2001). Introduction: Social accounting, reporting and auditing: beyond the rhetoric? *Business Ethics: A European Review, 10*(1), 4–8.

Porter, M. E., & Kramer, M. R. (2006, December). *Strategy and society: The link between competitive advantage and corporate social responsibility*. Harvard Business Review.
Robins, F. (2006). The challenge of TBL: A responsibility to whom? *Business and Society Review, 111*, 1–14.
Schadewitz, H., & Niskala, M. (2010). Communication via responsibility reporting and its effect on firm value in Finland. *Corporate Social Responsibility and Environmental Management, 17*, 96–106.
Scott, W. R. (1987). The adolescence of institutional theory. *Administrative Science Quarterly, 32*(4), 493–511. Accessed August 25, 2017, from https://doi.org/10.2307/2392880
Searcy, C., & Buslovich, R. (2014). Corporate perspectives on the development and use of sustainability reports. *Journal of Business Ethics, 121*(2), 146–169.
Seele, P., & Lock, I. (2015). Instrumental and/or deliberative? A typology of CSR communication tools. *Journal of Business Ethics, 131*(2), 401–414.
Selznik, P. (1957). *Leadership in administration*. New York: Harper & Row.
Serafeim, G. (2014). *Integrated reporting and investor clientele* (Working paper). Harvard University, Cambridge, MA.
Sherif, S. F. (2015). The role of higher education institutions in propagating corporate social responsibility case study: Universities in the Middle East. *International Journal of Education and Research, 3*(1), 217–226.
Simnett, R., Vanstraelen, A., & Chua, W. F. (2009). Assurance on sustainability reports: An international comparison. *Accounting Review, 84*(3), 937–967.
Skouloudis, A., Evangelinos, K., & Kourmousis, F. (2009). Development of an evaluation methodology for triple bottom line reports using international standards on reporting. *Environmental Management, 44*, 298–311.
Solomon, J. F., & Solomon, A. (2006). Private social, ethical and environmental disclosure. *Accounting, Auditing & Accountability Journal, 19*(4), 564–591.
Stacchezzini, R., Melloni, G., & Lai, A. (2016). Sustainability management and reporting: The role of integrated reporting for communicating corporate sustainability management. *Journal of Cleaner Production, 136*, 102–110.
Steurer, R., & Konrad, A. (2009). Business – Society relations in Central-Eastern and Western Europe: How those who lead in sustainability reporting bridge the gap in corporate (social) responsibility. *Scandinavian Journal of Management, 25*, 23–36.
Strumińska-Kutra, M., & Woźniczko, J. (2010). Empiryczne wymiary społecznej odpowiedzialności przedsiębiorstw – perspektywa instytucjonalna. *Problemy Zarządzania, 8*(2), 71–89.
Thomson, I. (2015). 'But does sustainability need capitalism or an integrated report' a commentary on 'The international integrated reporting council: A story of failure' by Flower, J. *Critical Perspectives on Accounting, 27*, 18–22.
Thorne, L., Mahoney, L., & Manetti, G. (2014). Motivations for issuing standalone CSR reports: A survey of Canadian firms. *Accounting Auditing & Accountability Journal, 27*(4), 686–714.
Tolbert, P., & Zucker, L. (1983). Institutional sources of change in the formal structure of organizations: The diffusion of civil service reform, 1880–1935. Articles and Chapters.
Tregidga, H. M., & Milne, M. J. (2006). From sustainable management to sustainable development: A longitudinal analysis of a leading New Zealand environmental reporter. *Business Strategy and the Environment, 15*(4), 219–241.
Tschopp, D., & Huefner, R. (2015). Comparing the evolution of CSR reporting to that of financial reporting. *Journal of Business Ethics, 127*, 565–577.
van Bommel, K. (2014). Towards a legitimate compromise? An exploration of integrated reporting in the Netherlands. *Accounting Auditing & Accountability Journal, 27*(7), 1157–1189.
Vermalen, T. (2009). Maximizing shareholder value. An ethical responsibility? In C. Smith & G. Lenssen (Eds.), *Mainstreaming corporate responsibility*. Chichester: Wiley.
Vogel, D. (2005). Is there market for virtue? The business case for corporate social responsibility. *California Management Review, 47*(4), 19–49.

Waris, A., Frynas, J., & Mahmood, Z. (2017). Determinants of corporate social responsibility (CSR) disclosure in developed and developing countries: A literature review. *Corporate Social Responsibility and Environmental Management, 24*(4), 273–294.

Wild, S., & van Staden, C. (2013). Integrated reporting: Initial analysis of early reporters – An institutional theory approach, paper presented at the seventh Asia pacific interdisciplinary research in accounting conference (APIRA), Kobe, Japan, 26–28 July, 2013. (Conference Contribution – Full Conference Paper).

Xu, S., & Liu, D. (2017). Do financial markets care about corporate social responsibility disclosure? Further evidence from China. *Australian Accounting Review, 34*, forthcoming.

Zimara, V., & Eidam, S. (2015). The benefits of social sustainability reporting for companies and stakeholders – Evidence from the German chemical industry. *Journal of Business Chemistry, 12*(3), 85–103.

Part VI
Social Aspects of CSR

Personal Social Responsibility and Its Impact to Consumerism

Marta Karwacka

The idea of corporate social responsibility has been promoted in Poland since the first decade of the twenty-first century. For a dozen of years the number of companies that implement individual projects related to social and environmental responsibility has grown steadily, or some of them have decided to develop complex and serious CSR strategies. In the meantime, a number of important organizations and individuals have also emerged to promote this idea in Polish society. Unfortunately many CSR projects were outsourced to PR or marketing department, which weren't prepared enough to carry out/develop projects with social impact. Low level of social and environmental awareness, the lack of a strategy and the lack of connection of social activities with the business profile, quickly began to raise many doubts and reduce trust to business. Today, fortunately, many companies realize that being responsible means also being honest, that's why some of them invest in real dialogue process with local communities, build cross-sector partnerships or start to cooperate with NGOs. The necessity to implement such activities is also due to the presence of group of responsible consumers. More and more people are beginning to communicate their individual social responsibility, and some of them become and play a role of opinion leaders, sometimes setting up their own businesses reflecting their personal beliefs and values. The need for integrity is reflected in consumer decisions and the lifestyle of individuals, who care about social and environmental issues. Personal social responsibility, manifested by consumer activity, blogging, creating responsible start-ups is becoming important power that drives the changes that are taking place in the modern economy.

In 1989 Polish society began its transition to democracy, and the Polish economy started undergoing transformation, slowly embracing the rules of free market capitalism. After living under real socialism for several decades, Polish people, who had

M. Karwacka (✉)
Toruń, Poland

A. Długopolska-Mikonowicz et al. (eds.), *Corporate Social Responsibility in Poland*, CSR, Sustainability, Ethics & Governance,
https://doi.org/10.1007/978-3-030-00440-8_15

long yearned for an open world and freedom in all aspects of life, including the sphere of economy, finally witnessed a totally different reality. The change was abrupt and nobody had had a chance to prepare for it. The development of free market economy in the 90s presented many Polish people with new challenges. Some of them quickly managed to adapt to the new reality (they are the so called winners of the transition), while others were not able to handle the situation equally well. In fact, in many families this failure to adapt to free market economy can be seen up to this day and the reason for that lies in the processes of intergenerational transmission. Even today there is still considerable public debate about the so called losers of the post-1989 transformation and about the processes which not only led to exposing and escalating social issues, but also made it impossible to effectively overcome these problems. What is certain, however, is the fact that we have become consumers. The new reality presented society with an opportunity (and for those with career ambitions even a necessity) to acquire new material status symbols. According to Krzysztof Pietrowicz, "the situation in the pre-1989 Poland was completely different. There were status symbols, but they were not as omnipresent as they are today, or they were simply different."[1]

The experience of living under socialism surely had an influence on how consumerism and consumer activities were to be perceived in Poland in years to come. There is no doubt that certain behaviours, fears and habits were passed on to the next generations. Those who could still remember empty shelves in shops, lack of even the most essential goods and having little or no product choice find it quite amusing now to recollect the times when the client had absolutely nothing to say.

In the last decade of the twentieth century the Polish people had to face many challenges that the new political system presented them with, but as consumers they wanted and still want to have infinite consumer opportunities and to be able to use them to the full. The Polish people were eager to discover the world, which up to that moment they had only seen from behind the Iron Curtain. Our society still has an insatiable hunger for new possibilities and always wants to have a wide choice of options. There is also no doubt that we wished to gain more and more status symbols by buying goods available on the market, which is typical of stratified societies. According to Tomasz Szlendak and Arkadiusz Karwacki, "material aspects of consumption are visible symbols of economic and social status in highly stratified cultures, and individuals who do not own them are considered socially worthless and unsuccessful."[2] The rapidly developing market in Poland quickly managed to shape consumer behaviour. Interestingly enough, for many years Polish consumers tended not to ask questions and not to think about the quality of products on offer, which means that they constituted a perfect example of consumer society, i.e. one which, as Zygmunt Bauman argues, "(...) has little need for mass industrial labor and conscript

[1]Pietrowicz, K. "Nasze gadżety codzienne albo o płynności symboli statusu" in *Na pokaz. O konsumeryzmie w kapitalizmie bez kapitału*, eds. Szlendak, T., Pietrowicz, K., p. 72.

[2]Szlendak, T., Karwacki, A., "Koncepcja poziomicy – cudowne lekarstwo czy utopijna terapia?", in: Studia Socjologiczne, 1/2010, p. 41.

armies; instead it needs to engage its members in their capacity as consumers. The way present-day society shapes its members is dictated first and foremost by the duty to play the role of the consumer."[3]

For over 30 years Polish society has been embracing consumer culture and, as has already been mentioned before, its appetite in this respect seems to be insatiable. One might then ask when their hunger will finally be satisfied and if it is indeed ever possible in the first place. Considering Erich Fromm's definition of consumerism as a state of never feeling fully satisfied, we might as well assume that our desires will never be quenched. According to Fromm, "in consumer society everybody is aware that their standard of life is nowhere near as good as it could be and that it is incomparably worse than the quality of life that wealthier individuals enjoy. This motivates people to work even harder and to have even more ambitions and desires, yet the standard of life they aspire to is somewhat similar to the horizon in that they will never able to reach it".[4]

Poland is a country in which consumption plays an extremely important social function[5] and in which social inequalities are clearly visible, but, similarly to what is happening in other countries, some of the Polish consumers have already decided to break free from consumer culture. By increasingly questioning hyperconsumerism, these individuals appear to add a few scratches to the hitherto perfect surface of the consumerist world. These are not cracks yet, but at least people in Poland are finally starting to talk about responsible consumer choices.

1 Personal Social Responsibility vs Corporate and Consumer Social Responsibility

As members of a global society, we all shape and influence the dynamics of global society. Our attitude to the environment and the way we treat other people are strongly linked to Personal Social Responsibility. (PSR). According to S. Lopez Davis PSR is "the way a person performs in his daily life as a member of society—and no only as a consumer—basing this decisions in a desire to minimze the negative impact and maximize the positive impacts on the social, environmental and economic in the long run".[6] Under the influence of decisions made by individuals, many social changes have taken place in history. PSR, understood as all human behaviour, cannot be reduced to a consumer role. However, today we can influence various processes to a large extent through consumption. To what extent are our decisions

[3]Bauman Z. Globalizacja, PIW, Warszawa, 2000, s. 95.

[4]Fromm E., Mieć czy być, Dom Wydawniczy Rebis, Poznań. 1995.

[5]Mróz, B, "Consumerism vs. sustainability: the emergence of new consumer trends in Poland", www.pol-int.org

[6]López Davis, S.,∗, Marín Rivesb, L. S. Ruiz de Maya, Introducing Personal Social Responsibility as a key element to upgrade CSR, Spanish journal of Marketing, 2017, s. 159.

influencing the development of sustainable consumption? In Poland, the issue of sustainable consumption is at the initial stage of development, however, we should consider it from the perspective of Personal Social Responsibility. When participating in the market of goods and services, each consumer is guided by a different sensitivity, grows up in a different environment, has his own interests, has different financial conditions, etc. Various circumstances lead us to participate in the consumer market in a very specific way.

Every change in society takes time, it can take several years to change attitudes, because it is a process that requires the spread of ideas initiated by individuals. What is needed, therefore, is people who initiate this change, but also people who pass it on to others. Malcolm Gladwell points out in his book "The tipping point: how little things can make a big difference" that in order to spread any information or idea on a large scale, we need not only visionaries of change, but also the so-called mowens, linkers and sellers. Each of these people is different in their personality. So, to change attitudes and behaviours, we need, firstly—people who collect information on a given subject, analyse and read; secondly—people who enjoy the trust of others, are social and disseminate that knowledge; and, thirdly—people who convince or dispel doubts if necessary. There is no doubt that both the initiators of change and the people who will follow and persuade others to follow change are needed in the debate on sustainable consumption.

The development of conditions conducive to sustainable consumption is based on Personal Social Responsibility, but it's worth to say, that different consumers behave differently. It can be assumed that just as at the beginning of the dissemination of new consumer behaviours and attitudes, people with a strong personal sense of responsibility for changes in the market are important, so at the later stage of development there is also the opinion and activity of people who react to changes taking place in companies. They are their reviewers, they watch closely. S. Lopez Davis points out that consumer behaviour can be divided into three categories: philanthropic, social and environmental impacts and ethical. Each of them is important in the context of developing sustainable consumption, but they differ in scope or level of consumer activity. The first group perceives consumption "from a humanitarian perspective, placing the consumer as a philantropist that seeks to help others, to care for other people or to contribute to fairness and solidarity".[7] The second group of consumers perceives their consumer choices from the perspective of their social and environmental impact, and through their attitudes and decisions they try to "contribute to the evolution of the social structures where they belong, maximazing the benefits and minimazing the harm".[8] The third group of ethical consumers has a very broad view of consumption and can be associated with both ethical investment, as well as consumer boycotts and fair trade purchases. Their activities "are detemined by moral principles and standards, ethical concerns, beliefs and values".[9]

[7]López Davis, S. op.cit 149.

[8]Ibidem, s154.

[9]Ibidem, s155.

Each of these groups influences the development of sustainable consumption, but also changes in the responsibility of companies all over the world.

2 Corporate Social Responsibility in Poland

Corporate social responsibility became the subject of debate in Poland when foreign-owned corporations started to enter the Polish market. Following in the footsteps of their parent companies, they began introducing CSR policies into their corporate strategies. It is thanks to these corporations that Polish consumers started to pay attention not so much to the role as to the concept of corporate responsibility for the environment. In the first couple of years since the idea of corporate social responsibility had been introduced in Poland it was not met with due appreciation. Undoubtedly this was because there were not enough experts who could quickly and responsibly introduce companies to the new concept. As time went by, however, there appeared a number of professional non-governmental organizations, such as Forum Odpowiedzialnego Biznesu (Responsible Business Forum), which started to spread knowledge about corporate social responsibility. According to Mirella Panek-Owsiańska, who has for many years been the President of Resopnsible Business Forum in Poland organization, "it was only after the transformation that the western understanding of corporate social responsibility, as defined in the corporate strategies of corporations opening their offices in Poland, contributed to the spread of this idea in our country. These major foreign-owned companies started to implement ethical codes, introduce environmental audits and get involved with local communities. They also initiated a debate about the role of business in society and about the difficulty of maintaining a balance between the economic, environmental, social and ethical aspects of their corporate activities."[10] It was extremely difficult at first because there was little awareness of these issues among consumers, employees and managers alike.

As time went by, however, the number of companies working in compliance with corporate social responsibility policies was gradually increasing, and the Polish people watched these changes with interest. Some of the activities undertaken were valuable, while others left a lot to be desired. There were a number of initiatives of questionable social or environmental value, which were still advertised under the 'corporate social responsibility' label. Most corporate social responsibility initiatives as far as businesses are concerned were mainly centered around marketing departments and PR agencies. Surely, corporate social responsibility policies *do* have an influence on how companies are perceived, but they definitely must not be reduced to serving as a PR tool only. Frequently what was labeled as corporate social responsibility was in fact only a social project, which consumers often interpreted

---

[10]Panek – Owsiańska, M.`CSR r(E)wolucja,[in:] Biznes, który zmienia świat, Forum Odpowiedzialnego Biznesu, Warszawa 2015, p. 14.

as nothing more than a cynical market game. In the end, Polish consumers started to seriously doubt that undertaking such social initiatives made any sense, and corporate social responsibility came to be more and more often considered as a PR tool used to create a more positive company image. It was, in fact, commonly thought, not only in Poland, but also elsewhere, that business has its own set of rules, and that it tends to turn a blind eye to unethical behaviours and incomplete truths. One might say that it is not necessarily wrong of people doing business to strive to multiply their profits, no matter the cost. The economic crisis only increased the lack of trust in business[11] and any initiatives labeled as corporate social responsibility. Here came a time when CSR was famously declared dead. A few years ago it was signalled that the "marketing" formula of CSR activities is over and that it is necessary to involve business in real solving social problems.. Even supporters of the concept themselves agreed back then that the role of business in society needed to be redefined—"Companies cannot continue to pretend to serve society while simultaneously acting against it. Neither can they continue to give shareholder's interest primacy above the interests of the public. No amount of investment in charitable causes or employee volunteering can change that fact. The purpose of a company will be to create shared value, where business and society achieve success together".[12] In fact, one cannot but notice that over the last couple of years companies have been changing their attitude towards social responsibility. The change has been an organic one and businesses are now increasingly willing to engage in a real dialogue and create lasting social partnerships.[13] Corporate social responsibility strategies are no longer the domain of marketing and sales departments, but they started to be implemented also in HR, logistics, manufacturing, etc. Given the situation on the job market, there are a number of initiatives now for the benefit of employees, who are more and more willing to discuss the expectations they have of their workplace. More attention is also given to the issue of responsibility in supply chains, leadership and marketing. Undoubtedly, none of these changes would have happened if it were not for the increasingly attentive consumers and citizens who, unlike previous generations, are a lot less tolerant of situations in which values are disrespected and information is manipulated. On the Polish market there appeared a group of consumers whose personal values and sense of responsibility raised the bar on ethical standards for businesses as far as manufacturing, distribution and goods and services are concerned. This is where the concept of personal responsibility comes in. Personal social responsibility is a catalyst for important changes on the Polish market. It makes it possible for corporate social responsibility to evolve. Companies are forced

[11]2018 Edelman Trust Barometer. Global Report; 2009 Edelman Trust Barometer; 2010 Edelman Trust Barometer.

[12]Egan, D, CSR is dead, long live social enterprise, The Gourdian, Aug 2011, https://www.theguardian.com/social-enterprise-network/2011/aug/09/shared-value-csr-social-enterprise

[13]Karwacka, M. Siła Współpracy. Relacje przedsiębiorstw i organizacji pozarządowych w kontekście społecznej odpowiedzialności biznesu, Wyd. NOMOS, Kraków 2016.

to try even harder to convince conscious customers that they act responsibly and that they care about and share their values.

3 Consumer Awakening

Paradoxically enough, the economic crisis and the growing scepticism about corporate social responsibility in Poland not only sparked off a debate concerning the role of business in society, but also encouraged Polish people to juxtapose their personal values with marketing campaigns and companies' social reports. Additionally, individuals' heightened sense of personal social responsibility contributed to strengthening the position of non-governmental organizations and movements fighting for a better quality of life. Generally speaking, in recent years the third sector in Poland (mostly the few powerful non-governmental organizations influential enough to exert pressure on public and market entities) has considerably gained strength, and Poland's accession to the European Union was undoubtedly a contributing factor in this respect. Thanks to running projects financed with EU public funds, non-governmental organizations were able to gain more expertise not only in the area of management, but also in cooperating and sharing knowledge and experiences with similar organizations in Europe. The consumer awakening might also to a large extent have been connected with generational change. Numerous studies have shown that Millenials find non-material values important. As stated in the latest global report, *The 2017 Deloitte Millennial Survey,*[14] Millenials, unlike previous generations, are not so much intent on striking a balance between their private and professional lives; rather, they wish to integrate their values in both spheres. Based on their own system of values, they look for an appropriate employer, whom they also expect to engage in addressing global social issues. Adherence to one's personal values at work is also reflected on the market. The young want consistency. They are critical consumers, who closely examine individual brands, openly criticise unethical business practices and more frequently oppose unbridled consumerism. More and more Polish people are starting to embrace the slow life philosophy (slow life, slow food, slow fashion), and non-governmental organizations such as Fundacja Kupuj Odpowiedzialnie (Buy Responsibly Foundation) are becoming increasingly influential. The popularity of blogs advocating the value of minimalism indicates that consumers are no longer uncritical of manufacturers' practices and products available on the market.

Polish critical consumers start to manifest their sense of responsibility, thus expressing their concern about their own health and the natural environment. The concept of personal social responsibility is relatively new in Poland, yet it already

[14]The 2017 Deloitte Millennial Survey Apprehensive millennials: seeking stability and opportunities in an uncertain world https://www2.deloitte.com/content/dam/Deloitte/global/Documents/About-Deloitte/gx-deloitte-millennial-survey-2017-executive-summary.pdf

manifests itself in consumers' attitudes. S. López Davisa, L. Marín Rivesb and S. Ruiz de Maya, define "Personal Social Responsibility (PSR) as the way a person performs in his daily life as a member of society—and not only as a consumer—basing his decisions in a desire to minimize the negative impacts and maximize the positive impacts on the social, economic and environmental in the long run."[15]

Personal social responsibility is thus a multi-faceted issue and it constitutes a sort of coherent system in which all values are integrated. Our personal moral compass and the conditions in which we operate influence consumer behaviour. PSR does not have to coincide with consumer responsibility, but as the market for sustainable products develops and consumer awareness grows, PSR can play an increasingly important role.

Undoubtedly, it also plays a crucial role in shaping ethical consumerism in Poland. Owing to the increasingly overt and unhampered display of non-material values, there is a growing number of ethical consumers, or at least consumers who are suspicious of mass-produced items and remain critical of products which disrupt our interaction with the natural world and affect social relations through exploitation and ill-treatment of employees at the manufacturing stage. Opposed to such practices, these consumers are on the lookout for alternative solutions.

Symbolic consumer awakening in Poland happened at the time of the Clean Clothes Campaign. Shortly after the Rana Plaza collapse, which occurred in Bangladesh in April 2013, claiming the lives of almost 1200 people, it turned out that one of the clothing brands that sourced from the factory was Cropp, which is managed by the Polish company LPP. The information about the tragedy and the poor safety and working conditions in most Asian factories quickly made the news all over the world. Both the media and non-governmental organizations, including Clean Clothes Polska, demanded that the companies sourcing from the Rana Plaza factory should be held responsible for the tragedy and insisted that appropriate steps should be taken to prevent such accidents from happening in the future. Experienced brands instantly responded to the appeal and decided to conduct additional inspections in factories and to allocate a sum of money to install fire protection systems in the buildings. They also donated one million dollars to the families of the victims. As a result, in May 2013 as many as 70 clothing companies (including H&M, Inditex, Esprit, Tesco, Lidl and Marks & Spencer) co-signed an accord in which they agreed to create a fund to finance safety inspectors in Bangladesh factories. Unlike most brands, LPP did not sign the accord. The Clean Clothes Polska organization launched a consumer campaign aiming to motivate LPP to sign the accord and pay due damages. As part of the campaign, over 2000 consumers signed a petition, which they later submitted to the LPP board. In response to the campaign, LPP released a statement in which they categorically denied any responsibility for the Rana Plaza collapse, claiming that they had no control over their subcontractors. The statement caused yet another wave of petitions, this time calling for the creation of

[15]López Davis, S.,∗, Marín Rivesb, L. S. Ruiz de Maya, Introducing Personal Social Responsibility as a key element to upgrade CSR, Spanish journal of Marketing, 2017.

better working conditions in Bangladesh factories. At around the same time the poor working conditions in subcontracted factories of Polish brands started to be openly debated in the Polish fashion industry and they were bitterly criticized by well-known fashion journalists. The issue was also thoroughly explored by Polish journalists, who provided heart-breaking coverage from the site of the collapse and talked to the factory employees and the families of the victims. Leading Polish newspapers published articles with dramatic headlines such as "Piekło szwaczek. Polska firma nie czuje się winna" (Sewing hell. Polish company denies responsibility); "Bunt odzieżowych wyrobników w Bangladeszu" (Bangladesh garment factory labourers on strike), "Wyzyskiwani pracownicy z Bangladeszu strajkują. Pracują po 19 godzin" (Exploited Bangladeshi workers on strike. Working 19 hours a day.). It has often been claimed that the protest of Polish consumers did not change or mean anything. I do agree that not too many people took part in the boycott campaign, but at the same time it must be said that thanks to consumers' actions after 6 months since the Rana Plaza tragedy LPP did finally sign the Accord on Fire and Building Safety in Bangladesh. The first consumer campaign in Poland was thus a successful one.[16] Several years on from the collapse, the company now employs corporate social responsibility experts and conducts thorough checks in factories it sources from.

4 Responsibility of Polish Consumer

Consumer responsibility or conscious consumption has been the subject of discussion in Poland since relatively recently and so far, it has only been of interest to a small group of people. Undoubtedly, the matter is strongly connected with the issue of corporate social responsibility. In my opinion, personal responsibility for the environment is key to developing a full-fledged CSR strategy. By ethical consumption I mean applying our personal values to our market behaviours and consumer choices, which we make based not only on the price of a given product, but also on its quality and social or environmental impact. A responsible consumer is a person who carefully considers his or her expenses and consciously reflects on the type of lifestyle he or she leads and on the impact it has on the environment.[17] It is also

[16]Compiled from the following sources: http://cleanclothes.pl/petycja37_godna_placa_dla_wszystkich_.html; http://swiat.newsweek.pl/lpp-potwierdza%2D%2Dprodukowalismy-odziez-cropp-w-rana-plaza,104690,1,1.html; http://wyborcza.pl/1,75248,14818998,Gdanskie_LPP_zaplaci_za_poprawe_warunkow_pracy_w_Bangladeszu.html, http://gosiaboy.com/post/62536468505/blogerzy-kontra-lpp-kontra-zlo-calego-swiata-o-co; http://natemat.pl/74795,piec-miesiecy-po-katastrofie-w-bangladeszu-wlasciciel-croppa-wciaz-niemal-nie-poprawil-bezpieczenstwa-w-fabrykach, http://michalzaczynski.com/2013/09/27/w-roli-ofiary/

[17]Koszewska, M. Wpływ wrażliwości ekologicznej i społecznej na sposób zachowania konsumentów na rynku tekstylno odzieżowym, Handel Wewnętrzny May-June 2012 Volume 3, p. 9.

someone who believes that his or her individual decision can make a difference. Responsible consumers are those who express their views and manifest their values by how they spend their money. As revealed in the Barometr CSR study, the number of such consumers is still growing. The authors who conducted the study distinguished six types of consumers based on their attitude towards corporate social responsibility[18]:

1. **satisfied realists**—people who are highly aware of what corporate social responsibility is and who are thus the most open-minded about the initiatives that companies undertake to reduce their impact on the environment. They firmly believe that business bears responsibility for the environment and they expect companies to engage in activities meant to benefit society. In comparison with other groups, satisfied realists quite frequently choose brands engaged in social initiatives and they suggest making similar choices to others.
2. **sensitive intellectuals**—people who have a positive attitude to the issue of corporate social responsibility and who know quite a lot about it. They tend to acknowledge the multidimensional nature of CSR and understand the mechanisms behind it. Sensitive intellectuals also to a large extent expect businesses to launch and run campaigns showcasing their social commitment.
3. **eco-pragmatists**—people who have little knowledge about corporate social responsibility and who are quite suspicious of companies' real intentions behind CSR policies. They do not approve of businesses benefiting from corporate social responsibility initiatives. Eco-pragmatists are slow to trust companies and believe that the latter should always attempt to prove that they have good intentions. Such people *do* take action to help protect the environment, but they can hardly be defined as responsible consumers.
4. **passive posers**—people who expect a lot from businesses, but they are not ready to be responsible consumers themselves. They believe that they know what corporate social responsibility is about, but in practice it turns out they do not in fact understand it. Passive posers seem to confuse corporate social responsibility with philanthropy and they are sceptical of communicating companies' commitment.
5. **excluded complainers**—people who do not understand the idea of corporate social responsibility and do not even want to learn anything about it. At the same time, they believe that because companies have huge financial resources, they should share them with others. When making consumer choices, excluded complainers do not stop to think about the social and environmental impact of products they decide to buy.
6. **corpo-egocentrics**—people who do not believe that their actions might have any real influence on the environment. They *do* notice problems, but they refuse to believe that by changing their own habits they could indeed make a difference.

[18]Barometr CSR study; http://csrconsulting.pl/wp-content/uploads/2016/11/BAROMETR-CSR-2016-wybrane-wyniki.pdf

The 2016 Barometr CSR study showed that eco-pragmatists were the biggest group (23%). 21% of consumers were satisfied realists, 7% were sensitive intellectuals and 15% were passive posers. The groups of complainers and corpo-egocentrics each constituted 12% of consumers. Compared with the 2013 Barometr CSR study, the number of eco-pragmatists and satisfied realists has increased (15% in 2013 compared to 21% in 2016), while the number of sensitive intellectuals, passive posers and corpo-egocentrics has decreased over the last couple of years.

Taking these study results under consideration, one might thus assume that corporal social and environmental responsibility is becoming more and more important to Polish consumers.

5 How to Wear Fair: Personal Social Responsibility While Shopping

My professional interests have lain in the issue of corporate social responsibility for over 10 years. I am a researcher and an advisor and I work with companies, non-governmental organizations and universities. In my professional life I endorse the same values that matter to me in my personal life. The ability to integrate professional and personal lives is really important for employees, organizations and the environment. One day I started to wonder if I also take my personal values into consideration while shopping.

I define personal social responsibility as adherence to the same values in various fields of our lives, both professional and personal. Ethical consumption is thus a derivative of personal social responsibility. It is about aiming to maintain coherence as far our consumer behaviours are concerned. Taking into consideration growing consumer awareness and the knowledge we now have about how food influences our health, it is quite understandable that consumers are paying more and more attention to food products. In this case personal social responsibility is more about personal health benefits and about taking care of oneself and those closest to us. We are then ready to spend our money and devote our time to finding the best products and we do not hesitate to criticize unethical business practices. But what if these benefits are not that obvious? I have asked myself this question in the context of the clothing industry. Although we dress everyday and we buy new clothes quite frequently, we rarely reflect on the impact that our shopping choices have on the environment and on others. We are very well aware that the clothes we buy are made in Asian factories, but we also think that the consequences of our actions are not immediate, and in fact, we do not believe that our choices might ever influence global brands to change their policies.

The experience of writing my How to Wear Fair? blog for almost 2 years has taught me that dressing consciously and responsibly is indeed possible, but the average customer must face a number of challenges along the way. First of all, it is quite difficult to precisely define sustainable fashion. Each and every one of us endorses different values—some people are committed to animal welfare, while others are more concerned about the environment or about the wages of garment factory workers. There are also those who wish to support local businesses, as well as those who try to keep their shopping needs to the minimum. At the moment buying clothes responsibly is a rather time-consuming and onerous task. One of the reasons for that is the fact that slow fashion is still a relatively new concept in Poland, which means that consumers have yet to learn more about supply chains or the mechanisms behind the clothing industry. Also, fast fashion has much more to offer in terms of choice and variety of products than slow fashion. In spite of these challenges, however, the idea of ethical consumption in the clothing industry will be gaining popularity in years to come. It is only enough to see how much interest the issue excites among consumers, journalists and even designers themselves. There are also more and more fashion brands in Poland which aim to satisfy the needs of ethical consumers. These companies offer products which are 100% vegan (Oak Bags), follow the principles of circular economy (Piel, Argentum and BoardThing) or are transparent in terms of margins and manufacturing costs (Elamanty). There is also an increasing number of initiatives aiming to help consumers who are looking for alternatives to fast fashion products, so it seems the trend is developing.

Taking into consideration the extent to which the clothing industry is closely watched by responsible consumers, non-governmental organizations and the media all over the world, it might be assumed that, paradoxically enough, it is this very sector that could become one of the most fertile grounds for implementing corporate social responsibility policies along the whole supply chain. The entire clothing industry is currently beset with numerous problems, but in the years to come companies must face these challenges and try to implement sustainable fashion practices.

References

2009 Edelman Trust Barometer.

2010 Edelman Trust Barometer.

2018 Edelman Trust Barometer. Global Report.

Barometr CSR study. (2016). http://csrconsulting.pl/wp-content/uploads/2016/11/BAROMETR-CSR-2016-wybrane-wyniki.pdf

Bauman Z., Globalizacja, PIW., Warszawa, 2000.

Egan, D. (2011, Aug). CSR is dead, long live social enterprise. *The Gourdian*. https://www.theguardian.com/social-enterprise-network/2011/aug/09/shared-value-csr-social-enterprise

Fromm, E. (1995). Mieć czy być, Dom Wydawniczy Rebis, Poznań.

http://cleanclothes.pl/petycja37_godna_placa_dla_wszystkich_.html

http://gosiaboy.com/post/62536468505/blogerzy-kontra-lpp-kontra-zlo-calego-swiata-o-co

http://michalzaczynski.com/2013/09/27/w-roli-ofiary/

http://natemat.pl/74795,piec-miesiecy-po-katastrofie-w-bangladeszu-wlasciciel-croppa-wciaz-niemal-nie-poprawil-bezpieczenstwa-w-fabrykach

http://swiat.newsweek.pl/lpp-potwierdza–produkowalismy-odziez-cropp-w-rana-plaza,104690,1,1.html http://wyborcza.pl/1,75248,14818998,Gdanskie_LPP_zaplaci_za_poprawe_warunkow_pracy_w_Bangladeszu.html

Karwacka, M. Siła Współpracy. Relacje przedsiębiorstw z organizacjami pozarządowymi w kontekście społecznej odpowiedzialności biznesu, Wyd. NOMOS, Kraków 2016.

Koszewska. Wpływ wrażliwości ekologicznej i społecznej na sposób zachowania konsumentów na rynku tekstylno odzieżowym, Handel Wewnętrzny May–June 2012 Volume 3.

López Davisa, S., Marín Rivesb, L. S., & de Maya, R. (2017). Introducing personal social responsibility as a key element to upgrade CSR. *Spanish journal of Marketing, 21*, 146–163.

Mróz, B. (2010). Consumerism vs. sustainability: The emergence of new consumer trends in Poland. *International Journal of Economic Policy in Emerging Economies, 3*(1), 1–15 www.pol-int.org

Panek-Owsiańska, M. (2015). CSR r(E)wolucja. In Biznes (Ed.), *który zmienia świat*. Warszawa: Forum Odpowiedzialnego Biznesu.

Pietrowicz, K. (2004). Nasze gadżety codzienne albo o płynności symboli statusu. In T. Szlendak & K. Pietrowicz (Eds.), *Na pokaz. O konsumeryzmie w kapitalizmie bez kapitału* (p. 72).

Szlendak, T., & Karwacki, A. (2010). Koncepcja poziomicy – cudowne lekarstwo czy utopijna terapia? *Studia Socjologiczne, 1*, 41.

The 2017 Deloitte Millennial Survey Apprehensive millennials: Seeking stability and opportunities in an uncertain world. https://www2.deloitte.com/content/dam/Deloitte/global/Documents/About-Deloitte/gx-deloitte-millennial-survey-2017-executive-summary.pdf

Employees Motivation in CSR Projects: Case Study of WrOpenUp

Florian Urbański

1 Introduction

The economic growth of Poland in the last 27 years has been one of the major success stories during the transformation processes in Central and Eastern Europe since 1990 (Alsund 2013; Kirov and Delteil 2013; Kuitto 2016). This economic development over the two last decades also meant an increased involvement of international corporations within the Polish market (Polityka Insight 2017; Floyd 2016; Cieślik and Hagemejer 2014). This influence on the Polish market and the Polish society by international corporations can also be identified with Corporate Social Responsibility Projects (CSR-Projects), because those corporations has been identified as main drivers of CSR in Poland (Lakatos 2017; Dyczkowska et al. 2016). International corporations in Poland employ Polish as well as international employees (here called expats) in their subsidiaries (Poór et al. 2014). Local and international employees in these corporations form a culturally diverse workforce (Przytuła 2014; Rozkwitalska 2014). This leads to the question of whether this internal diversity is also reflected in CSR Projects led by these corporations and if their employee volunteers are likewise diverse in aspects such as gender, religion and, of course, nationality. The main objective of this paper is to investigate the intrinsic and extrinsic motivation of these volunteers. It is the goal of this paper to examine possible significant differences of motivation between gender, religion and nationality among the volunteering employees. The core argument is to find a substantial difference of motivation between Polish and international employees to join CSR Projects. The core question developed out of this thesis is the following: "How to motivate Polish and international employees to volunteer for CSR Projects." Therefore, the research questions are defined as:

F. Urbański (✉)
JUTECH Sp.z.o.o, Wrocław, Poland

A. Długopolska-Mikonowicz et al. (eds.), *Corporate Social Responsibility in Poland*, CSR, Sustainability, Ethics & Governance,
https://doi.org/10.1007/978-3-030-00440-8_16

- What types of motivation move people to volunteer?
- Do significant differences exist in motivation between genders?
- Do significant differences exist in motivation between nationalities?
- Do significant differences exist in motivation regarding religion?

The article describes the current state of CSR and volunteering in Poland by reviewing some of the recent literature (books, online and journal articles), and by focusing on projects run by international corporations. The second part examines a case study as a practical example. Based on the results of interviews and a questionnaire with Polish and international volunteers of the CSR Project called WrOpenUp, the article aims at answering the question to the core argument. The goal is to deliver a guideline to motivate Polish and international employees to join CSR Projects in Poland.

2 Corporate Social Responsibility

2.1 *Definition of CSR*

The term Corporate Social Responsibility (CSR) can be defined as a business concept for a voluntary integration of social and environmental concerns into business operations and into the interaction with stakeholders (UNIDO 2017; Chandler 2017), in order to achieve substantial growth and economic development for the company (Eweje and Bathurst 2017). Based on this definition, CSR can be also described in a broader sense as social and environmental obligations by a company which are taken additionally on top of the legal necessity to consider social expectations (Hackett 2016; Rasche et al. 2017). In further consequence, CSR also can be understood as a business model to include social, economic and environmental concerns into an organisation's corporate culture and strategy in order to provide a larger approach of business. (Hawkins 2006; Eweje and Bathurst 2017). The communication and cooperation between a company and its stakeholders is hereby the crucial element as a kind of a social dialogue with a wide range of groups such as shareholders, employees, suppliers, customers, local communities, management, citizen groups, competitors and the press and, in a much wider perspective, the society as a whole (Hackett 2016; Chandler 2017). Therefore, the covered activities of CSR reflect the diversified identity of stakeholders and their various interests and values such as: (1) Organizational Governance; (2) Human Rights; (3) Labour Practices; (4) The Environment; (5) Fair Operating Practices, (6) Consumer Issues, (7) Community involvement and Development (ISO 2010; Hansen 2017; Aluchna 2017). Regarding Hiß (2006) such fields of activities can be divided into three spheres of responsibility (Rajak and Dolan 2016). First the internal sphere contains activities which are defined by the responsibilities regarding the market and law in order of a responsible maximization of profits. Not just by following rules and laws but also by additional commitments with social hardship clause, or the integration of international agreements of labour- and environmental standards into the company's culture. In this case CSR means the renunciation of profit at the expense of others. Second the intermediate sphere summarizes all activities which are beyond law enactments and international agreements regarding the responsibility to the value creation chain. It involves self-commitments in standards of production and business behavior (e.g. code of conducts and their control via monitoring, seals of quality by independent institutions). Third the external sphere is defined by all activities which are beyond the other two spheres and are aimed at social effects outside of the company. One part the Corporate Giving as activities of donation and sponsoring of social or environmental events or organizations. The second part is the Corporate Volunteering scheme (or Employee Volunteering) which is defined as the exemption of employees for community volunteering. (Hiß 2006, 2009).

2.2 CSR in Poland

The concept of Corporate Social Responsibility is a relatively new phenomenon in the Polish society and on the Polish Market (Aluchna 2015b; Potocki 2015; Wołczek et al. 2016; Popowska 2015). In the beginning of the transformation process from 1990 until 1996, the concept of CSR was nearly unknown in the Polish society. These years were a period of market liberalization as well as an economical individualization. As a result it was also a time of low volunteering and inactivity of stakeholders in terms of social commitments and performance at all. (Aluchna 2015b; Potocki 2015; Jastrzębska 2016; Dyczkowska et al. 2016; Waniak-Michałak et al. 2016). With the end of this initial phase of the Polish transformation in 1997, this picture was about to change very slowly, driven by the ongoing development on the market, and lately also by the political and economic integration of Poland into the European Union in 2004. Those two factors did bring a change of perspectives to the stakeholders as well as to the Polish society and furthermore the necessary legal basis and institutional frameworks for CSR (Popowska 2015; Waniak-Michałak et al. 2016). Although 2007 studies showed an increasing interest in CSR within Polish companies, still a major barrier was identified: the lack of appropriate legal frameworks for CSR and less international interaction like with forums of the United Nation Organisation. (Macuda et al. 2015; Popowska 2016; Wołczek 2015). The major developments regarding CSR in Poland took place in 2009 when the Polish government introduced a special CSR team under direct control of the Polish Prime Minister (Ministerstwo Gospodarki RP 2010), and their conclusions have been integrated into the new Guidelines of the National Development Strategy 2020 (Gruszecka-Tieśluk 2015; Grzybek 2014; Wołczek 2015). Furthermore, the RESPECT Index portfolio was established by the Warsaw Stock exchange (Lipiec 2016); an index to identify listed companies with responsible and sustainable behaviours and etiquettes (Potocki 2015; Aluchna 2015a; Jastrzębska 2016). What is also quite interesting is the fact that differences between Polish-owned companies and international corporations still exist in terms of their CSR performance (Wołczek

2015; Aluchna 2015a). A survey by Kuraszko (2007) showed still a significant difference between Polish-owned companies and international corporations in terms of CSR Strategies: 40% of companies have no CSR strategy at all (mainly Polish businesses), 45% of companies with uncoordinated CSR Projects (mainly international corporations and large Polish companies) and only 15% of companies have included a CSR Strategy into the overall company strategy (mainly large international corporations) (Aluchna 2015a; Potocki 2015; Jastrzębska 2016). A similar conclusion may be drawn from the latest Polish CSR Ranking of CSR 2017 by Deloitte (and paired up with company information by Bloomberg www.bloomsberg.com). Among the nine companies of the first level (Diamond Level) are six international corporations (foreign-owned) and only three large Polish corporations (Polish-owned). On the second level (Platinum Level) of 14 companies are 10 international Corporations (foreign-owned) corporations compared to only 4 large Polish corporations (Polish-owned). Meanwhile on the third level (Cristal Level) of 39 companies are 25 international corporations listed compared to just 14 large Polish Companies (Polish-owned) (Ranking 2017).

3 CSR and Volunteering

3.1 Employee Volunteering

One of the key elements of activities in Corporate Social Responsibility is employee volunteering (Barkay 2012)—also known as corporate volunteering or community volunteering (Aluchna 2015b). Especially in projects of community volunteering (Uyan-Atay 2013; Pedersen 2015; Bsoul 2014). Employee volunteering can be understood as *"any formal organized company support for employee and retirees who wish to volunteer their time and skills in service to the community"* (Wild 1993). The key characteristic of this particular CSR activity is defined by the condition that private companies provide resources of personnel and competence to external projects of social or environmental concerns (Tschirhart 2005; Tschirhart and St.Claire 2008; Michels 2016). Employees are directly involved in the activities and usually volunteer during their working hours or even beyond their work time (De Gilder et al. 2005; Michels 2016). Moreover, CSR activities are a form of direct cooperation between the company and its employees to integrate the principles of corporate social responsibility into their strategy and behaviour. Furthermore, it is a public demonstration of responsibility, transparency and accountability to the society. It is not only to be seen as merely yet another way of empowering employees and activating values of labour for society but it is also perceived as the highest form of involvement by stakeholders into the strategy of corporate social responsibility (Pedersen 2015; Aluchna 2015b). Due the fact that employees are involved into decision making, they become a strategic stakeholder (Herzig 2004; Michels 2016). Employee volunteering is an instrument to achieve social goals and a way of a social dialogue between stakeholders (Herzig 2004) and this evolving partnership between companies, non-governmental-organizations (NGO) and individuals creates a win-win situation as far as all involved partners are concerned (Aluchna 2015b; Pedersen 2015; Michels 2016). Regarding the scope, length, company policy and the attitude of employees, all volunteering programs can be different and diverse (Aluchna 2015b; Uyan-Atay 2013). The range of volunteering programs is wide and may contain a broad variety from poverty reduction, support to disabled persons, educational projects as well as charity programs (Peterson 2004a, b; Aluchna 2015b; Uyan-Atay 2013).

3.2 Volunteering in Poland

Volunteering in general is a new phenomenon in Poland due the fact that volunteering did not exist as an organised strategy in Poland before the political and economic changes of the late 1980s (Potocki 2015; Szymańska-Palaczyk 2016; Kosewska 2015). Due to the transformation of Poland and its integration into the European Union, volunteering has become a social factor (Szymańska-Palaczyk 2016; Kosewska 2015) and in the year 2003 Polish law had to cover and define voluntarism (The Chancellery of the Prime Minister 2003) and renew it an amendment in 2010 (The Chancellery of the Prime Minister 2010; Potocki 2015). Therefore, studies that concern employee volunteering in Poland in particular are still very rare (Aluchna 2015b). Nonetheless, the report on employee volunteering in 2011 showed numbers which may in fact point to the development of CSR in Poland (CSRInfo 2011): in 2010 only 6% of Poles volunteered, yet this figure rose to 11% in the following year. At present, 60% of Polish people consider volunteering as valuable and interesting. Another interesting finding was the fact that employee volunteering is a popular activity mainly among Polish subsidiaries of foreign-owned companies and corporations. Nevertheless, the level of participation is just about 6% up to 20% of the company's workforce (Aluchna 2015b; Centrum Wolontariatu 2008; Centrum Wolontariatu 2013). These results seem on par with other studies in the last 7 years which have shown Polish citizens' low commitment to social activities within the Polish society regarding general volunteering (Przewłocka 2011; Jacobson and Korołczuk 2017). Studies like the European Values Study (EVS 2017), the Eurobarometer (GHK 2010), CBOS Foundation (CBS 2017) and the European Social Survey (ESS 2017) showed that only 13–20% of Polish adults had been volunteering (Aluchna 2015b; Charycka 2016). The difference is visible if this number is compared with countries of very high level of over 40% like Austria, Netherlands, Sweden and UK or even countries with high level of 30–39% like Denmark, Finland, Germany and Luxembourg. (GHK 2010; Potocki 2015). Interestingly a complete different picture was given by institute Klon/Jawor: in 2015, even 61% of the Poles spent their time on social work in a NGO. Much more than compared to the prior years: (1) 2006: 40%; (2) 2008: 44%; (3) 2010: 50% (4) 2012: 55% (Adamiak et al. 2015). The low level of participation and interest in general and employee volunteering can be explained by low income levels, and the struggle of individuals when it comes to economical necessities when these are compared to Western standards. The level of volunteering in Poland correlates highly with level education and the amount of income (Potocki 2015; Szymańska-Palaczyk 2016; Kosewska 2015). Nevertheless, this seems to be the main influence to the slow development of employee volunteering in Poland, all studies also showed an increasing interest in volunteering in Poland (Potocki 2015; Aluchna 2015b).

4 Diversity and Diversity Management

Diversity as *"a range of many people or things that are very different from each other"* (Oxford Learner's Dictionary 2017) can be defined in sociology more concrete as differences of individuals (Tirana 2017; Zein 2015) in *"social categories like race, ethnicity, religion, gender, sexuality, socio-economic status, nationality and citizenship, parental status, body size and ability, and age and experience"* (SOC Washington 2017; McPherson 2018). Regarding the requirement of business management diversity can be defined much more precise as *"the representation of people of different group identities in the same organizational social system."* (Cox and Smolinski 1994) and even more concrete about diversity in terms cultural system as *"a system of values, behavioural norms, goal priorities and preferred work styles which distinguish one group from another."* (Thomas 2017; Tirana 2017). The theory of diversity management (DiM) is about acceptance and the positive esteem of this diversity regarding demographic variables within a company or an organisation in order to support the business success. The concept of diversity management is described by Aretz and Hansen (2003) as a basic attitude and according to Schulz (2009) this includes openness to multi-cultural diversity (Stangel-Meseke 2016; Hansen 2017). Furthermore, Quindt (2013) as well as Vedder and Krause (2014) described the overlapping contents of CSR and DiM regarding their orientation to company culture and its economic performance. Both management concepts contain demographic, economical and ethical aspects and are based on internal and external strategies to reach an added economic value for the company. In addition, both concepts have in common the point of responsibility for society regarding the equality of opportunities and fairness integrated in a management concept guided by values and norms in (1) Organizational Governance; (2) Human Rights; (3) Labour Practices; (4) The Environment; (5) Fair Operating Practices, (6) Consumer Issues, (7) Community involvement and Development. Therefore, both concepts have to be seen as equal conceptions with its mutual independencies and synergy effects (ISO 2010; Stangel-Meseke 2016; Hansen 2017). The basis of a diversity management within an organisation is the measurement of diversity (PeopleFluent 2015; Thomas 2017). Especially in question of human resource management and recruiting a successful diversity strategy is a question of being an attractive employer to well-educated and highly skilled employees (PeopleFluent 2015; Tirana 2017). The same applies to CSR-projects and to NGOs (Hansen 2017; Bhatta et al. 2015). The first and the easiest measurement is the headcount ("how much?") and the next level the percentage ("How much compared to the whole group") regarding demographic variables which is a kind of descriptive statistic (Thomas 2017).

5 Motivation and Volunteering

5.1 Motivation: Intrinsic and Extrinsic

Among other definitions (Shah and Gardner 2008), motivation can be understood as *"those psychological processes that cause the arousal, direction, and persistence of voluntary action that are goal directed"* (Mitchel 1982). Additionally, it can be described as *"the forces within us that activate our behaviour and direct it toward one goal rather than another"* (Phares and Chaplin 1997) In this meaning, it is a theoretical construct to explain human behaviour and the root cause for people's actions, desires, and needs (Dweck and Elliot 2017) and in a short sentence *"to be motivated means to be moved to do something"* (Deci and Ryan, 2002, 2017). The theory of self-determination by Deci (1971) describes human motivation by individually inherent tendencies to growth and psychological needs. It distinguishes human behaviour caused by external influence and outside interferences and caused by self-motivation and self-determination. This presupposes the true option toward choice and the sense of freedom to transform the will into their own actions (Guay et al. 2000; Dweck and Elliot 2017).

In light of this, theory motivation can be divided into three different types (Deci and Ryan 1985). First intrinsic motivation, by definition, means performing an activity for its inherent satisfaction as a self-desired discovery of new possibilities and new challenges as well as to observe one's own capacity and the ability to gain new knowledge (Deci 1971; Deci and Ryan 2002, 2017). It is the internal drive of interest and the enjoyment of the undertaking itself which is the main factor of this internal motivation which occurs in the individual and not by an external trigger or the mere wish of consideration (Beswick 2017; Deci and Ryan 2017). Second the extrinsic motivation is caused by influences from outside of the individual and is focused on the outcome beyond the task itself (Deci 1971; Deci and Ryan 2002, 2017). These influences are external regulations which occur when a behaviour is aimed to achieve rewards or to avoid sanctions or rather negative consequences (Deci and Ryan 2017; Dweck and Elliot 2017). According the sub theory of Organismic Integration Theory (OIT) (Deci and Ryan 1985), the extrinsic motivation can be divided in respect to the autonomy into four different types: (1) External regulation: behaviour as root cause in fulfilling of an external pressure or in expectation of an external reward; (2) Introjected regulation: behaviour as root cause in internalization of external expectations into self-imposed pressure to avoid feelings of guilt or lack of self-esteem (3) Identified regulation: behaviour as root cause in personal acceptance of external goals or regulation as personally important with focus on an instrumental value instead of the task itself. (4) Integrated Regulation: behaviour as root cause of assimilation of identified regulation in to own personal value (Deci and Ryan 2002; Beswick 2017). The third type amotivation is the lack of interest in the task and its outcomes and by this definition a complete opposite of motivation. (Deci 1971; Deci and Ryan 2017). Due motivation is a psychological construct, it cannot be directly observed and recorded but only determined by measuring indirect factors such as cognitive functions (e.g. perception, learning, remembering, assumptions), affective reactions (e.g. subjective experience of feelings and emotions) and behavioural responses (e.g. performance, action and reaction) or physiological reactions (e.g. brain activity) (Touré-Tillery and Fishbach 2014). The concepts of SDT and OIT regarding the different kinds of motivation also build a basis to measure motivation because it is focused on the "Why" of a behaviour (Deci and Ryan 1985; Dweck and Elliot 2017). On this postulate the developed Situational Motivation Scale by Grano et al. (2008) is based on cognitive and affective factors by simply asking about the root cause: "Why are you currently engaged in this activity?"

5.2 *Motivation Behind Volunteering*

Different models describe possible motivation in individuals to volunteer. The unidimensional model postulates that different motives (e.g. religion, reputation) form just one unidimensional scale by overlapping motives into one category. The two-factor model suggests the existence two kinds of motivation: (1) Altruism (concern and responsibility for others) and (2) Egoism (concern for themselves). The three-dimensional model divides motivation into three categories: (1) altruistic motives (concern for others), (2) material motives (concern for rewards) and (3) social motives (concern for society). Several multi-dimensional models describes multiple categories of motives and according to Clary et al. (1998) exactly six functions of motives: (1) Values (e.g. Altruism, Humanism), (2) Understanding (e.g. Will to learn, gathering new skills and knowledge), (3) Social (e.g. social interaction, meet new people), (4) Carrer (e.g. gain career experience), (5) Protective (e.g. reducing guilt towards the less fortunate), (6) Enhancement (e.g. self-development and growth) (Papadakis 2004; Widjaja 2010). According to Finkelstein (2009) these six functions can be categorized by intrinsic an extrinsic motivation by which the functions of values, understanding, social, enhancement and protective as the internal motives (intrinsic) and the career function as the external motive (extrinsic) (Widjaja 2010). Similar model was described by Gidron (1978, 1984) in which extrinsic motives like external incentives and injunctive social norms as well as intrinsic motives like inherent value, interest and enjoyment are identified as volunteering motives (Geiser et al. 2014). According to the motivation to volunteer scale by Grano et al. (2008) the research indicates a much higher influence of intrinsic factors instead of extrinsic factors for volunteering motivation (Geiser et al. 2014). The Question of different types of motivation and higher frequency regarding gender, national differences as well as, cultural, political, religious and social factors in volunteering is a controversial topic. Some studies found differences in intrinsic and extrinsic motivation between genders and nationality, other studies postulate that findings regarding such differences in volunteering motivation are not consistent. But in general studies indicates a higher percentage of women volunteering then men (Greely 1997; Geiser et al. 2014; Widjaja 2010).

6 The Case Study WrOpenUp

6.1 WrOpenUP

The CSR-Project WrOpenUp is a foundation based in the City of Wroclaw in Poland, supported by local subsidiaries of multinational corporations. The project was realized first in 2012 and meanwhile the project gets supported by Credit Suisse, Google, HP, HPE, IBM, UBS, NOKIA and Volvo. (Gazeta Wyborcza 2017; Tarasiewicz 2017). It is the aim of the project to support and develop openness towards cultural diversity in the City of Wroclaw. Therefore, WrOpenUp organizes workshops with pupils from local schools in Wroclaw to develop cross-cultural and intercultural competencies as well as teaching social intelligence, new media literacy, design mindset, cognitive load management and virtual collaboration. Every other year this is additionally complimented by a competition in which students are tasked with developing their own school projects about diversity. (Gazeta Wyborcza 2017; Tarasiewicz 2017). The crucial element here is employee volunteering and the volunteers are recruited out of the supporting companies. Based on the skills and the cultural diversity of those volunteers, the workshops are organized and facilitated by them (Tarasewicz-Gryt and Długopolska-Mikonowicz A (2017).

6.2 Methodology and Questions

In order to pursue the research goals a questionnaire was developed based on the self-determination-theory (SDT) (Deci 1971), Organismic Integration Theory (OIT) (Deci and Ryan 1985) and the theory of multi-dimensional motivation of volunteering as its categorization in intrinsic and extrinsic motivation (Finkelstein 2009). The questions where adapted to the case study and based on the research by Guay et al. (2000)—Situational Motivation Scale (SIMS)—and the research by Grano et al. (2008)—Motivation to Volunteer Scale (MVS). The survey asked in the first part about social aspects (age, gender, nationality, income etc.). and in a

Table 1 Questions regarding motivation

Motivation "Why have you been got engaged as a volunteer in WrOpenUp?"	
Strongly agree (4); Agree (3); Neither agree nor disagree (2); Disagree (1); Strongly disagree (0)	
1 Intrinsic motivation (Because...) A. I like to do good and positive things for society B. I Like to gather new experience and knowledge C. I like to socialize and meet new people D. I like to develop myself and grow E. I think volunteering is fun and pleasant	**2 External regulation (Because...)** A. In my company I am supposed to volunteer B. I get free time as compensation C. The recognition is good for my professional career D. My company provides financial benefits for volunteering E. I can learn new skills for my professional career
3 Introjected regulation (Because...) A. In my company I am expected to volunteer B. In my company others volunteer as well C. I feel responsible for my company D. I feel responsible to others E. I feel responsible for society	**4 Identified regulation (Because...)** A. I think WrOpenUp is important B. I think WrOpenUp is good for others C. I think WrOpenUp is good for the students D. I think WrOpenUp is good for Wroclaw E. I think WrOpenUp is good for society
5 Integrated regulation (Because...) A. I like to support the goals of WrOpenUp B. I share the same goals as WrOpenUp C. I like to support openness and tolerance in society D. I like to teach children openness and tolerance E. I like to teach children soft skills and knowledge	**6 Amotivation (Because...)** A. Maybe there is a reason to volunteer, but I do not know it B. I volunteer but I am not sure if it is really worth it C. I do not see any point in participating in volunteering D. I volunteer but I am not sure this is a good activity E. I do not know anymore why I volunteered

second part about their companies' relation to their volunteering work (encouraging, benefits, compensation for volunteering). Third part of the questionnaire was developed to identify the different types of motivation (see Table 1.) by focusing on the main question „Why have you been engaged as a volunteer in WrOpenUp?"

To measure the different types of motivation, for each of the six types of motivation a group of 5 questions has been developed. Each block was based on a four point likert-scala which in average showed the tendency of motivation. The answers have been analysed by nationality (Poles towards expats), gender (female towards male) and religion (religious towards secular people), the answers of all five questions got added and divided by amount of each group. The data of the survey was collected from online questionnaire which was composed of open-ended and closed-ended questions. The survey was send to $n = 150$ former and current volunteers since the project started in 2012. In total $n = 36$ answers (24%) formed the data for analysis. The Data was evaluated by descriptive statistics to answer the research questions.

7 Results

7.1 Diversity of Volunteers

The sample consisted of n = 36 volunteers of the CSR Project WrOpenUp. Of these participants, 55.6% where female and 44.4% were male. The majority (55.6%) where Polish and the minority (44.4%) comprised expats from different countries (See Table 2). Regarding religious beliefs 55.6% claimed to be Christian, 13.9% Islamic, 11.1% Agnostics and 19.4% Atheists, which were combined for the following analysis to the groups of religious people (69.4%) and secular people (39.6%).

The Question concerning the monthly cross income (see Table 3) was answered by 86.11% with over 4501 PLN and by 13.89% with between 2001–4500 PLN. With 47.2% the age group of 31–40 years of age was the majority towards 38.9% of 25–30 years of age and 13.9% of 41–50 years of age.

The answers regarding the question about the highest reached education level showed 63.9% with a master's degree, 27.8% with bachelor's degree and 2.8% each of high school with diploma, high school without diploma as well as other (see Table 4).

Table 2 Demographic diversity 1

Nationality			Beliefs/Religion		
	Number	Percentage		Number	Percentage
Polish	20	55.6	Christianity[a]	20	55.6
Indian	2	5.6	Islam[a]	5	13.9
Morocan	2	5.6	Agnosticism[b]	4	11.1
Italian	2	5.6	Atheism[b]	7	19.4
Pakistani	2	5.6			
Brazilian	2	5.6	Religious[a]	25	69.4
Belarusian	1	2.8	Secular[b]	11	30.6
Ukrainian	1	2.8			
Costa Rica	1	2.8	Gender		
Filipino	1	2.8		Number	Percentage
Turkish	1	2.8	Female	20	55.6
Spanish	1	2.8	Male	16	44.4

n = 36
[a]"Christianity" and "Islam" combined to "Religious"
[b]"Agnosticism and Atheism" combined to "Secular"

Table 3 Demographic diversity 2

Monthly cross income			Age groups		
	No.	%		No.	%
Above 4501 PLN	31	86.11	31–40 years	17	47.2
2001–4500 PLN	5	13.89	41–50 years	5	13.9
Under 2000 PLN	0	0	25–30 years	14	38.9

n = 36; No. = Number; % = Percentage

Table 4 Demographic diversity 2

Education	Number	Percentage
Some high school, no diploma (or equivalent)	1	2.8
High school graduate, diploma (or equivalent)	1	2.8
Bachelor's degree (or equivalent)	10	27.8
Master's degree (or equivalent)	23	63.9
Other	1	2.8

n = 36

Table 5 Relationship companies to volunteering

Q: Encourage			Q: Appreciation		
"Does your company encourage you to volunteer?"			"Does your company appreciate your volunteering (e.g. by feedback-tools)?"		
	No.	%		No.	%
Yes	22	61.1	Yes	21	58.3
No	11	30.6	No	13	36.1
I do not know	3	8.3	I do not know	2	5.6
Q: Compensation			Q: Benefits		
"Does your company provide compensation for volunteering time?"			"Does your company provide any financial benefits for volunteering?"		
	No.	%		No.	%
Yes	3	8.3	Yes	3	8.3
No	32	88.9	No	33	91.7
I do not know	1	2.8	I do not know	0	0.0

n = 36; No. = Number; % = Percentage

7.2 Company Results

In order to analyse possible factors extrinsic motivation out of the volunteer's company, the participants got asked about their companies' relationship regarding their volunteering effort (Table 5). The majority of 61% approved that their company does encourage them to volunteer. Also a majority of 58.3% confirmed that their company does appreciate their volunteering effort. A majority of 88.9% denied the question about possible time compensation and also a majority of 91.7% negate regarding the question of beneficial rewards for volunteering.

7.3 Motivation Scale

In order to measure the types of motivation, the participants were asked "Why have you been engaging as a volunteer in WrOpenUp? ". The response options where

30 different answers combined to 6 blocks in order of the six motivation types. The participants could express their agreement to these answers by a four point Likert-Skale.

The first comparison was between females (F) and males (M) (see Table 6). Intrinsic motivation showed in both samples a strong tendency (3.64 F/3.66 M). Clearly lower was the agreement to the block of external regulation (1.31 F/1.53 M). The answers about introjected regulation showed just a bit higher tendency (1.69 F/ 1.88 M). Again, a high score in both samples could be found in the field of identified regulation (3.59 F/3.66 M). Also, strong consent was visible in questions about integrated regulation with (3.59 F/3.73 M). The answers about amotivation showed a very low tendency (0.42 F/0.30 M). In summary, the results do not indicate any significant difference in motivation between genders.

The second comparison was regarding nationalities and it was simplified by comparing Polish participants (PL) with internationals Expats (EX) (See Table 7.). Both groups showed a strong tendency of intrinsic motivation (3.72 PL/3.56 EX). In both samples, the answers regarding external regulation scored significantly low (1.56 PL/1.21 EX). The introjected regulation scored here just a bit higher (1.78 PL/1.76 EX). The answers for identified regulation in the showed again higher tendency (3.61 PL/3.64 EX). Also, a high score was shown in the field of integrated regulation (3.67 PL/3.63 EX). The answers about amotivation showed the lowest agreement (0.33 PL/0.41 EX). In summary, the results do not indicate any significant

Table 6 Gender results

Gender				
	Female (*m*)	Male (*m*)	Difference (*m*)	Total (*m*)
1. Intrinsic motivation	3.64	3.66	0.02	3.65
2. External regulation	1.31	1.53	0.22	1.41
3. Introjected regulation	1.69	1.88	0.19	1.77
4. Identified regulation	3.59	3.66	0.07	3.62
5. Integrated regulation	3.59	3.73	0.14	3.65
6. Amotivaiton	0.42	0.30	0.12	0.37

m = mean

Table 7 National results

Nationality				
	Polish (*m*)	Expat (*m*)	Difference (*m*)	Total (*m*)
1. Intrinsic motivation	3.72	3.56	0.16	3.65
2. External regulation	1.56	1.21	0.35	1.41
3. Introjected regulation	1.78	1.76	0.02	1.77
4. Identified regulation	3.61	3.64	0.03	3.62
5. Integrated regulation	3.67	3.63	0.04	3.65
6. Amotivaiton	0.33	0.41	0.08	0.37

m = mean

Table 8 Gender results

Beliefs				
	Religion (*m*)	Secular (*m*)	Difference (*m*)	Total (*m*)
1. Intrinsic motivation	3.64	3.67	0.03	3.65
2. External regulation	1.29	1.67	0.38	1.41
3. Introjected regulation	1.74	1.85	0.12	1.77
4. Identified regulation	3.61	3.65	0.05	3.62
5. Integrated regulation	3.66	3.64	0.02	3.65
6. Amotivaiton	0.40	0.29	0.11	0.37

m = mean

difference in motivation between nationalities or at least between Polish and international volunteers.

The third comparison (See Table 8) was between the group of religious people (RL) and non-religious people (Secular: SL). Also in this point both samples showed a high tendency to intrinsic motivation (3.64 RL/3.67 SL). The external regulation had a lower tendency (1.29 RL/1.69 SL). Also, the group of questions about introjected regulation scored relatively low (1.74 RL/1.85 SL). With a higher tendency again scored the field of identified regulation (3.61 RL/3.64 SL). In addition, the field of integrated regulation had a very high acceptance (3.66 RL/3.64 SL). Likewise, in this comparison the answers about amotivation showed lowest acceptance (0.40 RL/0.29 SL). In summary, the results do not indicate any significant difference in motivation between religious and non-religious people.

8 Conclusion

8.1 Interpretation

The results showed a higher participation of women compared to men in volunteering, what corresponds to several other studies about volunteering. Also, the findings regarding the income, the age and the educational level of the participants corresponds to surveys about volunteering in Poland. The majority of volunteers in this case study were well educated young and middle-aged people of a relatively high income. Most of the volunteers were religious people and among them the Christians made up the majority, which may be caused by the fact of a Polish majority among the volunteers. Results indicate the importance of an encouraging atmosphere and culture of appreciation of volunteering within the involved companies of CSR Projects as internal factors of motivation. On the other hand, the survey indicates a less impact incentives like possible financial benefits or time compensation for volunteering time. The results of the survey indicate a dominance of intrinsic motivation to volunteer which leads to the assumption that the volunteers in the case studies are more driven by factors of values, understanding, social and enhancement. But also, the indication of the importance of identified motivation as well as the integrated regulation can be identified as major factors of motivation. This assumes a high identification as well as a strong integration of the projects' values and goals by the participating volunteers. As a last indication the results show a low influence of external regulation as well as of introjected regulation on motivation to volunteer. This allows the assumption that questions of career and a protective attitude plays a minor role. The postulated significant difference in motivation between Polish and international volunteers could not be found. Likewise, a substantial difference in motivation between genders as well as between religious and non-religious people could not be identified.

8.2 Recommendations

Some general conclusions can be drawn from the results in order the question, how to motivate in general employees for CSR Projects. Due the fact intrinsic motivation is one of the main factors for volunteering, companies should first focus on these very personal values in recruitment of volunteers. It is important to point out the personal chances such as (1) making a positive impact (2) new experience and knowledge; (3) socializing with new people; (4) personal development and growth; (5) fun and inner joy. These are the main triggers that will activate the already existing inner self-desired discovery. Of course, the external factors of identified regulation and integrated regulation must also be considered in communication with employees in order to recruit volunteers. Therefore, it is important to communicate clearly the values and the goals of the CSR Project. This will be the secondary trigger because those employees who will identify with the values and goals will be motivated to join. As soon as a volunteer is an active participant in the project, his motivation of identified regulation can develop external factor: the integrated regulation. But all this requires an encouraging atmosphere and culture of appreciation within the companies. This can be done by company internal feedback processes (e.g. digital feedback-tools) or feedback on external social media platforms as a sign of appreciation. As the final conclusion, it is important to point out that the most important value of community volunteering as a part of CSR lays within the employees as potential volunteers itself. It is important to activate their inner motivation in the first place.

References

Alsund, A. (2013). *How capitalism was built: The transformation of Central and Eastern Europe, Russia, the caucasus, and Central Asia* (2nd ed.). New York: Cambridge University Press.

Aluchna, M. (2015b). Employee volunteering as an element of corporate social responsibility: The evidence from Polish listed companies. In S. O. Idowu et al. (Eds.), *Corporate social responsibility and governance, CSR, sustainability, ethics and governance*. Cham: Springer. Accessed August 20, 2017. https://doi.org/10.1007/978-3-319-10909-1_6.

Aluchna, M. (2015a). Corporate social responsibility in Poland: From the perspective of listed companies. In S. O. Idowu et al. (Eds.), *Corporate social responsibility in Europe CSR, sustainability, ethics and governance* (pp. 271–289). Cham: Springer.

Aluchna, M. (2017). Chapter 3 form ideas to impact: CSR in the eyes of top managers in Poland. In S. O. Idowu et al. (Eds.), *Stages of corporate social responsibility in Europe CSR, sustainability, ethics and governance* (pp. 47–63). Cham: Springer.

Aretz, H.J., & Hansen, K. (2003). Erfolgreiches Management von Diversity. Die multikulturelle Organisation als Strategie zur Verbesserung einer nachhaltigen Wettbewerbsfähigkeit. *Zeitschrift für Personalforschung / German Journal of Research in Human Resource Management,* 17. Jahrg., H. 1, 9–36.

Barkay, T. (2012). Employee volunteering: Soul, body and CSR. *Social Responsibility Journal, 8* (1), 48–62.

Beswick, D. (2017). *Cognitive motivation: From curiosity to identity, purpose and meaning*. Cambridge: Cambridge University Press.

Bhatta, T.R., Buruiana, S.C., Hasan, M. & Løwe, J.M. (2015). *Managing cultural diversity in NGOs*. Accessed September 10, 2017, from core.ac.uk/download/pdf/43031159.pdf.

Bsoul, M. (2014). Wolontariat pracowniczy jako przejaw działań z zakresu społecznie odpowiedzialnego biznesu (CSR), Cracow university of economics. *Rynek-Społeczeństwo-Kultura, 3*(11), 10–15. bibliogr. 10 poz.

CBS. (2017). Accessed August 30, 2017, from www.cbos.pl

Centrum Wolontariatu (2008). *I. Ogólnopolskie badania wolontariatu pracowniczego - raport*. Warszawa: Centrum Wolontariatu.

Centrum Wolontariatu. (2013). *II. Ogólnopolskie badania wolontariatu pracowniczego - Wolontariat pracowniczy w największych firmach w Polsce*. Warszawa: Stowarzyszenie Centrum Wolontariatu w Warszawie.

Chandler, D. (2017). *Strategic corporate social responsibility: Sustainable value creation*. London: Sage.

Charycka, B. (2016). *Wolontariat. Mocny w teorii, słaby w praktyce*. Accessed September 10, 2017, from http://fakty.ngo.pl/wiadomosc/2005671.html

Cieślik, A. & Hagemejer, J. (2014). Multinational enterprises, absorptive capacity and export spillovers: Evidence from Polish firm-level data. *Review of Development Economics, 18*(4), 709–726. 18p.

Clary, E.G., Snyder, M., Ridge, R.D., Copeland, J., Stukas, A. A., Haugen, J. & Miene, P. (1998). Understanding and assessing the motivations of volunteers: A functional approach. *Journal of Personality and Social Psychology, 74*, 1516–1530.

Cox, T.J. & Smolinski, C. (1994). *Managing diversity and glass ceiling initiatives as national economic imperatives*.

CSRInfo. (2011). *Wolontariat pracowniczy w Polsce. Przewodnik dla biznesu* [Employee volunteering in Poland. The guidance for business]. Warszawa: CSRInfo.

De Gilder, D., Schuyt, TNM & Breedijk, M. (2005). Effects of an employee volunteering program on the work force: The ABN-AMRO case. *Journal of Business Ethics, 61*(2), 143–152.

Deci, E.L. (1971). Effects of externally mediated rewards on intrinsic motivation. *Journal of Personality and Social Psychology, 18*, 105–115.

Deci, E.L. & Ryan, R.M. (2017). *Self-determination theory: Basic psychological needs in motivation*. London: The Guildford Press.

Deci, E.L. & Ryan, R.M. (1985). *Intrinsic motivation and selfdetermination in human behavior*. New York: Plenum.

Deci, E.L., & Ryan, R.M. (2002). Self-determination theory and the facilitation of intrinsic motivation, social development, and well-being. *American Psychologist, 55*, 68–78. https://doi.org/10.1037/0003-066x.55.1.68.

Dweck, C.S. & Elliot, A.J. (2017). *Handbook of competence and motivation*. Theory and application (2nd ed.). London: The Guildford Press.
Dyczkowska, J., Krasodomska, J. & Michalak, J. (2016). CSR in Poland: Institutional context, legal framework and voluntary initiatives. *Accounting and Management Information Systems, 15*(2), 206–254. 49p.
ESS. (2017). Accessed September 10, 2017, from www.europeansocialsurvey.org
EVS. (2017). Accessed September 10, 2017, from http://www.europeanvaluesstudy.eu
Eweje, G. & Bathurst, R. (2017). *CSR, sustainability, and leadership*. New York: Routledge.
Finkelstein, M. A. (2009). Intrinsic vs. extrinsic motivational orientations and the volunteer process. *Personality and Individual Differences, 46*(5–6), 653–658.
Floyd, D. (2016). Local sourcing practices of foreign owned multinational subsidiaries operatin in Poland. In Millar C.C J.M. and Choi G.C (Eds.), *International business: Emerging issues and emerging markets*. London: Palgrave Macmillan.
Gazeta Wyborcza. (2017). *Gala tegorocznego WrOpenUp w Kinie Nowe Horyzonty. Zwycięzcy już są znani*. Wrocław: Gazeta Wyborcza Wroclaw. 10 czerwca.
Geiser, G., Okun, M. A. & Grano, C. (2014). Who is motivated to volunteer? A latent profile analysis linking volunteer motivation to frequency of volunteering. *Psychological Test and Assessment Modeling, 56*(1), 3–24.
GHK (2010) Volunteering in the European Union. Brussels: GHK.
Gidron, B. (1978). Volunteer work and its rewards. *Volunteer Administration, 11*, 18–32.
Gidron, B. (1984). Predictors of retention and turnover among service volunteer workers. *Journal of Social Service Research, 8*, 1–16.
Grano, C., Lucidi, F., Zelli, A., & Violani, C. (2008). Motives and determinants of volunteering in older adults: An integrated model. *International Journal of Aging and Human Development, 67*, 305–326.
Greely, A. (1997). The other civic American: Religion and social capital. *The American Prospect, 32*, 68–73.
Gruszecka-Tieśluk, A. (2015). What's what . . .CSR in research in 2015. In *Responsible business in Poland 2015. Good practices*. Warsaw: Forum Odpowiedzialnego Biznesu.
Grzybek, M. (2014). CSR research in 2014. In *Responsible business in Poland 2014. Good practices*. Warsaw: Forum Odpowiedzialnego Biznesu.
Guay, F., Vallerand, R. J. & Blanchard, C. (2000). On the assessment of Situational Intrinsic and Extrinsic Motivation Scale (SIMS). *Motivation and Emotion, 24*(3), 175–213.
Hackett, C. (2016). *Development in an Era of capital control: Embedding corporate social responsibility within a transnational regulatory framework*. Palgrave socio-legal studies. Basingstoke: Palgrave Macmillan.
Hansen, K. (2017). CSR und diversity management. In *Management-Reihe corporate social responsibility*. Heidelberg: Springer-Verlag GmbH.
Hawkins, D. E. (2006). *Balancing tomorrow's sustainability and today's profitability*. London: Palgrave Macmillan.
Herzig, C. (2004). *Corporate volunteering in Germany: Survey and empirical evidence*. Lüneburg: Centre for Sustainability Management, Universität Lüneburg.
Hiß, S. (2006). *Warum übernehmen Unternehmen gesellschaftliche Verantwortung -Ein soziologischer Erklärungsversuch*. Frankfurt am Main: Campus.
Hiß, S. (2009). From implicit to explicit corporate social responsibility – Institutional change as a fight for Myths. *Business Ethics Quarterly, Special Issue on "The Changing Role of Business in a Global Society: New Challenges and Responsibilities", 19*(3), 433–451.
ISO. (2010, November). *Die DIN ISO 26000 "Leitfaden zur gesellschaftlichen Verantwortung von Organisationen"*. Bonn: Bundesministerium für Arbeit und Soziales.
Jastrzębska, E. (2016). *Zarządzanie różnorodnością jako element CSR - dobre praktyki i korzyści* [The diversity management as an element of CSR – Good practices and benefits]. Kraków: Cracow University of Economics.

Kirov, N. & Delteil, V. (2013). *Labour and social transformation in Central and Eastern Europe: Europeanization and beyond*. Cambridge: Cambridge University Press.
Adamiak, P., Charycka, B. & Gumkowska, M. (2015). *POLSKIE ORGANIZACJE POZARZĄDOWE 2015*. Warsaw: Stowarzyszenia Klon/Jawor.
Jacobson, K. and Korołczuk, E. (2017). *Civil society revisited: Lessons from Poland*. New York: Berghahn.
Kosewska, B. (2015). Voluntary activity of Polish people and its motives in recent years: New volunteering, De Gruyter open. *Practice and Theory in Systems of Education, 10*(4), 357–362.
Kuraszko, I. (2007). *Społeczna odpowiedzialność biznesu w Polsce*. Warszawa: Wstępna analiza in Odpowiedzialny biznes w Polsce 2007 [Responsible business in Poland 2007].
Kuitto, K. (2016). *Post-communist welfare states in European context patterns of welfare policies in Central and Eastern Europe*. Cheltenham: Edward Elgar.
Lakatos, Z. (2017). *Corporate social performance in emerging markets: Sustainable leadership in an interdependent world*. New York: Gower Publishing.
Macuda, M., Matuszak, Ł. & Różańska, E. (2015). The concept of CSR in accounting theory and practice in Poland: An empirical study. *Zeszyty Teoretyczne Rachunkowości, 84*(140), 115–138.
McPherson, S. (2018, January 12). *8 Corporate Social Responsibility (CSR) trends to look for in 2018*. Forbes.
Michels, S. (2016). *Employee Volunteering als Win-Win-Konstellation - Ergebnisse zweier quasi-experimenteller Studien*. Berlin: Springer Fachmedien Wiesbaden.
Ministerstwo Gospodarki RP [Ministry of Economy of the Republic of Poland]. (2010). *CSR Społeczna Odpowiedzialnosść Biznesu w Polsce* [CSR Corporate Social Responsibility in Poland]. National Reform Program Europe 2020.
Papadakis, K. (2004). *Understanding volunteers' motivations*. Recreation and Leisure Studies Master's Theses.
Paper 1 Oxford Learner's Dictionary. (2017): *Definition of diversity*. Accessed August 8, 2017, from http://www.oxfordlearnersdictionaries.com/definition/english/diversity?q=Diversity
Pedersen, E. R. G. (2015). *Corporate social responsibility*. London: Sage.
PeopleFluent. (2015). *Measuring diversity for success – How and why organizations use strategic diversity measurement*. A PeopleFluent-White Paper.
Peterson, D. (2004a). Recruitment strategies for encouraging participation in corporate volunteer programs. *Business Ethics, 49*(4), 371–386.
Peterson, D. (2004b). Benefits of participation in corporate volunteer programs: Employees' perceptions. *Personnel Review, 33*(6), 615–627.
Phares, E.J. & Chaplin, F.W. (1997). *Introduction to personality* (4th ed.). London: Pearson.
Polityka Insight. (2017). *Research: The impact of foreign direct investment – Contribution to the Polish economy in the past quarter century*. Accessed June 22, 2017, from https://www.spcc.pl/images/file/events_projects/2017/inwestycje_zagraniczne_EN_01_03.pdf
Poór, J., Karoliny, Z., Dobrai, K., Slavic, A., Kerekes, K., Farkas, F. and Engle, A.D. (2014) Factors influencing human resource management solutions at subsidiaries of multinational companies in Central and Eastern Europe. *Journal of East-West Business, 20* (2), 93–119. 27p. / April–June
Popowska, M. (2015). Corporate social responsibility in Poland. From theory to practice. In S. O. Idowu et al. (Eds.), *Corporate social responsibility in Europe*. New York: United Sustainable Diversity.
Popowska, M. (2016). Shaping new generations of managers and consumers: CSR implementation and higher education system in Poland. In D. Turker et al. (Eds.), *Social responsibility education across Europe*. CSR, sustainability, ethics and governance. Cham: Springer.
Potocki, T. (2015). Coroporate social responsibility in Poland: From theory to practice. In S. O. Idowu et al. (Eds.), *Corporate social responsibility in Europe*. CSR, sustainability, ethics and governance. Cham: Springer.
Przewłocka, J. (2011). *Zaangaz'owanie społeczne Polak_ow w roki 2010: Wolontariat, Filantropia, 1% Raport z badan*. Warszawa: Stowarzyszenie Klon/Jawor.

Przytuła, S. (2014). WPŁYW CZYNNIKÓW KULTUROWYCH NA PRAKTYKĘ ZARZĄDZANIA EKSPATAMI W POLSKICH FILIACH - WYNIKI BADAŃ EMPIRYCZNYCH (Vol. 349, pp. 337–350). Wrocław: Prace Naukowe Uniwersytetu Ekonomicznego we Wroclawiu. 14p.

Quindt, K. (2013). *Diversity Management im Kontext von Business Case und Corporate Social Responsibility. Unveröffentlichter Forschungsbericht an der Business Information and Technology School GmbH (BiTS) Iserlohn*. Iserlohn: Masterstudiengang Business Psychology.

Rajak, D. & Dolan, C. (2016). *The anthropology of corporate social responsibility*. New York: Bergham.

Ranking. (2017). XI Ranking odpowiedzialnych firm 2017 in DODATEK DO D ZIENNIKA GAZETY PR AWNEJ NR 91, piątek–niedziela, 12–14 maja 2017.

Rasche, A., Morsing, M., & Moon, J. (2017). *Corporate social responsibility: Strategy, communication, governance*. Cambridge: Cambridge University Press.

Rozkwitalska, M. (2014). Negative and positive aspects of cross-cultural interactions: A case on multinational subsidiaries in Poland. *Engineering Economics, 25*(2), 231–240.

Schulz, A. (2009). *Strategisches Diversitätsmanagement. Unternehmensführung im Zeitalter der kulturellen Vielfalt*. Wiesbaden: Gabler GWV Facherlage GmbH.

Shah, J.Y. & Gardner, W.L. (2008). *Handbook of motivation science*. New York: Guilford Press.

Stangel-Meseke, M. (2016). Diversity management – Fluch oder Segen für Unternehmen. Ein kritischer Diskurs. In Genkova, P. and Ringeisen, P. (Eds.), *Handbuch Diversity Kompetenz*. Wiesbaden: Springer Reference Psychologie. https://doi.org/10.1007/978-3-658-08594-0_10.

SOC Washington. (2017). https://soc.washington.edu/diversity

Szymańska-Palaczyk, A. (2016). Wolontariat w Polsce. Przykład projektu Polska Cyfrowa Równych Szans [Volunteering in Poland: The digital Poland of equal opportunities project (PCRS)]. *Poznan University of Economics, Ruch Prawniczy, Ekonomiczny i Socjologiczny, 1*, 255–271.

Tarasewicz-Gryt, O., & Długopolska-Mikonowicz, A. (2017). Model współpracy międzysektorowej w projekcie społecznej odpowiedzialności. Studium przypadku WrOpenUp. *Marketing i Rynek, 11*, 53–63.

Tarasiewicz (2017) WrOpenUp case study. http://akademiaodpowiedzialnosci.pl/baza-wiedzy?id=23andt=WrOpenUp_Case_Study__Aneta_Dlugopolska_Mikonowicz_i_Oliwia_Tarasewicz_Gryt. Accessed 06 August 2017.

The Chancellery of the Prime Minister. (2003). Law on public benefits activity and volunteerism from 23 April 2003.

The Chancellery of the Prime Minister. (2010). Act of law of April 24th 2003 on public benefit and volunteer work – Consolidated text after last amendment on January 22nd 2010.

Thomas, T.A. (2017). Diversity as a strategy. In Reiche, S.B., Stahl G.K, Mendenhall, M.E. and Oddou, R.G. (Eds.), *Readings and cases in international human resource management*. New York: Routledge.

Tirana, M. (2017). *Managing diversity in organizations: A global perspective*. New York: Routledge.

Touré-Tillery, M. and Fishbach, A. (2014). How to measure motivation: A guide for the experimental social psychologist. *Social and Personality Psychology Compass, 8*(7), 328–341. https://doi.org/10.1111/spc3.12110.

Tschirhart, M. & St.Claire, L. (2008). Fine lines: Design and implementation challenges in employee volunteer programms. In M. Liao-Troth (Ed.), *Challangesin volunteer management. A volume in research in public management* (pp. 205–225). Charlotte, NC: Information Age Publishing.

Tschirhart, M. (2005). Employee volunteers. In J. L. Brudney (Ed.), *Emerging areas of volunteering* (pp. 13–29). Indianapolis, IN: ARNOVA.

UNIDO. (2017). United Nations Industrial Development Organization. Accessed June 22, 2017, from http://www.unido.org/csr/o72054.html

Uyan-Atay, B. (2013). *Corporate community involvement: A visible face of CSR in practice.* Farnham: Gower.

Vedder, G. & Krause, F. (2014). Corporate social responsibility and diversity management – Eine Win-Win-Situation. In K. Hansen (Ed.), *CSR und Diversity Management. Erfolgreiche Vielfalt in Organisationen* (pp. 57–75). Springer Gabler: Berlin.

Waniak-Michałak, H., Macuda, M. and Krasodomska. J. (2016). Corporate social responsibility and accounting in Poland: A literature review. *Accounting and Management Information Systems* [Contabilitate si Informatica de Gestiune] *15*(2), 255–303. 49p.

Lipiec, J. (2016). Warsaw stock exchange. *Social Responsibility Journal, 6*(4), 611–626.

Widjaja, E. (2010). *Motivation behind volunteerism.* CMC senior theses. Paper 4. Accessed August 10, 2017, from http://scholarship.claremont.edu/cmc_theses/4

Wild, C. (1993). *Corporate volunteer programs: Benefits to business.* New York: Conference Board Report.

Wołczek, P., Adam, S. & Szymański, R. (2016). *Strategiczny CSR.* Warszawa: Texter.

Wołczek, P. (2015). Development of the CSR concept in Poland – Progress or stagnation? *Research Papers of the Wroclaw University of Economics* [Prace Naukowe Uniwersytetu Ekonomicznego we Wroclawiu] *387*, 200–214. 15p.

Zein, O. (2015). *Culture and project management: Managing diversity in multicultural projects* (p. 20). London: Routledge.

Social Media for Corporate Social Responsibility Strategy Creation and Communication in Poland

Justyna Szumniak-Samolej

1 Introduction

Companies are elements of an increasingly complex global system. Its organizational structure is becoming a network, whose logic exerts more and more influence on the shape and functioning of modern societies and economies. Communication networks, especially the Internet, are changing many aspects of people's lives. The Internet is being used in an ever-widening scope by representatives of all generations all over the world, with the most particular in that regard being the habits and behaviors of young people who grew up in the digital environment (Generation Y, Generation Z). The development of informational-communicative technologies allows them to create connections and relations which had been previously unavailable. In the world of network connections, users have access to new interactive communication tools, create online content, connect with a social community, talk about everything and undertake joint actions. They are open, active, and willing to cooperate with companies (Tapscott 2009). At the same time, they monitor their activities and, when necessary, are ready to initiate a boycott. Without a doubt, the Internet has made it possible for consumers and other stakeholders to react more quickly and with greater force.

It would seem that these observed changes cannot help but have an influence on corporate social responsibility (CSR). This is all the more so, because this is a concept which has been a controversial topic, due to its frequent use as a public relations tool serving solely to improve a company's image, rather than a real intent to introduce social and ecological change (Filek 2004, p. 55; Hope 2007, p. 162).

J. Szumniak-Samolej (✉)
Department of Management Theory, Warsaw School of Economics, Warsaw, Poland
e-mail: justyna.szumniak@sgh.waw.pl

A. Długopolska-Mikonowicz et al. (eds.), *Corporate Social Responsibility in Poland*, CSR, Sustainability, Ethics & Governance,
https://doi.org/10.1007/978-3-030-00440-8_17

The goal of this chapter is to chart the possibilities that social media offer in terms of creating strategy and communicating corporate social responsibility activities, as well as to analyze the use of these tools in the area of CSR in Poland. Additionally, the author will attempt to point out potential future areas for change in corporate social responsibility actions caused by the further development of the Internet.

These are the research questions which the author wanted to find answers to in this chapter:

- RQ1. What meaning do social media have in the creation and communication of CSR actions?
- RQ2. Do the corporate leaders of CSR in Poland use social media in the area of corporate social responsibility?
- RQ3. Will the further development of the Internet (Web 4.0) have an influence on the concept of corporate social responsibility, and if so, what kind?

The study was conducted in three steps. Firstly, a literature review was conducted to identify the fundamental themes of the research. Secondly, the author identified potential areas of utilizing social media in the creation and communication of CSR strategies as well as its related benefits and risks. Thirdly, the range of utilizing social media in the creation and communication of CSR activities in Poland was analyzed.

2 Technological and Social Changes in the Corporate Realm

Until recently, companies' communication with their environment generally transpired in one direction. Companies could control the flow of information and the degree to which they were disseminated. The dynamic development of ICT technology, in particular Web 2.0 and social media based on this technology, did away with this static order.

Web 2.0 is defined in different ways. However, the common point of these various approaches is that they focus on the most important feature which differentiates Web 2.0 from the previous generation of Internet (Web 1.0): the increasing role of users in the creation of the global network (Mazurek 2008, p. 97). Unlike Web 1.0, where one-way communication prevailed, the Web 2.0 environment allows for the sort of communication, where users' relationships have a wide variety of forms and content. Each user can simultaneously be a sender and recipient of a message. Information can be sent by and addressed to many individual participants who can choose the information they are interested in (Burnett and Marshall 2003, p. 47).

Therefore, Web 2.0 is not only a new technological solution, but also above all a new philosophy of how the Internet works, based on openness, freedom, equality, creativity and cooperation. Because of this, societal changes are an important consequence of this technological development. Many authors argue that the Internet makes possible the emergence of new societal forms and a change in existing types of social interaction (Christakis and Fowler 2011, p. 263; Barney 2008, p. 54; Shirky 2008, p. 106; Filiciak et al. 2010, p. 37; Li and Bernoff 2008, p. 9).

Modern communication technologies support immediate, interdependent and non-hierarchical group forms of activity, consumption and creativity. A significant amount of users' practices depends on bypassing traditional hierarchies and methods of formalized activities. As a result, what we are dealing with is a peculiar de-institutionalization where the role of the individual and groups has been strengthened, while formal institutions have been weakened (Filiciak et al. 2010, pp. 148–149). As a result, we now speak of such social phenomena as commons-based peer production (Benkler 2008, pp. 75–143), peer production (Tapscott and Williams 2008, pp. 106–111) and digital activism (Brodock et al. 2009).

3 The Meaning of Social Media for Corporate Social Responsibility

The concept of corporate social responsibility has had numerous definitions, yet there is no agreement amongst theoreticians and practitioners as to which of them is the single correct one. Currently, one of the most popular definitions of CSR in Poland is the one proposed in the standard ISO 26000 from 2010 (ISO 2017): "The responsibility of an organization for the impacts of its decision and activities on society and the environment, through transparency and ethical behavior that: contributes to sustainable development, including health and welfare of society takes into account the expectations of stakeholders; is in compliance with applicable law and consistent with international norms of behavior; is integrated throughout the organization and practiced in its relationships."

In spite of their noble assumptions, CSR activities carried out by corporations often come under criticism (Szumniak-Samolej 2013, pp. 45–53; Szumniak-Samolej 2015, p. 611). We can point out two essential problems with implementation of this concept in practice:

1. Instrumental use of "CSR programs" as smokescreens for unethical behaviors (Morsing and Vallentin 2006, p. 248; Filek 2004, p. 55; Hope 2007, p. 162).
2. Taking action in the area of CSR in good faith though incompetently (on-off and random actions in the field of social involvement, lack of CSR strategy and integrating it with the business strategy of the company, not realizing the real business challenges, lack of cooperation with key stakeholders) (Bachnik 2011; Coombs and Holladay 2015, p. 86; Elkington and Zeitz 2014, p. 38).

Visser (2011) also claims that the current model of corporate social responsibility is outdated. It is not only ineffective but even destructive for economies, communities and the environment.

It would seem that social phenomena observed on the Internet, like the aforementioned peer production or digital activism, should change the current approach of companies to corporate social responsibility in both of the above areas.

In theory, social media force organizations to be transparent. Stakeholders are connected, so it is much easier for them to verify information coming from corporations by reaching out to alternative sources and their partner networks. They also have new ways of exerting pressure on organizations (Coombs and Holladay 2015, p. 87; Płoszajski 2012, p. 15). Transparency is also one of the key elements in CSR activities. The more transparent CSR activities are, the more credible they are to stakeholders (European Commission 2009, p. 32). In addition, dishonest companies will not open up to a real dialogue, which will be quickly noticed in social media because dialogue is essential to them. Transparency and its associated dialogue are therefore necessary to convince stakeholders of their position, gain their trust, and lay the groundwork for cooperation (Tapscott and Ticoll 2003; Zumer Sustainability Strategy Consulting 2011, p. 3).

Besides transparency, another important dimension of CSR is involving stakeholders in creating and implementing a social engagement strategy (Morsing and Schultz 2006; Tapscott and Ticoll 2003; Zadek et al. 2003; Laszlo 2008). So far, however, there have been few examples of real dialogue, cooperation, partnership and shared development (Morsing and Schultz 2006). Possible reasons for this state of affairs include companies not perceiving stakeholders as partners, lack of access to knowledge about CSR, and lack of an appropriate technological infrastructure, tools, or collaboration platforms (Szumniak-Samolej 2013, p. 242). Theoretically, social media have potential to bring about change in these areas.

Theoretical analysis suggests that the development of information and communications technology (ICT) as well as, changes in organizations' environment, force the transformation of existing, not always the best CSR practices and create conditions for the existence of a new quality of corporate social responsibility, in which organizations take advantage of ICT tools as well as new cooperation models in the process of creation and implementation of CSR activities (Visser 2008; Peborgh and Odiseo Team 2008; Hammer et al. 2009; Bitan 2008; Weber 2009, p. 18; Tapscott and Ticoll 2003; Pohle and Hittner 2008).

The author of this chapter has identified two potential areas for change, due to the development of social media in the area of managing corporate social responsibility. The first is external to the organization and involves stakeholders' potential control of companies' activities and the social pressure exerted through the new media. On the other hand, the second one is about the abilities of companies to use social media and new cooperation models in the process of creating and realizing CSR strategies. The outlined areas for change are presented in Table 1.

It would seem that social media paradoxically will enable us to return to theoretical sources of CSR notions, i.e. ethical decision-making and business operations, taking into account the impact the company has on people and the environment, and dialogue and cooperation with stakeholders.

The benefits of such a change in approach to corporate social responsibility can, as it seems, impact both companies as well as their stakeholders and, on a wider

Table 1 Corporate social responsibility in the age of social media

1. Control/social pressure through new media:
• Verification of ethicality of business activities • Enforcement of transparency of CSR initiatives
2. Use of Web 2.0 tools and new cooperation models by organizations in the process of creating and implementing CSR strategies:
• Communicating CSR activities: – Reporting in real time – Interaction, dialogue with stakeholders • Creating a community • Corporate cooperation with stakeholders – Activating stakeholders in CSR activities – Creating shared strategy for corporate social responsibility – Developing CSR innovations (processes, products, services, business models)

Source: Szumniak-Samolej 2013, p. 244

scale, entire societies and the environment. Some of the benefits for stakeholders and societies and/or the environment (Szumniak-Samolej 2013, pp. 249–251) may include (i) impediment for companies to lie and hide uncomfortable facts in social media; (ii) responding to the real needs of the market and society (thanks to collaborative methods such as crowdsourcing or cooperation, one can look for social and environmental issues that really need urgent attention and create projects that will solve them more effectively And their form will be more attractive to the participants); (iii) an increase in the number, variety and innovation of CSR initiatives.

Whereas, some of the benefits for companies may include (Szumniak-Samolej 2013, pp. 249–251): (i) faster and easier dialogue with stakeholders; (ii) greater transparency and credibility of CSR activities; (iii) utilizing the activity and creativity of (mainly young) people on the Internet and their willingness to engage in innovative CSR solutions; (iv) using the expert knowledge of external stakeholders (NGOs and international organizations, research centers, universities, independent experts, other companies) to create innovative solutions; (v) better social perception of CSR activities created together with stakeholders; (vi) receiving feedback from stakeholders; (vii) avoiding crises—creating an early warning system providing information on abnormalities. Businesses that react quickly and are open to cooperation can transform emerging distressing signals or pressures from the market into opportunities to create common solutions that benefit all parties; (viii) promoting the company as well as the project using the information diffusion mechanism on the Internet (e.g. viral marketing); (ix) lower costs of creating and promoting CSR projects.

Of course, the use of social media in creating and communicating CSR strategies is also linked to potential threats. However, it seems that the negative aspects in this area are above all threats to companies and their existing practices (e.g. unwanted effects of sharing information or the practical aspects of engaging with stakeholders), not to stakeholders or society.

Among other things, managers will have the dilemma of how much information to share, which parts of it are necessary, and which ones should be kept confidential. Additionally, the increasing demand for information from stakeholders may exceed the company's ability to deliver it (Pohle and Hittner 2008, p. 9). The practical challenges of engaging with stakeholders in social media can also be the following: How to attract, interest and engage stakeholders to work together? How to be sure that the solutions they propose are really the best? What tools and collaboration platforms should be used?

Every company must answer these questions on their own, because the scope of stakeholder engagement and the tools used are partially determined by such factors as the scale of the company's impact on the environment, the industry and specificity of the products and services offered, stakeholders' characteristics and the current level of trust or lack of trust in the company, product or industry (Epstein 2008, p. 178).

4 The Use of Social Media for CSR Activities in Poland

Below you will find examples of the use of social media by companies and organizations operating in Poland in the areas described in the previous section, i.e. control and social pressure by way of new media and companies using social media and new collaborative models in the creation and implementation of CSR strategies (communication, community building, stakeholder engagement).

4.1 Control and Social Pressure by Way of New Media

In this section exemplary social media initiatives, whose goal it was to enable control and emphasize social community, were selected and analyzed by the author.

A good example of many years of active control and social pressure on businesses is "Buy Responsibly!", a portal operated by the Buy Responsibly Foundation, which provides information about companies and products and encourages responsible consumption (Buy Responsibly 2017). Although the site is not built on Web 2.0 tools, the project has profiles on many social network sites, including the extensive "Buy Responsibly" Facebook page (59,528 likes, 57,941 followers) (Buy Responsibly Facebook Page 2017). Regular posts and calls for civic action are usually met with an exuberant response in the form of comments, shares and declarations of participation.

Another interesting project is Mintu.me which is aimed at "people who care about the world around them" and giving them "the opportunity to work together" (Mintu.me 2017). In practice, the site allows users to organize crowd funding actions in the area of ecology and sustainable development, buy and sell ecological and handmade objects, and co-author a blog on ecology and social issues. Although the site has been in operation for several years, it is not developing too dynamically. The shopping section is the most active. The portal also has its own Facebook page, which regularly presents interesting news on sustainable development and sharing economy from Poland and the rest of the world. There is not much discussion, but there is a group of people who click "like" under posts. The site has 3173 likes and 3030 followers (Mintu.me Facebook Page 2017).

The Facebook page "Black and White Shopping List", on which the authors recommend which products to buy or avoid because of their ingredients, is well-maintained. It is liked by over 130,000 people, which is also the number of observers. Comments and discussions usually appear under posts (Black and White Shopping List Facebook Page 2017).

Interestingly, many of the projects exposing irresponsible corporate actions began operating in 2011/2012 but quickly ceased to exist. One example of this is "Irresponsible Business", a website designed to boost stakeholder awareness and provide reliable information and data on irresponsible business practices and activities. For this purpose, a database of bad practices was created on the portal with information contributed by its authors as well as Internet users. On its Facebook page, you will find posts published between February and April 2012 (Irresponsible Business Facebook Page 2017). The website also does not work.

Another example is the "Hyper-Exploitation" portal, which enables social monitoring of trade. Its purpose was to create a map of the cases of employee exploitation in hypermarkets in Poland. Both employees and customers who experienced or witnessed a marketer's mistreatment could report a problem on the portal and propose a solution to the problem. The site was created in July 2011. It is difficult to find the precise date of its completion, but it probably lasted several months. The last post in the "Hyper-Exploitation" discussion group on the social networking site nk.pl appeared in February 2012 (Hyper-Exploitation Nk.pl Group 2017).

4.2 *The Use of Web 2.0 Tools and New Collaborative Models by Businesses in the Process of Creating and Implementing CSR Strategies*

To see how companies operating in Poland use social media to communicate and create CSR programs, the author traced the actions of three companies that have been recognized as CSR leaders in Poland by the "2017 Ranking of Responsible Companies". The Ranking of Responsible Companies is a long-running recognition of the largest companies operating in Poland, rated for their quality of management of corporate social responsibility (Ranking of Responsible Companies 2017a).

According to the "2017 Ranking of Responsible Companies", the three most responsible companies in Poland are Coca-Cola HBC Poland, Schenker Sp. z o.o. (DB Schenker), and Polpharma; the first two scored 100% of all possible points while the third got 98%. Runners-up include Orange Polska, Grupa Lotos, Kompania

Piwowarska, Bank Zachodni WBK, DANONE Group and Tesco Polska (Ranking of Responsible Companies 2017b).

4.2.1 Coca-Cola HBC Polska

The latest available social report from Coca-Cola HBC Poland is the 2015 Corporate Social Responsibility Report (Coca-Cola HBC Poland CSR Report 2017). It is very sketchy and does not include any information about the company's use of social media in CSR, except for the information buried in the volunteer video "We Create Sustainable Future Right Now" page on Facebook. As of August 24, 2017, this page has 45,842 likes and 45,626 followers (Sustainable Future Facebook Page 2017). According to the site's description, "it brings together people who are passionate about ecology, physical activity, healthy lifestyles, helping others, and pro-social activities. If you believe that changes in the world start with us and the best way to predict the future is to create it, then we encourage you to jointly create a sustainable future right now. Share what you do and think. Find inspiration, exchange experiences, and see how many people share your values." There are general posts about the environment and sustainability, motivational posts, and posts promoting an active lifestyle, environmental behaviors, and social reports. From time to time, there is a brief account of Coca-Cola's CSR activities, such as a video report from one of the employee volunteering actions, a corporate eco-picnic report, or information about a sustainable development project that Coca-Cola HBC has planned for 2020. Despite the authors encouraging page visitors to be more active and contribute to the creation of the site, most of the posts have no comments and very few likes. Among the few initiatives involving stakeholders and instigating their discussion is one in which the company declares to plant one tree for every like under a specific post. In 2017, this action lasted 2 weeks and resulted in 15,000 trees being planted.

In addition, Coca Cola in Poland has a general Facebook page (https://www.facebook.com/cocacolapoland), as well as an Instagram account (https://www.instagram.com/cocacolapolska/), a Twitter profile (https://twitter.com/cocacolapolska) and YouTube channel (https://www.youtube.com/user/cocacola). Its Facebook page contains marketing posts, as does its Instagram account. On its Twitter profile, CSR posts appear every once in awhile (for example, information about donating to help victims of storms). Its YouTube channel is an international channel for The Coca Cola Company, which publishes a variety of videos from around the world.

4.2.2 Schenker Sp. z o.o. (DB Schenker)

The most up-to-date social report on the DB Schenker website is the Responsibility Report for the years 2015–2016 (DB Schenker 2017a). The report was prepared according to GRI standards and verified by an outside company. Furthermore, the company conducts a dialogue with stakeholders based on the AA 1000 standard.

In the context of online tools, it is worth noting that Internet users are allowed to evaluate the report in a short online survey (DB Schenker 2017b).

As far as using social media in its dialogue, it is targeted to a single group: potential employees. Since December 2016, the company runs a "Careers" page on Facebook (DB Schenker Facebook Page 2017). Posts concern the company's life, including CSR initiatives such as Schenker's volunteer projects, information on the company's social involvement, as well as CSR accolades and recognitions. However, there are not many comments.

An interesting solution is the newest technology used by the company to communicate with subcontractors. In 2016, the DB Schenker Group invested 25 million dollars to create Drive4Schenker, which was made in co-operation with the American online transportation platform uShip. Drive4Schenker is an Internet marketplace offering partial and full truck transport. A carrier who logs onto the platform has access to all the transportation offers from DB Schenker branches across Europe. In no time at all, the carrier can choose the routes and loads he or she is interested in and propose a price. The platform offers various search options, e.g. by location or cargo size. With Drive4Schenker, drivers have access to all the data they need on their mobile phones. Such automation of information flow improves quality of service and eliminates so-called 'empty courses' in Europe (DB Schenker 2017c).

EcoTransIT is another technological solution in the spirit of sustainable development: an environmental impact calculator for various modes of transport, which is available on the DB Schenker website (EcoTransIT 2017).

4.2.3 Polpharma

The latest available social report is Polpharma's Social Responsibility Report 2013–2014 (Polpharma 2017). Its complexity and thoroughness make it stand out. The report was prepared according to guidelines of the G4 Core Global Reporting Initiative and was subjected to external verification. However, for the management, measurement and reporting of social involvement, the company used the London Benchmarking Group (LBG) methodology.

The company however has very few CSR activities which utilize social media. Polpharma does not even have pages and channels on the most popular social media sites (except YouTube). This does not mean though that it does not use ICT tools.

The company initiates the creation of IT tools in the form of databases and applications for pharmacists and physicians. They include, for example: (i) the DrWidget Drug Database for desktop computers and mobile applications. This is a comprehensive database of more than 17,000 descriptions of medicinal products, dietary supplements and dermocosmetics with continuous updates (DrWidget 2017). DrWidget also has a Facebook page with 1394 likes and 1338 followers (DrWidget Facebook Page 2017). Posts appear regularly and quite often, but there are virtually no comments from Internet users, except for a few notifications about technical issues or deficiencies in the database; (ii) cardiological calculators—a mobile application that includes subsidiary regimens for patients with cardiovascular disease;

(iii) pulmonary algorithms—an analogous application that includes subsidiary regimens for patients with respiratory diseases.

An interesting initiative is also Polpharma.tv, a video portal with dedicated access for doctors, pharmacists and patients, featuring videos on the medical practice, medications and preventive treatments. Users can rate videos, share them on social media sites and add comments. The service was launched in late 2016. In the section for patients where no login is required, there are currently nine videos without any comments. There is also a live channel on the portal, where the company reports online on important events. During interviews, virtual participants can send questions to the interviewer through a moderated chat window.

Since December 2013, the company also runs a Youtube channel titled "People Helping People" (People Helping People 2017) which features video coverage from employee volunteering initiatives as well as promotional corporate movies, aimed at business partners among others. There are no comments under most of the films.

5 Wrap-up and Conclusions

According to the report *Social Media Sustainability Index* (cyclically published since 2011), which documents the scope of social media use in communicating CSR and sustainability development activities, the number of companies that use these technologies is growing every year (120 companies in 2011, 176 in 2012, 230 in 2013, 273 in 2014, 283 in 2015) (Yeomans 2016, p. 10). The number of social reports shared via social media is also increasing regularly (p. 13). The most popular social channels in 2015 were Twitter (the number one choice amongst companies that wanted to communicate sustainability), Linkedin, Facebook, and YouTube (p. 15). According to this report, the top five companies communicating CSR through social media in 2015 were General Electric Co., Unilever PLC, Patagonia, Sainsbury's, and General Mills (p. 29).

Despite the upward trend, there are still no groundbreaking CSR initiatives in social media. As Kent and Taylor (2016, p. 60) write, "Social media's strength as a relationship-building tool is not realized as CSR activities are often communicated unidirectionally." They describe using social media for CSR as "one-way, monological, *Homo Economicus* based practices".

Similar conclusions can be drawn by looking at above described CSR leaders' involvement in social media in Poland. It is difficult to acknowledge that these companies use the potential of this technology to create and communicate CSR activities. Besides random examples and one-off, not-very-creative actions on social media, the biggest emphasis is on traditional dialogue methods (meetings, surveys), social reporting, working with partner organizations, employee volunteering, and communication through a Web 1.0 web site.

Social media and Web 2.0 tools are the landscape of the modern internet with which most users probably still identify. Of course, the technology behind the development of the Internet is constantly evolving. Increasing global network resources require better tools to deal with the enormous amounts of data. Progress towards solutions that will maximize users' ease of use of information gathered in a global network, based on technologies such as artificial intelligence (AI), will carry the Internet into its next stage of development which is Web 3.0. (Sarowski 2017, p. 37). This is not the final stage, as the next step is Web 4.0. However, for now, the concept of Web 4.0 is irregular and liable to change (Almeida 2017, p. 7044). The key role in the adoption and implementation of Web 4.0 will be probably played by such applications and technologies as social networks, Internet of Things, Big Data, AI and M2M (Almeida 2017, p. 7044).

Some authors (Khoo 2010; Perera et al. 2014; White 2015) equate Web 4.0 with the Internet of Things (IoT) concept meaning the integration of the network with almost all types of devices (Sarowski 2017, p. 37). The revolutionary element here is a new type of communication involving the connecting of people with objects as well as connecting the objects themselves (Kwiatkowska 2014, p. 63). Cisco, as one of the technology developers behind the development of Web 4.0 solutions, speaks about The Internet of Everything and defines it as "the intelligent connection of people, processes, data, and things on the network" (Cisco 2017).

This is where the question arises, what is the significance of Web 4.0 and the further development of ICT for the concept of CSR? Roscoe (2015a) argues that "CR initiatives are hampered by four key challenges: poor traceability, supply chain inefficiencies, environmental sustainability, and impact measurement." The solution to each of these problems may be the IoE (Roscoe 2015b, p. 2). Some say that the treatment of "CSR as a goal, with IoT as the means" creates breakthroughs in productivity and efficiency (Triple Pundit 2015). Some sample applications include (Holler et al. 2014, pp. 21–22):

(i) more effective use of natural resources; use of IoT to increase yield and productivity, and reduce supply chain losses;
(ii) IoT as a technology to help provide assisted living for older or disabled persons and to reduce the cost of healthcare and patient care systems;
(iii) using technology to reduce the impact of human activities on the environment;
(iv) monitoring of product life cycle and supply chain (including production and recycling conditions).

While attempting to point out the potential direction of shared development CSR and IoE as well as promising areas for further research, one can again quote Cisco: "The true impact of the Internet of Everything will be measured not by the number of connections, but by the extent they are harnessed to benefit humanity" (Cisco 2017).

References

Almeida, F. (2017). Concept and dimensions of Web 4.0. *International Journal of Computers and Technology, 16*(7), 7040–7046.

Bachnik, K. (2011). A revolution in CSR (interview with W. Visser) [Rewolucja w CSR (wywiad z W. Visserem)]. *Harvard Business Review Polska. "Responsible Business" Supplement.*

Accessed September 30, 2012, from http://odpowiedzialnybiznes.pl/artykuly/rewolucja-w-csr-wywiad-z-w-visserem/

Barney, D. (2008). *The network society* [Społeczeństwo sieci]. Warsaw: Wydawnictwo Sic!

Benkler, Y. (2008). *The wealth of networks: How social production transforms markets and freedom* [Bogactwo sieci. Jak produkcja społeczna zmienia rynki i wolność.]. Warsaw: Wydawnictwa Akademickie i Profesjonalne.

Bitan, R. (2008). Web 2.0 and CSR – Getting the most out of virtual communications. *CSR@Intel*. Accessed December 5, 2008, from http://blogs.intel.com/csr/2008/11/web_20_and_csr_-_getting_the_m.php

Black and White Shopping List Facebook Page. (2017). Accessed August 30, 2017, from https://www.facebook.com/pg/CzarnaiBialaListaZakupow/about/?ref=page_internal

Brodock, K., Joyce, M., & Zaeck T. (2009, July). *Digital activism survey report 2009*. DigiActive Research Series. Accessed August 25, 2017, from https://akgul.bilkent.edu.tr/DigitalActivisim-SurveyReport2009.pdf

Burnett, R., & Marshall, P. D. (2003). *Web theory: An introduction*. London: Routledge.

Buy Responsibly. (2017). Accessed August 28, 2017, from http://www.ekonsument.pl

Buy Responsibly Facebook Page. (2017). Accessed August 28, 2017, from https://www.facebook.com/kupujodpowiedzialnie/

Christakis, N. A., & Fowler, J. H. (2011). *Connected* [W sieci]. Sopot: Smak Słowa.

Cisco. (2017). Accessed September 1, 2017, from http://csr.cisco.com/pages/internet-of-everything-social-good

Coca-Cola HBC Poland CSR Report. (2017). Accessed August 31, 2017, from http://2015.raportcsr-cchellenic.pl/pl/index

Coombs, W. T., & Holladay, S. J. (2015, March 30). How activists shape CSR: Insights from internet contagion and contingency theories. In *Corporate social responsibility in the digital age* (pp. 85–87). Bingley: Emerald Group.

DB Schenker. (2017a). Accessed August 25, 2017, from http://www.dbschenker-csr.pl

DB Schenker. (2017b). Accessed August 25, 2017, from http://www.dbschenker-csr.pl/ankieta.html

DB Schenker. (2017c). Accessed August 25, 2017, from http://www.dbschenker.pl/log-pl-pl/start/drive4schenker_.html

DB Schenker Facebook Page. (2017). Accessed August 25, 2017, from https://www.facebook.com/DBSchenkerPLKariera/

DrWidget. (2017). Accessed August 28, 2017, from http://drwidget.pl/o-serwisie.html

DrWidget Facebook Page. (2017). Accessed August 28, 2017, from https://pl-pl.facebook.com/DrWidget/

EcoTransIT. (2017). Accessed August 25, 2017, from http://www.ecotransit.org/calculation.en.html

Elkington, J., & Zeitz, J. (2014). *The breakthrough challenge: 10 ways to connect today's profits with tomorrow's bottom line*. San Francisco: Jossey-Bass.

Epstein, M. J. (2008). *Making sustainability work*. Sheffield: Greenleaf Publishing.

European Commission. (2009). *Towards greater corporate responsibility: Conclusions of EU-funded research*. Luxembourg, EUR-OP.

Filek, J. (2004). *An introduction to business ethics* [Wprowadzenie do etyki biznesu]. Cracow: Akademia Ekonomiczna w Krakowie.

Filiciak, M., Danielewicz, M., Halawa, M., Mazurek, P., & Nowotny, A. (2010, January). Youth and media: New media and participation in culture [Młodzi i media. Nowe media a uczestnictwo w kulturze]. In *Report of the center for popular culture research*. Warsaw: SWPS University of Social Sciences and Humanities.

Hammer, A., Sadowski, M., Hester, K. F., Kiewiet de Jonge, E., & Shah, J. (2009, May). *Enhancing stakeholder engagement through Web 2.0*. London: SustainAbility.

Holler, J., Tsiatsis, V., Mulligan, C., Avesand, S., Karnouskos, S., & Boyle, D. (2014). *From machine-to-machine to the internet of things: Introduction to a new age of intelligence*. Amsterdam: Academic Press.

Hope, E. (2007). Responsibility in public relations activities: Ethical principles [Odpowiedzialność w działaniach public relations – zasady etyczne]. In Olędzki, J., & Tworzydło, D. (Eds), *Public relations: Social significance and direction of development* [Public relations. Znaczenie społeczne i kierunki rozwoju]. Warszawa: Wydawnictwo Naukowe PWN.

https://polpharma.tv

https://twitter.com/cocacolapolska

https://www.facebook.com/cocacolapoland

https://www.instagram.com/cocacolapolska/

https://www.youtube.com/user/cocacola

Hyper-Exploitation Nk.pl Group. (2017). Accessed August 30, 2017, from http://nk.pl/grupa/208751/hiperwyzysk?page=1

Irresponsible Business Facebook Page. (2017). Accessed August 30, 2017, from https://pl-pl.facebook.com/Nieodpowiedzialny-Biznes-308567729179574/

ISO. (2017). Accessed August 24, 2017, from http://iso26000.info/iso26000/

Kent, M. L., & Taylor, M. (2016). From homo economicus to homo dialogicus: Rethinking social media use in CSR communication. *Public Relations Review, 42*(1), 60–67.

Khoo, B. (2010). *Rfid- from tracking to the internet of things: A review of developments.* Proceedings of the 2010 EEE/ACM International Conference on Green Computing and Communications, Washington, USA, 533–538.

Kwiatkowska, E. M. (2014). The development of the internet of things: Opportunities and threats [Rozwój Internetu rzeczy – szanse i zagrożenia]. *Antimonopoly and Regulatory Online Quarterly* [Internetowy Kwartalnik Antymonopolowy i Regulacyjny], *8*(3), 60–70. http://ikar.wz.uw.edu.pl/numery/22/pdf/60.pdf.

Laszlo, Ch. (2008). *The sustainable development company* [Firma zrównoważonego rozwoju]. Warsaw: Studio Emka.

Li, C., & Bernoff, J. (2008). *Groundswell: Winning in a world transformed by social technologies.* Boston: Harvard Business Press.

Mazurek, G. (2008). *Blogs and virtual communities – Their use in marketing* [Blogi i wirtualne społeczności – wykorzystanie w marketingu]. Warsaw: Wolters Kluwer Polska.

Mintu.me. (2017). Accessed August 28, 2017, from https://mintu.me

Mintu.me Facebook Page. (2017). Accessed August 30, 2017, from https://pl-pl.facebook.com/www.mintu.me/

Morsing, M., & Schultz, M. (2006, October). Corporate social responsibility communications: Stakeholder information, response and involvement strategies. *Business Ethics: A European Review, 15*(4), 323–338.

Morsing, M., & Vallentin, S. (2006). CSR and stakeholder involvement: The challenge of organizational integration. In A. Kakabadse & M. Morsing (Eds.), *Corporate social responsibility: Reconciling aspiration with application.* Basingstoke: Palgrave Macmillan.

Peborgh, E., & Odiseo Team. (2008). *Sustainability 2.0.* Odiseo: Buenos Aires.

People Helping People. (2017). Accessed August 28, 2017, from https://www.youtube.com/channel/UCw0kZLNYbPU3-osaC0DlYsg/videos

Perera, C., Zaslavsky, A., Christen, P., & Georgakopoulos, D. (2014). IEEE. *Communications Surveys & Tutorials, 16*(1), 414–454.

Płoszajski, P. (2012). A movable goal: The need to redefine the theory and practice of corporate social responsibility in the new economy [Ruchomy cel: o konieczności redefinicji teorii i praktyki społecznej odpowiedzialności przedsiębiorstw w warunkach nowej gospodarki]. *Corporate social responsibility in the new economy* [Społeczna odpowiedzialność biznesu w nowej gospodarce]. Warsaw: Wydawnictwo OpenLinks.

Pohle, G., & Hittner, J. (2008). *Attaining sustainable growth through corporate social responsibility.* New York: IBM Institute for Business Value. IBM Corporation.

Polpharma. (2017). Accessed August 28, 2017, from https://www.polpharma.pl/aktualnosci-firma/raport-spoleczny-grupy-polpharma

Ranking of Responsible Companies. (2017a). Accessed August 31, 2017, from http://rankingodpowiedzialnychfirm.pl/o-nas/

Ranking of Responsible Companies. (2017b). Accessed August 31, 2017, from http://rankingodpowiedzialnychfirm.pl/2017/05/xi-ranking-odpowiedzialnych-firm-wyniki/

Roscoe, A. (2015a, April 10). *How the internet of everything can jumpstart CSR*. Accessed September 1, 2017, from https://alexaroscoe.com/2015/04/10/how-the-internet-of-everything-can-jumpstart-corporate-responsibility/

Roscoe, A. (2015b). *Corporate responsibility and the internet of everything*. Submission to the 2015 Robert Davies Commemorative Essay Award. Accessed September 1, 2017, from https://www.sbs.ox.ac.uk/sites/default/files/Skoll_Centre/Images/internet-of-everything-alexa-roscoe-winning-essay.pdf

Sarowski, Ł. (2017). From Web 1.0 internet to Web 4.0 internet: Evolution of communication space forms in the global network [Od Internetu Web 1.0 do Internetu Web 4.0 – ewolucja form przestrzeni komunikacyjnych w globalnej sieci]. *Social Treatises* [Rozprawy Społeczne], *11*(1), 32–39.

Shirky, C. (2008). *Here comes everybody: The power of organizing without organizations*. London: Penguin Books.

Sustainable Future Facebook Page. (2017). Accessed August 24, 2017, from https://www.facebook.com/ZrownowazonaPrzyszlosc/

Szumniak-Samolej, J. (2013). *Responsible business in the network economy* [Odpowiedzialność biznesu w gospodarce sieciowe]. Warsaw: Poltext.

Szumniak-Samolej, J. (2015, December). Innovative business models and initiatives based on the social and/or ecological mission. *China-USA Business Review, 14*(12), 609–620.

Tapscott, D. (2009). *Grown up digital*. New York: McGraw-Hill.

Tapscott, D., & Ticoll, D. (2003). *The naked corporation*. New York: Free Press.

Tapscott, D., & Williams, A. D. (2008). *Wikinomics: How mass collaboration changes everything* [Wikinomia. O globalnej współpracy, która zmienia wszystko]. Warsaw: Wydawnictwa Akademickie i Profesjonalne.

Triple Pundit. (2015, October 8). *IoT & CSR: Better together*. Accessed September 1, 2017, from http://www.triplepundit.com/podium/internet-of-things-csr-better-together/

Visser, W. (2008). CSR 2.0: The new era of corporate sustainability and responsibility. *CSR Inspiration Series*, No. 1.

Visser, W. (2011). *The age of responsibility: CSR 2.0 and the new DNA of business*. London: Wiley.

Weber, L. (2009). *Marketing to the social web*. New Jersey: Wiley.

White, B. (2015). Discovering the future of the web. *Journal of Computing and Information Technology, 23*(1), 87–93.

Yeomans, M. (2016). *The 6th annual social media sustainability index*. Sustainly. Accessed August 31, 2017, from https://sustainly.com/sites/default/files/The%206th%20Annual%20Social%20Media%20Sustainability%20Index.pdf

Zadek, S., Sabapathy, J., Dossing, H., & Swift, T. (2003, January). *Responsible competitiveness: Corporate responsibility clusters in action*. London: The Copenhagen Centre & Account Ability.

Zumer Sustainability Strategy Consulting. (2011). *Sustainability 2.0: Current trends at the confluence of social media and CSR*. San Francisco, CA: SB Insights.

Part VII
Outlook on CSR in Poland and Vicegrad

Strategic CSR in Poland

Przemysław Wołczek

1 Introduction: Purpose of the Article and Research Method

The contemporary turbulent environment, in which global conditions of development are changing dynamically, causes that the achievement of the goals set by companies becomes a very demanding challenge. In order to meet this challenge, companies try to pursue their goals based on the development strategy. That's because a properly developed and implemented strategy often determines the success of a given company. However, in the twenty-first century, it should be remembered to take into account the requirements that are set for companies in the strategy development and implementation process by the concept of corporate social responsibility or corporate sustainability and responsibility (CSR).

The importance of a strategic approach to CSR in terms of building the competitiveness of a company and its relations with stakeholders is emphasized, inter alia, by Rok (2004), Porter and Kramer (2006), Adamczyk (2009), Wołczek (2010, 2011a) and Gligor-Cimpoieru and Munteanu (2015). These authors argue that a strategic approach to CSR is an opportunity for companies to achieve social and economic benefits from the socially responsible actions.

The aim of this paper is to assess and analyse the status of strategic CSR in Poland. The article assesses, whether the legal and political background, the social context, and the actions undertaken by companies are sufficient to classify them as approaching the idea of CSR strategically.

P. Wołczek (✉)
Department of Strategic Management, Wroclaw University of Economics, Wrocław, Poland
e-mail: przemyslaw.wolczek@ue.wroc.pl

A. Długopolska-Mikonowicz et al. (eds.), *Corporate Social Responsibility in Poland*, CSR, Sustainability, Ethics & Governance,
https://doi.org/10.1007/978-3-030-00440-8_18

The assessment was conducted by means of the desk research method, i.e. analysing the existing data (reports, analyses, databases, etc.), and the method of analysing the national and foreign literature discussing corporate social responsibility.

The structure of the article was subordinated to the purpose of the article. In the first part sought answer to the question: What is strategic CSR? For this purpose, the literature of the subject in the context of defining the place of CSR concept in enterprise strategy was analyzed. In the second part the considerations on how far the concept of CSR in Poland has a strategic dimension was presented. The considerations concerned three areas: legal and political context, civil society context, enterprises context. In the third part there are presented five examples of companies utilising a strategic approach to CSR.

2 What Is Strategic CSR?

Discussing strategic CSR should not commence without familiarising oneself with the term "strategy", which will assist in understanding the actual meaning of strategic activities undertaken by companies. The management theory does not feature any singular definition of the term "strategy", which stems from the fact that "strategy" is a "live" category—it changes over time (Penc 1999, p. 143). This is why the source literature features varying definitions of strategy, depending on the author.[1] The exact definition of strategy notwithstanding, its crucial role in how the companies direct their actions is unquestionable (Moszkowicz 2000, p. 175). One

[1]See, among others: A. D. Chandler jr., *Strategy and Structure* (Cambridge, Mass.: The MIT Press, 1962, p. 13; as cited in: H. Koontz, C. O'Donnell, H. Weihrich, *Management* (eight edition), McGraw-Hill Book Co., New York, 1984, p. 107; H.A. Simon, *Działanie administracji*, PWN, Warszawa 1976, p. 163, as cited in: *Zarządzanie. Teoria i praktyka,* A. K. Koźmiński, W. Piotrowski, et al. Wyd. Naukowe PWN, Warsaw 1995, p. 123.; W.F. Glueck, *Business Policy and Strategic Management* (third edition), McGraw-Hill Book Co., New York, 1980, p. 9; L.W. Rue, P. G. Holland, *Strategic Management, Concepts and Experiences,* McGraw-Hill Book Co., New York 1986, pp. 4–5; *Opening up the definition of strategy,* In: Quinn J. B., Mintzberg H., James R. M. (edit.), *The Strategy Process,* Prentice-Hall, Englewood Cliffs, New Jersey 1988, p. 14, as cited in: T.J. Watson, *W poszukiwaniu doskonałego zarządzania*, Wyd. Naukowe PWN, Warsaw 2001, p. 34; J. Jeżak, *Zarządzanie strategiczne przedsiębiorstwem,* Wyd. Uniwersytetu Łódzkiego, Łódź 1990, p.14; A. A. Thompson Jr., A. J. Strickland III, *Strategic Management. Concepts and Cases,* (seventh edition) Irwin Homewood, Boston 1993, pp. 6–7; M. Romanowska, *Zarządzanie strategiczne firmą,* Wyd. Centrum Informacji Menedżera, Warsaw 1995, p. 2; H. Kreikebaum, *Strategiczne planowanie w przedsiębiorstwie,* Wyd. Naukowe PWN, Warszawa 1996, p. 26; G. Hamel, C.K. Prahalad, *Przewaga konkurencyjna jutra,* Business Press, Warsaw 1999, p. 20; M. Moszkowicz, *Strategia przedsiębiorstwa okresu przemian*, PWE, Warsaw 2000, p. 34; K. Obłój, *Strategia organizacji*, PWE, Warsaw 2001, p. 9; *Zarządzanie strategiczne. Koncepcje—metody,* (fifth edition), R. Krupski et al., Wyd. AE Wrocław, Wrocław 2003., p. 16; M.E. Porter, *Czym jest strategia?* Harvard Business Review, July–August 2005, p. 169; G. Johnson, K. Scholes, R. Whittington, *Podstawy strategii,* PWE, Warsaw 2010, p. 21, A. S. Huff, S. W. Floyd, H. D. Sherman, S. Terjesen, *Zarządzanie strategiczne,* Oficyna a Wolters Kluwer business, Warsaw 2011, p. 19.

should simultaneously underline that each company, regardless of its size, should utilise some sort of a strategy—a central, integrated, external concept on how to achieve its goals (Hambrick and Fredrickson 2001), direct its actions, divide its resources in the most effective manner, coordinate decisions made by different units (Grant 2011, p. 15), and consisting of several fundamental elements that cooperate and complement one another (Ignacy and Wołczek 2014, p. 69). It could be therefore stated that strategy is an element that defines the direction for all initiatives and actions undertaken in a company. This is why it is crucial to define what actual aspects, trends, and concepts have been included in a company's strategy.

The current situation in the business world clearly shows that CSR concepts play a significant role, e.g. in the investor decision-making processes. This can be proved by, e.g. the birth of the Socially Responsible Investing (SRI) concept, which made its appearance in the 1980s in the UK (Murawski 2011). Another concept more and more often mentioned is the idea of Sustainable and Responsible Investing (also SRI), defined as an investment approach taking into consideration the ESG (Environmental, Social, and Governance) factors within the realm of portfolio selection and management (Bloomberg 2017, p. 3). Socially responsible investment has been gaining in popularity since 1990s and this trend still continues to this day. This can be confirmed e.g. by the research conducted by The Global Sustainable Investment Alliance, according to which the worldwide SRI market in 2016 was worth c. USD 23 trillion. This market is dominated by Europe, whose share of the value is assessed at c. USD 12 trillion. This is followed by the United States, valued at USD 8.7 trillion. Within the realm of relative categories, the socially responsible investing currently constitutes 26% of all worldwide professionally managed assets. In the European market, the share of responsible investment strategies already constitutes 53% of all professionally managed assets. SRI is therefore becoming the primary force in the global financial markets (Bloomberg 2017, pp. 3–9).

On the other hand, the research conducted by Deutsche Bank's climate change advisory department shows that skilful management of ESG factors results in tangible profits for a company, primarily within the realm of lower capital expenses and better financial results (DB Climate Change Advisors 2012).[2]

[2]The department conducted a comprehensive analysis of research focusing on the connection between ESG performance of companies and, among other items, the costs of capital acquisition and their financial results. In order to promote the validity of the analysis, only the documents that met the minimum requirements of academic discipline have been selected (i.e. those that had been published in known scientific journals). The early stages identified over 100 works from the last 15 years, of which the analysis focused on 56 papers, as well as 2 literature reviews and 4 meta-analyses. As for capital expenses, the results of all 19 analysed works focusing on the relationship between the company's CSR rating (scoring the company's methods of ESG management) and capital expenses (both external—loans and bonds, and internal—stock) indicate that entities with high CSR rating experience lower (ex-ante) capital expenses. This indicates that the market perceives the companies that better manage ESG factors as less risky than others and rewards them for it suitably. It has also been revealed that good management of ESG factors is positively correlated with financial results of the company. The vast majority of the 36 works showed that companies with high CSR ratings are characterised by above average financial results, and their

Taking the above into consideration, one could surmise that if companies wish to achieve and retain some degree of competitive advantage, they should base their strategy on the principles of CSR. This is also affected by the noticeable worldwide trend, especially present in highly developed countries, of increased expectations and requirements of investors with regards to companies. Companies are expected to not only deliver their respective products and services, but also to engage in solving complex social issues (Adamczyk 2009, p. 9; Eggers and Macmillan 2013). This perception of the desired role of companies gives rise to the notion that in the coming years the success of the company will be determined by a skilfully designed and applied business strategy based on the CSR principles (Wołczek 2010, 2011a, b).

What is strategic CSR, then? In order to answer such a basic question, one first needs to define the role of CSR in the company's strategy, because it would seem that CSR on a strategic level may only be applicable once CSR becomes the foundation for the company's strategic outlook. This, in turn, suggests that CSR should constitute the central element of the company's vision, mission, strategy, and operations (Wołczek 2010).

The strategic approach to CSR is underlined by e.g. the European Commission. In their statement from 2011 they focused on and emphasised the role of CSR in a strategic outlook. The document states: „*A strategic approach to CSR is increasingly important to the competitiveness of enterprises. It can bring benefits in terms of risk management, cost savings, access to capital, customer relationships, human resource management, and innovation capacity*" (European Commission 2011, p. 3).

The significance of strategic CSR within the realm of company's competitiveness and its relationship with stakeholders has already been underlined by e.g. M. E. Porter and M. R. Kramer. These authors claimed that the key to the company's ability to achieve social and economic benefits from socially responsible activities is the strategic approach to CSR. Such an approach assumes undertaking a low number of initiatives, aiming to bring about social and economic benefits promoting the company (Porter and Kramer 2006). Porter and Kramer believe that a company's social engagement brings about best results in areas where the company's interests and the society's needs overlap (Świerżewski 2006, p. 3). This fusion allows one to achieve the state of shared value, in which both the company and its stakeholders will benefit from the company's social engagement.

On the other hand, as per R. W. Griffin, one could surmise that a company characterised by a strategic outlook on CSR is a company that assumes an active attitude towards this concept (Griffin 2005, p. 123). This attitude can be defined as the company perceiving itself as a "citizen" in the society in which it functions and actively seeking ways to improve the general social well-being. An active attitude

---

stock is doing better than the respective market indices. Most investors perceive ESG as medium- and long-term opportunity (3–5 and 5–10 years, respectively).

Source: DB Climate Change Advisors (2012) Sustainable Investing. Establishing Long-Term Value and Performance. Deutsche Bank AG, Frankfurt am Main

https://www.db.com/cr/en/docs/Sustainable_Investing_2012.pdf. Accessed 20 August 2017.

means consciously and willingly accepting responsibility for the social and environmental context in which it functions, as well as undertaking actions aiming at mutual benefits to both the company and its stakeholders.

Additional references to the concept of strategic CSR can be found in the work of A. Filek concerning the stages of constructing corporate social responsibility, in which the author analyses the types, stages, and levels of responsibility (Filek 2006). Strategic CSR is only applicable starting with the stage four i.e. the "developed CSR". This stage is characterised by social engagement aiming at creating social relations with stakeholders and maintaining balanced development. The company's responsibility is a positive factor that focuses on its respective value that it can generate for its stakeholders. It can also be described as a conscious responsibility aiming at integrating some of the societal goals into the company's activity. The most advanced stage of constructing social responsibility, however, is the fifth stage, i.e. "advanced CSR". This stage is characterised by the company's obligation to improve the quality of life of all members of its society.

The importance of strategic CSR is also defined in the concept of CSR 2.0, proposed by W. Visser. This author claims that the currently prevalent model of CSR (so-called CSR 1.0) has failed, because any changes introduced by the companies on its basis are restrained by three key constraints: limited scale, peripherality, and economic ineffectiveness (Visser 2012). It can be said that CSR 1.0 has its roots in the limited motivation of companies to reform their basic areas of activity to include the implications of the concept of responsible business and balanced development. As a result, the only CSR-related undertakings that are dominating in the current market have a very limited influence on the way the company functions and its strategy. The solution to this issue, according to Visser, is CSR 2.0—a systematic concept of business social responsibility. It entails assuming true responsibility for the company's actions, including its failures. The systematic approach to CSR may necessitate modifications of current business models, in order for the company to assume true and complete responsibility.

P. Ratajczak proposes an interesting approach to analysing company strategies and the concept of CSR. The author concluded that the level of cooperation between the CSR activities and the company's strategy has an influence on the company's value for its owners and stakeholders. The author assumed two scales: value for the company's owners and for the company's stakeholders. He generated three types of socially responsible activities: non-strategic CSR, strategic CSR, and business CSR (Ratajczak 2014). According to Ratajczak, business CSR stands for an approach constituting the basis for the company's strategy and resulting from its mission and vision. The company functioning based on the business CSR model aspires to solve selected social issues in a long term. Business CSR, when compared to non-strategic and strategic CSR, allows one to generate the highest possible value for both the owners of the company and its stakeholders.

The analysis of above propositions relating to the role and scope of CSR in corporate strategy allows one to conclude that strategic social responsibility begins when the company realises the scope of the role it plays for its stakeholders and willingly begins integrating into its activity aims, whose realisation targets fulfilling at least a certain degree of the stakeholders' needs. Any socially responsible action undertaken by a company should at least follow the general strategy of the company, but the best possible scenario entails the CSR concept becoming the foundation, upon which the overall strategy of the company is built. Only such an approach allows one to generate benefit for both parties—the company itself and its stakeholders.

3 The Image of the Strategic CSR in Poland

This point of the analysis is devoted to measuring the strategic aspect of CSR in Poland. This will include three areas: legal and political context, civil society context, enterprises context.

3.1 Legal and Political Context

This section is devoted to analysing institutional solutions assisting in the development of CSR within the legal and political environment on a national level.

In order to ascertain a level of integrity and coordination of activities undertaken by various levels of government administration within the realm of CSR, according to the disposition no. 23 of the Minister of Development dated April 11th 2016, a Corporate Social Responsibility Committee was established, with an idea of being a supplementary organ of the Minister of Development (Zarządzenie nr 23, 2016). Previous CSR committees, first established for the Prime Minister, and subsequently for the Minister of Economy, are now therefore seeing a successor. The Committee is tasked with: (1) maintaining dialogue, exchanging experiences between public administrative bodies, the business, social partners, as well as NGOs and research institutions, with regards to the ideas of balanced development and corporate social responsibility; (2) popularising the idea of social solidarity and responsible business conduct; (3) devising recommendations on directions of CSR implementation with regards to the *Strategy for Responsible Development*; (4) maintaining the relationship between the competitiveness of economy, environmental protection and quality of life, as per the *Plan for Responsible Development* and the *Strategy for Responsible Development.*

The fact that such a body has been established suggests that the Polish government is intensifying its actions within the realm of CSR and is realising that this concept can only be properly developed with the government's assistance—support from the business and social organisations may not be sufficient. It can therefore be surmised that there exists in Poland a governmental entity leading the management of CSR issues, which may significantly simplify the work of enhancing the socially responsible attitudes within the Polish market.

Another important issue is the matter of a national strategy of sustainable development and/or strategy of CSR. In turn, with reference to this issue, it should be noted that *Strategy for Sustainable Development of Poland by 2025* (Strategia zrównoważonego… 1999) was developed as early as in 1999. The main objective set forth in this document was to maintain a GDP growth at a level of approx. 5% a year and at the same time increase fourfold the efficiency of the use of raw materials, fuels and natural resources. It was also assumed that in the 25-year time horizon of the strategy implementation there will be taken primarily the actions that would serve reaching a sustainable development model by Poland. These kind of actions may be exemplified by the *Strategy for Responsible Development until 2020 (potentially until 2030)* instituted by the government on February 14th 2017. As per the work's introductory segment: The Strategy "*is a key Polish document within the realm of medium- and long-term foreign economic policy. […] The Strategy, aimed at a responsible and unified development, has a goal of liberation of entrepreneurship, inventiveness, and productivity. Coherence between social programs, spending tens of billions PLN for equalising social opportunities, and economic development and proper financing of both economic and social projects, are all key elements of the Strategy's success*" (Strategia 2017, pp. 2–4). Unfortunately, a national CSR strategy in Poland is still non-existent. The Corporate Social Responsibility Committee, acting on behalf of the Ministry of Development, has, however, announced the introduction of a National Plan for CSR.

Another important institutional aspect influencing the promotion of CSR are the "green" and "social" public procurements. As far as the strategy for public procurements goes, assuming a strategy containing environmental and social elements, it may be surmised that the Polish Public Procurement Office is active within this realm. Since 2007, four documents concerning this issue have been published (Krajowy Plan Działań … 2007, 2010, 2013, 2017). Sadly, when analysing Polish "achievements" within the realm of "green" or "social" public procurements, one has to conclude that both these contract types constitute only a tiny amount of the entire population (Table 1).

Such low percentages involved in the green and social pubic procurements are probably caused by the low level of knowledge of the ordering parties with respect to such contracts types. This is why it is crucial to institute intensive educational and promotional activities within the realm of social and green public procurements.

Table 1 The level of green or social public procurements per year

	2006	2009	2010	2011	2012	2013	2014	2015	2016 (plan)
Green public procurements	4%	10.5%	9%	12%	12%	12.1%	9.25%	11.4%	20%
Social public procurements	–	–	2.83%	2.04%	2.90%	4.50%	3.10%	4.08%	10%

Source: Krajowy Plan Działań w zakresie zrównoważonych zamówień publicznych na lata 2017–2020 (2017) Urząd Zamówień Publicznych, Warszawa, pp. 35–36

In turn, when considering the issue of specific provisions of the law, which would promote widespread adoption of good practices in the scope of CSR, it should be noted that such solutions can hardly be found in the Polish legislation. Certain incentives, mainly of a tax nature, can be found, *"however it can be considered that they are of inconsiderable importance in every-day business practice and do not really concern the fundamental economic, environmental, labour and social problems in the context of the CSR development"* (Przegląd zachęt… 2011, p. 5). The experts agree that the idea of CSR is quite novel in Poland and the Polish law is not entirely managing to keep up with it (Pawłowski 2016).

Taking the above into consideration, one could conclude that there are positive legal and political changes on an institutional level, aiming at systematic governmental actions cooperating with the private sector and NGOs in order to raise awareness and understanding of social, environmental, and ethical issues. This is exemplified e.g. by periodical meetings of the subcommittees of the Corporate Social Responsibility Committee. One can also notice how the ideas of sustainable and responsible development are mentioned in strategic governmental documentation. There is a distinct lack, however, of any actual changes within the scope of public procurements or legal solutions, which would encourage companies to engage in socially responsible activities. Without such long-term actions it will be difficult to achieve widespread support for the idea of CSR.

3.2 Civil Society Context

Within this context, the analysis shall focus on the ideas of independence (in terms of funding), NGOs, and involvement of the academic community and the media in the development and promotion of CSR in Poland.

Non-governmental organizations use different sources to finance their activities. Such sources may include membership fees, donations, subsidies from public and private funds, sponsorship, public fundraising events or incomes from a business activity. The surveys conducted by the Klon/Jawor Association in 2015 reveal that shifts in the budgets of Polish NGOs took place in the recent years (Kondycja sektora … 2016). It turns out that the national funds from the central budget are becoming less and less important (31% share in the revenues of the entire sector in 2007 and 15%—in 2014). However, taking into account an increase in the share of funds from budgets of local government units (from 12% in 2007 to 15% in 2014) and a growing share of EU funds (from 9% in 2007 to 23% in 2014), it should be noted that the total share of domestic and foreign public funds in the budget of the sector is still at a high level—in 2014 it was 55% (while in 2007–57%). Taking into account the above, it should be stated that functioning of NGOs in Poland is largely dependent on public funds, which can significantly influence the scope of activities undertaken by these organizations.

Referring to the involvement of the world of science in the development of the CSR concept in Poland, it should be noted that the currently leading Polish higher education

Table 2 Sample programmes and courses concerning the CSR concept offered by Polish higher education institutions (data for September 2017)

Programme/course	Organiser
CSR. The goals of sustainable development in the company strategy	Kozminski University
CSR manager	Collegium Civitas
Leader of corporate social responsibility	Wroclaw University of Economics
Corporate social responsibility (CSR)	Katowice University of Economics
Strategy of social responsibility in management	Krakow University of Economics
Strategic leadership for responsible management	Warsaw University of Technology
New consumer—psychological aspects of corporate social responsibility (CSR)	University of Warsaw
Strategy of corporate social responsibility	Warsaw University of Life Sciences

Source: Own study

Table 3 The total number of press articles associated with the subject of responsible business and related issues in the years 2008–2016

	Years								
	2008	2009	2010	2011	2012	2013	2014	2015	2016
Total number of press articles	829	1512	2248	2527	3342	5591	4702	6930	7504
Increase in relation to previous year (in %)	–	82%	49%	12%	32%	67%	−16%	47%	8%

Source: Own study based on: Raport (2017), p. 135; Raport (2016), p. 152; Raport (2015), p. 128; Raport (2014), p. 135; Raport (2013), p. 123; Raport (2012), p. 109; Raport (2011), p. 99; Raport (2010), p. 76; Raport (2009), p. 69

institutions offer programmes and courses concerning the CSR concept (Table 2). Most often they take the form of paid post-graduate studies. However, there is a lack of offers (e.g. fields of studies) for students of traditional studies. Some of the paid post-graduate studies, like "CSR. The goals of Sustainable Development in the company strategy"; "CSR Manager" or "Leader of Corporate Social Responsibility", are conducted by scientists who for many years have been involved in the promotion of the CSR concept with an active participation of CSR managers from companies most advanced in this scope in Poland and in cooperation with leading consulting firms, which proves that the level of professionalism in the approach to the CSR issues in higher education institutions is increasing.

The involvement of the media in promoting the CSR concept is the next area subjected to an analysis. Based on the data published by FOB (Responsible Business Forum), it can be stated that the involvement of the media in promoting the CSR concept in Poland has been growing steadily since 2008 (Table 3).

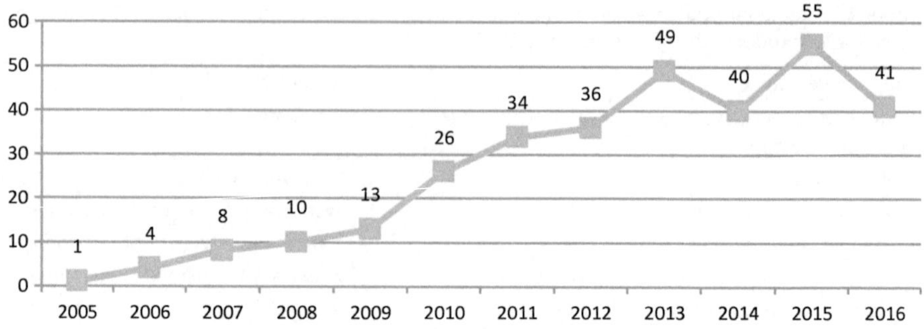

Fig. 1 The number of published CSR/social reports in Poland in years 2005–2016 (the data shows 317 reports published within the scope of 10 years (01.01.2005—31.12.2016), registered in the Report Registry (www.rejestrraportow.pl) managed by CSRInfo.). Source: Raportowanie niefinansowe w Polsce 2016, CSRInfo, 2016

3.3 Enterprises Context

This point shall discuss the ideas concerning CSR/social reporting and standards employed therein, as well as the structure of the RESPECT Index—the index of socially responsible companies of the Warsaw Stock Exchange.

As far as CSR reporting activities and standards used therein are concerned, one may conclude that social reporting is slowly rising in prominence in Poland (Fig. 1). Data for years 2005–2016 shows a significant increase in the numbers of published social reports.

Until the end of 2016, non-financial reporting was not obligatory in Poland. Within the last 5 years, however, i.e. 2012–2016, as much as 78% of all reports have been put together in accordance with the Global Reporting Initiative reporting standards. Within the entire summarised period, on other hand, this ratio nonetheless equals 67%, which puts Poland above the worldwide average. According to the GRI data, 60% of reporting entities in the world employ these standards (10 lat raportowania niefinansowego w Polsce). Trying to find an answer to the question, what kind of information are published in CSR/social reports, we may refer to the research results by Fijałkowska and Macuda (2017). Within this cited research, the Authors analyzed all the available CSR reports for the period of 2013–2015. The researched population embraced 98 CSR reports derived from the www.raportyspoleczne.pl website—all the reports from the 2016 (31), 2015 (37) and 2014 (30) editions. The results of the research reveal that, citing the Authors, "all the analyzed companies describe their CSR activities and actions, but the vast majority of them do not measure or give information about the impact, value, or monetary effects of their CSR activities. The analyzed reports are mainly narrative and include mostly descriptions of activities in a qualitative form. This kind of information is highly difficult to judge, control, and compare. It can be also easily manipulated and „colored" to attract and impress the reader". These not very optimistic conclusions are very important in the context of the fact that January 2017 brought about changes to the accounting act (making it in-line with the directive 2014/09/UE), forcing the biggest companies in Poland (c. 300 entities) to more widely disclose non-financial information. One could therefore expect that in 2018 Poland will witness a true revolution and within 1 year one can expect as many CSR/social reports as have been produced within the previous 10 years. It is to be hoped that the quality of information published in these reports will be significantly improved.

The perspective of a significant increase in published CSR/social reports is encouraging, but a question should be asked what percentage of the 100 largest companies publishes social reports. From the analysis of the first 100 companies in the ranking of 500 largest Polish companies organised by the "Rzeczpospolita" newspaper, it appears that in 2010 only 12 of these companies prepared CSR reports, of which only 4 prepared such reports in accordance with the GRI guidelines (Raportowanie CSR w Polsce 2010 (2011), pp. 7–8). In 2011, the number of companies included in this ranking that prepared reports according to GRI increased up to eight (5 lat raportowania odpowiedzialnego biznesu w Polsce). In turn, the studies conducted on a sample of the 120 largest Polish enterprises and financial institutions show that only 26 of them (less than 22%) publish social reports on their websites. Out of this, only 15 companies (12.5%) prepare reports using the GRI guidelines (Krzemień and Piskalski 2012). It can therefore be concluded that the process of creating CSR reports by the largest companies in Poland on the basis of international standards is very slow, but since 2010 it has accelerated.

At this point it is worth to take a closer look at the knowledge of the CSR concept among the companies operating in Poland. The results of the surveys conducted in late 2011 reveal that the concept of "corporate social responsibility" is known only to 31% of the companies operating in Poland, out of which large companies constitute 70%. In addition, 67% of the companies whose representatives declare the knowledge of the CSR concept carry out activities related to it. In the group of these companies, micro-entities (56%) are significantly less represented than others (approx. 80%). Among the companies that carry out CSR activities, 57% do this in accordance with specific standards and guidelines. ISO standards (ISO 14000 and ISO 26000) are most commonly used—54% of the companies choose one of them.[3]

The issue of the perception of CSR by entrepreneurs was also undertaken in the study "Społeczna odpowiedzialność biznesu: fakty i opinie. CSR oczami dużych i średnich firm w Polsce". The results of the survey reveal that 96% of Polish big and medium–sized companies believe that it is a business responsibility to respond to social and environmental challenges. On the other hand only 46% of the surveyed companies conduct activities related to CSR (most often these were very large enterprises—employing over 1000 employees—and with foreign capital). The vast

[3]Studies conducted by MillwardBrown SMG/KRC and PwC at the request of Polish Agency for Enterprise Development on a sample of 850 companies (including: large companies: 100, medium companies: 300, small companies: 250, micro companies: 200). More: *Ocena stanu...* (2011).

majority (78%) of respondents recognized that the significance of the CSR concept is important (63%—important, 15%—very important) in creating the company's strategy. But on the other hand, only 63% of companies involved in social issues include CSR in their strategies. Summarizing the results of the research, it can be stated that only 18% of the surveyed companies are socially mature organizations, which means that they approach the CSR concept comprehensively (among others they report non-financial data and take into account CSR in their strategies).[4]

As a final note, one should refer to the RESPECT Index, the oldest (created in 2009) index of socially responsible companies in Middle and Eastern Europe. The RESPECT Index is meant to be an index focusing on companies that fulfil the requirements of organisational order and order of information, investor relations and, above all, fulfil a range of environmental, social, and governance requirements (ESG). The index focuses on companies listed on the main stock of the Warsaw Stock Exchange (GPW).

It may therefore be worth exploring whether it actually includes companies with the highest level of ESG management. The ESG analysis of companies with regards to the RESPECT index, whose results should confirm their high level of social and environmental responsibility and above-average organisation and order, is based on the information extracted from the ASSET4 ESG database. This database contains detailed, objective, updated, and comparable ESG data. It also includes negative and adverse information on companies published in various global media (Thomson Reuters 2017a, b).

Currently the tenth edition of the RESPECT Index contains 24 companies.[5] The purpose of the analysis was to collect ESG data from the ASSET4 ESG database for all 24 companies. The results for twelve of these companies are presented in the Table 4. Lack of data for the other 12 companies in the ASSET4 ESG database makes it unable to assess their actual situation at this time. The data shown applies to the year 2015 (six companies), and to the year 2016 (also six companies) (this is the most up-to-date data as at December 31st 2017.

A detailed analysis of the data from Table 4 allows one to surmise that the best result within the category of Environmental Performance was achieved by ING Bank Śląski S.A. (93% in 2016). There should also be mentioned that high results within the same category were achieved by two resource companies: KGHM Polska Miedź S.A. (87% in 2015) and PKN Orlen S.A. (85% in 2015). This should not be a surprise for anyone, seeing as these entities operate in sectors with a significant influence on the environment. This is why a high degree of environmental management in these companies seems to be a natural consequence of the fact that any

[4]Studies conducted by KPMG and FOB (Responsible Business Forum) on a sample of 101 companies operating in Poland, including large and medium companies. The cited research was conducted using the CATI method in order to collect the opinions of the top management and employees responsible for CSR, as well as the activities that the surveyed companies undertake. More: *Społeczna odpowiedzialność biznesu: fakty a opinie* (2014).

[5]Data as at December 31st 2017. The tenth edition of the index was announced on December 14th 2016. Initially the index contained 25 companies, but as at June 8th 2017, in accordance with the decision of the Board, the company Pelion S. A. withdrew from the stock market.

Table 4 ESG results for RESPECT Index companies, based on the ASSET4 ESG database (data for December 2017)

Company	Year	Environmental performance (%)	Social performance (%)	Corporate governance performance (%)	Economic performance (%)	Overall performance (%)	Transparency[a]
PKN Orlen S.A.	2015	85	93	48	83	86	92/123
Bank Zachodni WBK S.A.	2015	65	78	61	94	83	91/123
ING Bank Śląski S.A.	2016	93	80	66	86	89	73/123
Bank millennium S.A.	2016	79	93	34	96	84	91/123
KGHM Polska Miedź S.A.	2015	87	87	31	57	72	96/123
Bank Pekao S.A.	2016	76	55	38	90	72	75/123
PZU S.A.	2015	38	71	56	64	61	75/123
Bank Handlowy w Warszawie S.A.	2016	57	71	19	76	59	71/123
PGNiG S.A.	2016	73	59	26	82	66	76/123
PGE S.A.	2015	36	34	50	51	39	98/123
Tauron Polska Energia S.A.	2016	54	20	32	36	29	86/123
Orange Polska S.A.	2015	85	91	52	90	88	91/123

Source: Own work

[a]Transparency—is the number of data points taken into consideration out of the maximum of 123

instances of negligence could result in irreparable environmental damage and, in turn, loss of reputation and investor trust. The worst scores in the "environment" category were PGE S.A. (36%) and PZU S.A. (38%).

In the Social Performance category, the best scores were held by: PKN Orlen S.A. (93%) and Bank Millennium S.A. (93%). The worst was Tauron Polska Energia S.A. (20%). Best results in the Corporate Governance Performance category could be seen in the cases of following companies: ING Bank Śląski S.A. (66%), Bank Zachodni WBK S.A. (61%) and PZU S.A. (56%). The worst was Bank Handlowy w Warszawie S.A. (19%). It may be worth mentioning at this point that the Corporate Governance Performance area is the one that sees least activity (according to the reported data) among the four areas analysed. In the Economic Performance category, the best scores were held by: Bank Millennium S.A. (96%) and Bank Zachodni WBK S.A. (94%). The worst was again held by Tauron Polska Energia S.A. (36%).

The analysis of the ESG scores of the RESPECT Index companies suggests very large discrepancies between the scores of various entities. Assuming, however, that the RESPECT Index should only contain entities that achieved at least 75% overall performance score, only 5 companies out of 12 analysed deserve to be placed in the index at all. The research suggests therefore that the procedure of appointing entities into the RESPECT Index is questionable, at best (more about functioning the RESPECT Index: Wołczek and Sikacz 2017; Sikacz and Wołczek 2017a, b, c, 2018).

4 Strategic CSR in Companies: Examples

When seeking examples of Polish companies utilising a strategic approach to CSR, the focus was on such entities that seek initiatives aiming at shared value, i.e. a state in which both the company and its stakeholders benefit from the company's social engagement.

The first example is a pharmaceutical company. Pharmaceutical companies are burdened with significant social responsibility, seeing as they are producing medicine, influencing people's health, and even their lives (Makowska 2010). Drug production, however, is not the only realm within which pharmaceutical companies can apply the CSR principles. Philanthropy is also such a sector. One example of strategic philanthropy, mentioned by Porter and Kramer (2006), constituting a strategic CSR approach, is the activity of the company Polpharma S.A. This company has focused on philanthropy activities within the scope of their usual business. An example of such activity is the creation of the Polpharma Scientific Foundation in 2001. Its main goal is financing scientific projects of substantial influence on the development of medicine and pharmaceutics. Since the Foundation's inception, almost 16 million PLN has been devoted to that goal, supplying almost 60 different research teams with appropriate funding.[6] It is almost certain that the work of these

[6]More on the foundation: https://www.polpharma.pl/fundacja/o-fundacji/jak-dzialamy/—Accessed 30 Aug 2017.

research teams will also benefit the development of Polpharma itself. This is therefore an example of a mutually beneficial activity: young researchers may develop their portfolio by means of their own research projects, while the company gains access to the results of these projects, as well as the ability to cooperate with teams of qualified researchers.

Another example is also a pharmaceutical company—the ADAMED S.A. Group ADAMED's strategic approach to CSR is exemplified by the ADAMED SmartUp project that the company launched in 2014.[7] This is a scientific-educational project aimed at high school students (15–19 years old), focusing on popularising sciences and natural studies, as well as supporting the passions and interests of particularly gifted young individuals. The project features first and foremost a scholarship scheme, offering participation in a science camp, individual educational consultation, and financial support for the best students. The scheme focuses on the following areas: chemistry and biochemistry, medicine and sciences, engineering and robotics, and physics and nanotechnology. The participants showing the highest potential, the ones with the highest degrees of creativity, passion for learning and motivation, may count on a scholarship for further scientific education, individually tailored to their needs. The scholarship fellows may also receive support necessary to continue their education, conduct scientific research, learn new languages, undergo traineeships or participate in other similar activities. So far, in scope of 3 editions of the project, 420 students participated in free of charge weekend scientific workshops, 9 students received ADAMED SmartUp scholarships, and almost 50 students won the project, which presented them with unique opportunities for further scientific development.

The final example of a large company with a strategic approach to CSR concept is ERGO Hestia S.A. This company is significant for the very wide array of socially responsible activities undertaken by its stakeholders. The last FOB report (Odpowiedzialny biznes 2017) pinpointed as many as 20 good CSR practices, which constituted the best score among all insurance companies. The company focuses on activities responsible towards its stakeholders within the scope of areas of CSR, listed in the ISO 26000 norm (Table 5).

On the other hand, there are also small companies, whose business core is based on the ideals of social responsibility—such as a recycling company Deko Eko. The company's ideals are based on the approach that *"respect for the natural environment, ethical conduct of manufacturers, and conscious consumer choices are all key elements in building a new quality in the market, whose foundation consists of care for nature and finite resources of the Earth, respect for another man's work and profession, and combining passion with business".*[8] The company has been conducting e.g. upcycling, which is described as the most innovative and creative method of recycling. Upcycling is all about bringing value to seemingly spent products and materials. Dekoeko.com is the largest sales and promotional platform in Poland, combining the designer and local business markets involved in upcycling,

[7]More on the project: https://adamedsmartup.pl/o-programie/—Accessed 10 Sep 2017.

[8]More on the company: http://dekoeko.com/—Accessed 10 Sep 2017.

Table 5 Examples of responsible activities undertaken by ERGO Hestia S.A.

CSR area as per ISO 26000	Example of activity
Community involvement and development	*Hestia's artistic journey* A national competition aimed at art students of all polish universities, established in 2002. The winners are rewarded with tickets for an artistic journey to the centres of modern art—New York and Valencia. The young artists, once they return, may prepare their own exposition in Warsaw's prestigious galleries, whereas the finalists will be able to show their work in an individual exhibition in ERGO Hestia's arts pavilion
Labour practises	*Kindergarten for children of ERGO Hestia's employees* The company is the first insurer in Poland to open a kindergarten for its employees' children, currently providing care for 72 children. Apart from the general playing areas, children also have access to e.g. a rehabilitation area, speech therapy area, or an artistic stage, supporting, among other activities, painting and clay modelling. The kindergarten is an integrated facility, it also accepts children with disabilities
Consumer issues	*Sign language services* The company aims at making its offer available to a variety of social groups. In this case, it is concerned with the deaf. Any deaf clients wishing to purchase an insurance policy, report damages, or make changes in their existing policy, may freely use the services of a sign language interpreter via the insurer's website. This service is supplied by the company Migam, which also happens to be a good example of a company with a strategic outlook on CSR[a]
Environment	*Eco-transport* The company has been monitoring carbon emissions for years and has consistently introduced projects aiming at their reduction. In 2016 Hestia began replacing the cars used by their mobile liquidators with hybrid cars, purchasing over 200 Toyota Auris hybrid vehicles—36% of the entire fleet. This has brought about reduction in carbon emissions by over 1/3. It's the most modern and eco-friendly vehicle fleet employed by insurance appraisers in Poland
Organisational governance	*Balanced development week—get your balance* An activity aimed at promoting the ideas and rules governing balanced development among the company's employees. Within its scope, the most important social and ecological projects of the company, as well as new initiatives, are being presented in practice

Source: Own work
[a]More on the company: http://migam.org/—Accessed 10 Sep 2017

ethical fashion, and eco-friendly design. Within its own "Brand Upcycling" project, the platform has begun cooperating with companies and organisations within the realm of devising innovative solutions to usage of waste, such as for utility, education, social issues, or reputation.

Another small company that aims at solving social issues with its business activity is Asante Bamboo Bikes. The idea behind this company was to conduct social activity in Africa. It focuses on selling handmade bamboo bicycles, whose frames are manufactured in Ghana, as part of the Yonso Project social initiative. A certain percentage of the profits from sales of the bicycles is being used to fund educational scholarships for young people in the Asante region (Szumniak-Samolej 2016, p. 11).

To summarise, one can find various examples of companies in Poland with a strategic approach to CSR. While medium and large companies employ single or perhaps several socially responsible strategic activities, there are small, newly-created companies that approach CSR in a more fundamental manner. It would therefore seem that the true revolution in the corporate approach to CSR—the transition from CSR 1.0 to CSR 2.0, as per W. Visser, is going to begin on the level of the small and micro-companies, not the big corporations. It is the founders of these small companies that truly see the potential of strategic approach to CSR.

5 Conclusion

The analysis and assessment of the status of strategic CSR in Poland carried out for the purposes of this paper suggests that we are on a very early stage of comprehension and practical implementation of this concept within the social and economic scope. Such assessment stems from the analysis of the legal and political circumstances, the social context, and the actions undertaken by various companies in the market.

One can pinpoint some positive legal and political changes on an institutional level, aiming at systematic governmental actions cooperating with the private sector and NGOs in order to raise awareness and understanding of social, environmental, and ethical issues. One can also notice how the ideas of sustainable and responsible development are mentioned in strategic governmental documentation. There is a distinct lack, however, of any actual changes within the scope of public procurement or legal solutions, which would encourage companies to engage in CSR activities.

The civil society context shows that NGO activity is significantly affected by public resources (both national and foreign), which limits the scope of initiatives undertaken by such organisations. The involvement of the academia and the media in the promotion of CSR in Poland, on the other hand, is gaining track and is becoming more and more professional. There is, however, a dearth of groundbreaking projects or activities aiming at raising awareness of the importance of strategic CSR for the social and economic development of the country.

The analysis of CSR-related activities undertaken by the companies themselves, on other hand, suggests that entities approaching the idea of CSR on a strategic level are still a distinct minority. Medium and large companies often feature only singular CSR activities, rarely reaching the level of being strategic. It would seem that more understanding of the impact of strategic CSR on the survivorship, development and competitiveness of companies can be found within the realm of micro and small companies, which more and more often assume the idea of CSR as the foundation and the core of their business activities. The confirmation of these presumptions may be the results of research published in February 2018 (Gangi et al. 2018). The authors, using a sample of 2480 firms from 51 countries covering the period 2010–2015, find that the favorable influence of CSR on the firms' capability of survivorship is more pronounced in SMEs than in large firms.

In conclusion it is worth considering the question: What should be the next steps to make CSR in Poland more strategic?

Trying to answer this question, it can be say that at **legal and political** level we should focus on:

- intensifying work of Corporate Social Responsibility Committee which integrates and coordinates activities undertaken by various levels of government administration within the realm of CSR,
- increasing the number of governmental actions conducted in cooperation with the private sector and NGOs in order to raise awareness and understanding of social, environmental, and ethical issues.
- implementing the idea of sustainable and responsible development mentioned in the strategic government document titled *Strategy for Responsible Development until 2020 (potentially until 2030).*
- increasing the scope of "green" and "social" public procurements.
- increasing the scope of legal solutions (law, taxes, financial incentives etc.), which would encourage companies to engage in CSR activities.

At **civil society** level we should focus on:

- increasing the scope of incentives and support for NGOs engaged in the development and promotion of CSR concept in Poland,
- increasing the number of fields of studies, programmes and courses concerning the CSR and sustainable development concept offered not only by Polish higher education institutions but also by education institutions on the primary and secondary level of education,

At **enterprises** level we should focus on:

- increasing the number of CSR/social reports published not only by large companies but also by medium and small ones,
- increasing the number of CSR/social reports prepared in accordance with appropriate reporting standards (for example GRI),
- increasing the quality of information published in CSR/social reports,
- increasing the knowledge about CSR concept among both managers and employees from enterprises of all sizes,
- increasing the number of companies for which CSR concept is a philosophy of doing business.
- rethinking the process of selecting companies to the RESPECT Index, so that only really responsible enterprises that meet the highest standards, could be part of this index.

References

10 lat raportowania niefinansowego w Polsce. Accessed August 20, 2017, from http://www.csrinfo.org/10-raportowania-niefinansowego-polsce/

5 lat raportowania odpowiedzialnego biznesu w Polsce. Accessed August 30, 2017, from http://www.csrinfo.org/wp-content/uploads/2016/03/5_lat_raportowania_odpowiedzialnego_biznesu_w_polsce_csrinfo.pdf

Adamczyk, J. (2009). *Społeczna odpowiedzialność przedsiębiorstw*. Warszawa: PWE.

Bloomberg. (2017). Global sustainable investment review. Accessed August 10, 2017, from http://www.gsi-alliance.org/wp-content/uploads/2017/03/GSIR_Review2016.F.pdf

Bowen, H. R. (1953). *Social responsibilities of the businessman*. New York: Harper and Brothers.

Carroll, A. B. (1979). A three-dimensional conceptual model of corporate performance. *Academy of Management Review, 4*(4), 497–505.

DB Climate Change Advisors. (2012). *Sustainable investing. Establishing long-term value and performance*. Frankfurt am Main: Deutsche Bank AG. Accessed August 20, 2017, from https://www.db.com/cr/en/docs/Sustainable_Investing_2012.pdf

Eggers, W. D., Macmillan, P. (2013, September). *Government alone can't solve society's biggest problems*. Harvard Business Review.

Epstein, E. M. (1987). The corporate social policy process: Beyond business ethics, corporate social responsibility, and corporate social responsiveness. *California Management Review, 29*(3), 99–114.

European Commision. (2011). *A renewed EU strategy 2011–14 for corporate social responsibility*. Brussels.

Fijałkowska, J., & Macuda, M. (2017). Value creation disclosure in CSR reports – Evidence from Poland. *Zeszyty Teoretyczne Rachunkowości, tom, 93*(149), 61–78.

Filek, J. (2006). *Społeczna odpowiedzialność biznesu. Tylko moda czy nowy model prowadzenia działalności gospodarczej?* Warszawa: UOKiK.

Gangi, F., Males, A., Monferra, S., & Mustilli, M. (2018). *Does corporate social responsibility help the survivorship of SMEs and large firms*? https://doi.org/10.1016/j.gfj.2018.01.006.

Gligor-Cimpoieru, D. C., & Munteanu, V. P. (2015). CSR benefits and costs in a strategic approach. *Annals of the University of Craiova, Economic Sciences Series, 1*, 90–99.

Grant, R. M. (2011). *Współczesna analiza strategii*. Warszawa: Oficyna a Wolters Kluwer business.

Griffin, R. W. (2005). *Podstawy zarządzania organizacjami*. Warszawa: Wydawnictwo Naukowe PWN.

Hambrick, D. C., & Fredrickson, J. W. (2001). Are you sure you have a strategy? *The Academy of Management Executive, 15*(4), 48–59.

Ignacy, J., & Wołczek, P. (2014). Forma strategii. In A. Kaleta (Ed.), *Ewolucja zarządzania strategicznego w trakcie rozwoju przedsiębiorstw*. Warszawa: Wydawnictwo C.H. Beck.

Kondycja sektora organizacji pozarządowych w Polsce 2015. (2016). Stowarzyszenie Klon/Jawor, Warszawa.

Krajowy Plan Działań w zakresie zielonych zamówień publicznych na lata 2007–2009. (2007). Urząd Zamówień Publicznych, Warszawa.

Krajowy Plan Działań w zakresie zrównoważonych zamówień publicznych na lata 2010–2012. (2010). Urząd Zamówień Publicznych, Warszawa.

Krajowy Plan Działań w zakresie zrównoważonych zamówień publicznych na lata 2013–2016. (2013). Urząd Zamówień Publicznych, Warszawa.

Krajowy Plan Działań w zakresie zrównoważonych zamówień publicznych na lata 2017–2020. (2017). Urząd Zamówień Publicznych, Warszawa.

Krzemień, M., & Piskalski, G. (2012). Największe polskie firmy a społeczna odpowiedzialność biznesu. Raport z badania 2012. Fundacja Centrum CSR.PL.

Makowska, M. (2010). Koncepcja odpowiedzialności społecznej firm farmaceutycznych. *Annales. Etyka w życiu gospodarczym, 13*(2), 95–104.

McGuire, J. (1963). *Business and society*. New York: McGraw-Hill.

Moszkowicz, M. (2000). *Strategia przedsiębiorstwa okresu przemian*. Warszawa: PWE.

Murawski, T. (2011). Społecznie odpowiedzialne inwestowanie. *Miesięcznik Kapitałowy, 2*, 59–60.

Ocena stanu wdrażania standardów społecznej odpowiedzialności biznesu. Raport. (2011). MillwardBrown SMG/KRC and PwC, Warszawa.

Pańków, W., et al. (2010). *Oblicza społecznej odpowiedzialności przedsiębiorstw*. Warszawa: Scholar.

Pawłowski, J. (2016). Biznes społecznie odpowiedzialny: Rząd zachęca, ale fiskus niekoniecznie. Gazeta Prawna.pl. Accessed August 20, 2017, from http://podatki.gazetaprawna.pl/artykuly/980148,biznes-spolecznie-odpowiedzialny-rzad-zacheca-ale-fiskus-niekoniecznie.html

Penc, J. (1999). *Strategie zarządzania. Perspektywiczne myślenie, systemowe działanie*. Warszawa: Placet.

Porter, M. E., & Kramer, M. R. (2006). Strategy and society: The link between competitive advantage and corporate social responsibility. *Harvard Business Review, 84*(12), 78–92.

Przegląd zachęt podatkowych w kontekście CSR w wybranych krajach europejskich. (2011). Accessed August 15, 2017, from http://odpowiedzialnybiznes.pl/public/files/Zachety_podatkowe_CSR_AccreoTaxand_2011.pdf

Raport. Odpowiedzialny biznes 2008. (2009). Forum Odpowiedzialnego Biznesu, Warszawa.

Raport. Odpowiedzialny biznes 2009. (2010). Forum Odpowiedzialnego Biznesu, Warszawa.

Raport. Odpowiedzialny biznes 2010. (2011). Forum Odpowiedzialnego Biznesu, Warszawa.

Raport. Odpowiedzialny biznes 2011. (2012). Forum Odpowiedzialnego Biznesu, Warszawa.

Raport. Odpowiedzialny biznes 2012. (2013). Forum Odpowiedzialnego Biznesu, Warszawa.

Raport. Odpowiedzialny biznes 2013. (2014). Forum Odpowiedzialnego Biznesu, Warszawa.

Raport. Odpowiedzialny biznes 2014. (2015). Forum Odpowiedzialnego Biznesu, Warszawa.

Raport. Odpowiedzialny biznes 2015. (2016). Forum Odpowiedzialnego Biznesu, Warszawa.

Raport. Odpowiedzialny biznes 2016. (2017). Forum Odpowiedzialnego Biznesu, Warszawa.

Raportowanie CSR w Polsce 2010. (2011). CSRinfo.

Ratajczak, P. (2014). Powiązanie działalności społecznie odpowiedzialnej ze strategią przedsiębiorstwa a wartość dla jego właścicieli oraz pozostałych interesariuszy. *Studia Oeconomica Posnaniensia, 2*(11(272)), 122–136.

Rok, B. (2004). *Odpowiedzialny biznes w nieodpowiedzialnym świecie. Akademia Rozwoju Filantropii w Polsce*. Warszawa: Forum Odpowiedzialnego Biznesu.

Sethi, S. P. (1975). Dimensions of corporate social performance: An analytical framework. *California Management Review, 17*(3), 58–64.

Sikacz, H., & Wołczek, P. (2017a). Analiza raportów informacji niefinansowych spółek z RESPECT Index. *Marketing I Rynek, 11*(CD), 540–552.

Sikacz, H., & Wołczek, P. (2017b). Analiza ESG spółek z indeksu RESPECT na podstawie bazy ASSET4 ESG. *Zeszyty Naukowe Szkoły Głównej Gospodarstwa Wiejskiego W Warszawie. Polityki Europejskie, Finanse I Marketing, 18*(67), 158–169. https://doi.org/10.22630/PEFIM.2017.18.67.31.

Sikacz, H., & Wołczek, P. (2017c). Analiza ESG spółek z indeksu RESPECT – podsumowanie badań. *Zeszyty Naukowe Szkoły Głównej Gospodarstwa Wiejskiego W Warszawie. Polityki Europejskie, Finanse I Marketing, 18*(67), 170–180. https://doi.org/10.22630/PEFIM.2017.18.67.32.

Sikacz, H., & Wołczek, P. (2018). Analysis and evaluation of the functioning of the RESPECT Index. *Management Sciences. Nauki o Zarządzaniu, 23*(3), 30–38. https://doi.org/10.15611/ms.2018.3.04.

Społeczna odpowiedzialność biznesu: fakty i opinie. CSR oczami dużych i średnich firm w Polsce. (2014). KPMG and Forum Odpowiedzialnego Biznesu, Warszawa.

Strategia na rzecz Odpowiedzialnego Rozwoju do roku 2020 (z perspektywą do 2030 r.). (2017). Warszawa.

Strategia Zrównoważonego Rozwoju Polski do 2025 roku. (1999). Warszawa. Accessed July 31, 2017, from http://www.access.zgwrp.org.pl/materialy/dokumenty/StrategiaZrownowazonegoRozwojuPolski/index1.html

Świerżewski, Ł. (2006). Czy zaangażowanie społeczne może mieć dla firmy wymiar strategiczny. In: Brief. Kiedy zaangażowanie społeczne ma dla firmy kluczowe znaczenie. Suplement promocyjny do Harvard Business Review Polska nr 7/8.

Szumniak-Samolej, J. (2016). *Rozmowy z liderami biznesu społecznego w Polsce*. Warszawa: Oficyna Wydawnicza Szkoła Główna Handlowa w Warszawie.

Thomson Reuters. (2017a). Thomson Reuters ESG scores for public companies, Thomson Reuter ESG score data fact sheet. Accessed August 30, 2017, from https://financial.thomsonreuters.com/content/dam/openweb/documents/pdf/financial/esg-scores-factsheet.pdf

Thomson Reuters. (2017b). Thomson Reuters ESG scores, methodology. Accessed August 30, 2017, from https://financial.thomsonreuters.com/content/dam/openweb/documents/pdf/financial/esg-scores-methodology.pdf

Visser, W. (2012). The future of CSR: Towards transformative CSR, or CSR 2.0. Kaleidoscope futures paper series, no. 1.

Wołczek, P. (2010). Społeczna odpowiedzialność biznesu w ujęciu strategicznym. *Prace Naukowe Uniwersytetu Ekonomicznego we Wrocławiu, 100*, 133–145.

Wołczek, P. (2011a). Strategia a CSR. *Prace Naukowe Uniwersytetu Ekonomicznego we Wrocławiu, 156*, 275–286.

Wołczek, P. (2011b). Społeczna odpowiedzialność biznesu źródłem przewagi strategicznej. In E. Urbanowska-Sojkin (Ed.), *Wybory strategiczne w przedsiębiorstwach. Rezultaty ekonomiczne, organizacyjne i społeczne* (pp. 321–331). Poznań: Zeszyty Naukowe nr 170, Wydawnictwo Uniwersytetu Ekonomicznego w Poznaniu.

Wołczek, P., & Sikacz, H. (2017). Stan raportowania informacji o stosowanej polityce różnorodności przez spółki z Respect Index. *Marketing i Rynek, 11*(CD), 707–720.

Zarządzenie nr 23 Ministra Rozwoju z dnia 11 kwietnia 2016 r. w sprawie powołania Zespołu do spraw Społecznej Odpowiedzialności Przedsiębiorstw.

Quo Vadis, Corporate Social Responsibility in Poland?

Institutional Perspectives and Challenges for Public Policy

Aldona Wiktorska-Święcka

1 Introduction

Corporate Social Responsibility (CSR), the concept which implies that companies voluntarily take social and environmental interests into account as a part of their business strategy and act on relationships with different stakeholders, contributes to public value. In this sense, it falls into one of the many definitions of Corporate Social Responsibility. Because of the lack of the clear and widely accepted way of understanding of this approach, the considerations in this chapter are perceived in line with the revised EU Strategy 2011–2014 on Corporate Social Responsibility. CSR is treated here as "the responsibility of enterprises for their impacts on society. Respect for applicable legislation, and for collective agreements between social partners, is a prerequisite for meeting that responsibility. To fully meet their corporate social responsibility, enterprises should have in place a process to integrate social, environmental, ethical, human rights and consumer concerns into their business operations and core strategy in close collaboration with their stakeholders, with the aim of: a) maximizing the creation of shared value for their owners/shareholders and for their other stakeholders and society at large; b) identifying, preventing and mitigating their possible adverse impacts" (COM 2011, 681). In this way, the CSR concept underlines three dimensions of entrepreneurial responsibility: a) on a micro scale, i.e. the entrepreneur's attitude toward him/herself and others, e.g. employees, contractors, and the attitude towards non-personal goods, e.g. to the environment; b) the meso dimension on the organization of business units, and c) the macro level, covering issues of the organization of the socio-economic system. The Corporate Social Responsibility treats the indicated dimensions in a holistic and

A. Wiktorska-Święcka (✉)
University of Wroclaw, Wrocław, Poland
e-mail: aldona.wiktorska-swiecka@uwr.edu.pl

A. Długopolska-Mikonowicz et al. (eds.), *Corporate Social Responsibility in Poland*, CSR, Sustainability, Ethics & Governance,
https://doi.org/10.1007/978-3-030-00440-8_19

integral way (Dylus 2002, 271). In the context of establishing public policies, the state interacts directly with the macro dimension, which consequence is the meso dimension. On the other hand, the macro dimension is an outcome of a CSR micro dimension. The state is directly responsible for creating conditions for the institutionalization of the law, which regulates the behavior of certain entities (individuals and organizations) and the behavior of the recipients of those activities, but also it affects collective bargaining between partners in the public space and the building of relationships. Administration in a democratic state should act as a intermediary in a dialogue for social and environmental issues, and its actions should focus on creating of favorable conditions for business and other market participants to make voluntary commitments for sustainable development. In addition, the administration should ensure effective coordination of activities by providing partners with a platform for exchanging knowledge and experience, and be an example of a socially responsible organization that respects sustainability standards. Lastly, the public administration can play a key role in disseminating of the idea of CSR in the process of creating of the public opinion by influencing public media and through the attitudes of political elites.

In this context, the aim of the chapter is the reflection on the future of legal dimension of CSR in Poland based on experiences before and after the EU accession and current institutional arrangements. The time perspective is not accidental: with the adoption of European legislation and standards (Śmigielska and Oczkowska 2017), but also with the opening up of markets, the transfer of know-how and people, the issue of CSR in the context of public policy has become not only fashionable but even necessary and sometimes mandatory. In addition, the reflection is related to the latest period after 2015 when, with the change of government in Poland, there was another change in the state vision and the revision of key concepts important for the implementation of the CSR approach (Horváth and Pütter 2017).

In the Polish context one can also observe the takeover and privatization of business discussions on the CSR subject (Wołodźko and Woźniak 2017; Kudlak and Ogórek 2017; Stawicka 2017). In one of the analyzes examining the implementation of the CSR concept into the practice of managing large enterprises in Poland, it is emphasized that "Polish public administration has not done much to involve the public in a real debate about the responsibility of Polish business. As a result, representatives of the third sector—trade unions, consumer organizations—are withdrawing from this area without seeing their role in it" (Piskalski 2015, 8).

Moreover, it is worthwhile to take into consideration that undertaking the issue of Corporate Social Responsibility in institutional terms poses a challenge because of its rooting in the normative dimension. Its narrowing to the institutional dimension does not simplify matters: Corporate Social Responsibility should be discussed in terms of a Socially Responsible State, a socially responsible law and socially responsible institutions. This is not the topic here. In this chapter an outline of the development of Polish public policy arrangements will be presented in the context of taking into account the CSR concept in two dimensions: a wide and a specific one. The wide approach includes initiatives which indirectly consider it as a basis for public policies. It is a vision in which the state and the business cooperate in different areas, social and economic partners are involved in public debate and co-decision processes, in which the public administration is striving to raise awareness and increase market participants' responsibility, as well as strengthening cooperation and cross-sectoral partnerships for values close to the social responsibility. The specific approach takes directly into account initiatives of the public administration related to the implementation of the CSR concept.

In order to better understand the essence of the institutionalization of CSR approach in Poland, both in a wide and a specific term, a brief reflection on the first period of creating the democratic state's vision after the 1989 systemic transformation is needed. It has consequences for current solutions as well as those that will be adopted in this regard in the future.

2 Identification of CSR Concept in Public Governance in Poland

Poland, after the political transition of the early 1990s, set a liberal market economy model. The retreat from prosocial solutions offered by companies was not only with the spirit of the new, and how universally perceived, better postcommunist era, but above all, it was a part of the free-market pragmatism, close to ideas of M. Friedman (Chryssides et al. 1999, 261). The intellectual debate about the belief that "the business of business is business" was focused mainly on economic terms rather than social ones. The first years of democratic experience in Poland are the time in which either views of the main reformers gathered around Professor Leszek Balcerowicz, the Vice-President of the Council of Ministers and Finance Minister in the government of Tadeusz Mazowiecki, the first post-communist government nor their actions did not relate to the current world trends that were taking place in Western Europe and the United States. They were the consequence of ever more violent globalization processes and brought about a change in the paradigm of the economy and society (e.g. the first initiatives for sustainable development). The voice of citizens and their organizations, increasingly calling for a departure from the classical liberal model for the social market economy, has grown stronger. This trend, however, was lagging behind in the Polish social-political realities, and even more in institutional practice. So, the first decade after the political transformation did not introduce any significant regulations, which fostered the idea of a socially responsible business. At that time, the solutions that today form the basis of CSR, for example tools in the area of social dialogue, did not fit in Western European standards. One can explain it with the fact that the corporate structures that were inherited after socialism were "not firmly entrenched in the subject matter of the political system and the ruling elites, prompted them to express their demanding demands through spontaneous protest actions destabilizing the economy [. . .]. The corporate structures thus consolidated, and their emphasis on the government increased" (Hausner 1992, 34). The social attitudes formed in the conditions of socialist corporatism, i.e. the maintenance of family-friendly systems and awareness, which did not exceed the level of corporate consciousness, were preserved. In 1994, views expressed by Professor Jerzy Hausner did not break into public consciousness and did not affect institutional practice, when he proposed institutionalized action as an optimal model for representing interests in inter-sectoral relationships. It would involve "the interaction of basic groups and organizations of interest in solving basic economic development problems which should be based on the legal and organizational independence of the social partners, who must be aware not only of their particular interests but also of their interdependence. It would be possible only as a result of social discourse. The foundation of this model was supposed not to be a social solidarity but a permanent agreement between the different actors that reasonably judged, that cooperation—despite the differences—was possible and would bring more benefits than the struggle for domination. The formation of such a model couldn't not in any case be the result of the implementation of any political model, but only the result of a long-term process of social interaction stimulated by specific institutional arrangements in the field of economy and politics" (Hausner 1995b, 324).

One of the reasons why the postulated model did not exist in practice and which could open up space for socially responsible business was the lack of a strong partner representing private entrepreneurship (Hausner 1995a, 312). The analysis of the institutional arrangements established at that time, which were supposed to stimulate the dialogue between social partners and which should be a prerequisite for the fulfillment of the obligations arising from the responsibility of the enterprises, is reflected in their superficiality (Hausner 2007). The reason may be that "the main bind of the institution is not law, [...] but the relationships between the actors involved in their actions, shaped by evolving norms and patterns of behavior" (Sroka 2003, 26–27), and in the discussed period there was no plan to create space for cooperation between the state and business sector. This absence of the "trilateral ideology" could have been the result, among others, of a deformed para-coalition agreement between political parties and interest groups (ibidem). These groups included both entrepreneurs who remain largely unrepresented today. The organizations that are associated with them even now retain their club and elite character (such as the Business Center Club) as well as non-governmental organizations, especially those active in favor of entrepreneurship. Although at the turn of the decades a Responsible Business Forum (Pol. Forum Odpowiedzialnego Rozwoju) was established, the organization, which was one of the first in Poland to begin to disseminate the issues of Corporate Social Responsibility, it did not break into social consciousness and did not influence the institutional arrangements that were being established at the time. An additional restriction on the creation of space for subsequent implementation of the CSR concept were social attitudes inherited from the previous system, including less favorable towards entrepreneurs (e.g., stereotypes which emphasized neglected attitude towards business activity:fli "An entrepreneur is a wealthy man, but his wealth results from greed and addiction to the need to get rich. The size of the entrepreneur's assets is a result of frauds and mothering, possibly a good birth and a happy coincidence. Regardless of which option we take, too much property must be covered by the exploitation of employees who have the characteristics of a crime; You do not think that a high level of wealth can be achieved in a fair way", Szreder 2006). Corruption and lack of respect for the basic values of capitalism remain unresolved (Corruption Perception Index 2016).

At the end of this part of the discussion it is worth mentioning that the period 1989–2004, in which institutions of a democratic state in Poland were formed after the political shift, and which did not take into account the base for institutionalizing the normative basis of CSR, was the time of the governments of representatives of different political views and ideologies. Thus, it cannot be left to say that the institutional reality, dominated by the liberal approach, was the cause of this deficit; liberals, but also socialists, conservatives, populists ruling at that time in Poland did not take care of such a vision of a state in which Corporate Social Responsibility became an element of public consciousness. The consequence was the superficiality of established dialogue bodies as well as the lack of institutional and legal regulations strengthening the implementation of the CSR concept into public policy as such. The political environment, whatever the ideology represented, did not create a space for stakeholder engagement and did not take any significant political initiative that would turn out to be a landmark moment and a paradigm shift for a state where Corporate Social Responsibility exists in both wide and specific terms.

3 Europeanization of CSR Concept in Institutional Practice in Poland

The Millennium Year 2000 was an important milestone in the international debate on the paradigms of doing business in a globalized world. The European Union, which first launched CSR operations in the mid-1990s, was actively involved in the ongoing discourse and business standards initiatives. This resulted, among others, in incorporation of the CSR standards into soft law in which companies take social and environmental aspects into account in their business activities on a voluntary basis. Poland remained passive in these initiatives, but with ever increasing mobility of people and ideas, and with ever more widespread ICT, more and more particular groups of stakeholders have opened up towards new ideas. These processes have been dynamic with the ongoing integration with EU structures and Europeanisation, and also with the transfer of values and/or standards from the European level to the national one. At the same time, it is worth noting that Europeanisation in the Polish public debate conducted since the late 1990s is regarded as a term used to describe adaptation processes taking place in all spheres of life—or even more widely—everywhere there has been an advanced act of qualitatively resonating parts of post-socialist reality to the standards known from Western Europe (Graziano and Vink 2007). Of course, much more complex views on Europeanization are known today (Börzel and Risse 2007; Radaelli 2006; Rosamund 2004). However, when implementing the concept of CSR, it can be argued that this process is clearly top-down: the level of the national administration is not the creator of any solutions, which then affects both bottom-up the European level and top-down the regional, subregional or local level. If any CSR initiatives were taken on a national level after 2004, they were more likely to respond to EU requirements and to be reactive and adaptable.

The specific CSR concept began to disseminate in Poland with the development of companies with mainly Western capital and initiatives undertaken by the Responsible Business Forum, as well as other organizations active in the field, such as the Social Communication Foundation (Pol. Fundacja Komunikacji Społecznej) and the Polish branch of Business Leaders Forum. During the pre-accession period many initiatives promoting CSR were organized, including numerous conferences, training sessions, seminars, and best practices exchanges. They were neither coordinated nor co-created by the public sector. It is only after 2004 one can state that the CSR concept starts to be transferred slowly and indirectly to public institutions operating on the cross-border of the market and the state, or important from the perspective of the social dialogue. In this respect, it should be mentioned that several State Treasury companies have adopted and implemented a CSR strategy over the last few years, and key social dialogue partners such as employers' organizations and trade unions have spread this approach among their organizations. One can also mention here "CSR Guide for Self-Evaluation for Companies", published by Global Compact in cooperation with PKPP Lewiatan, NSZZ Solidarność and Deloitte, which is a tool for promoting social responsibility standards. The government also takes initiatives that promote the issue of CSR. Here, one can call for training sessions organized by the Ministry of Labor and Social Policy from 2008 through its representatives from the Department of Dialogue and Social Partnership, which promote CSR among key representatives of social dialogue in Poland (representatives of various levels of administration, both public and local government, regional and regional chambers of commerce, associations active in the National Chamber of Commerce). These initiatives are, however, an acronym that is not a part of the integrated vision of a Corporate Social Responsibility promoted by the public administration at this time. In addition, it is worth pointing out that along with the progressive Europeanization of the CSR concept, an important role in its dissemination in the Polish institutional reality was played not only by the EU institutions. It is worth emphasizing the involvement of the United Nations Development Programme, which after 2000 launched the global campaign "Global Compact" for sustainable development. The determination and personal contribution of UN representatives to building relations with the Polish administration resulted among others in establishment of the Team for Corporate Social Responsibility (Zarządzenie 2009, 2014) at the Ministry of Economy. This fact begun to be a part of Poland's increased activity in the European Union after 2007 in this regard. The initiative to set up the mentioned CSR Team can be considered as a breakthrough in the institutionalization of CSR concept in specific terms, as it was the first attempt of an integrated approach to embedding this concept in national public policies. The goals of the CSR Team included: a) developing recommendations for coordinating the activities of various public administration bodies, b) analyzing and using experiences and disseminating of good practices in CSR from other countries, c) creating conditions for better communication and dialogue between administration, business, social partners and non-governmental organizations (www.mg.gov.pl). The CSR Team has become the most important platform for the exchange of knowledge and experience of a wide range of stakeholders representing governmental organizations, trade unions, NGOs (e.g. producer and consumer associations), employers' organizations, business clubs and consulting companies. While set up at the central level, the team directly addressed issues of institutionalization of CSR in Polish public policies. The result of its work was a series of activities promoting the idea of CSR, including tools such as the website, publications, events organized but above all the formulation of recommendations for the implementation of the assumptions of corporate social responsibility in Poland. They included:

- the need to identify CSR as part of strategic governance and competitive advantage;
- taking into account the social and environmental aspects of creating the state's economic policy;
- providing political support and leadership;
- cooperation with key stakeholders;
- strengthening the manifestations of voluntary initiatives in the supply chain (ibidem).

It is noteworthy that the conclusions of the work of the CSR Team were likely to be the foundation of an integrated institutionalization of CSR in Polish regulations and standards, as they were incorporated into key policy documents developed at that time, e.g. in the "National Reform Program for Europe 2020", "Strategy on Innovation and Efficiency", "Strategy on Energy Security and Environment", "Strategy for the Development of Human Capital" and the National Strategy "Efficient State".

In particular, the CSR Team joined activities related to the Polish Presidency of the European Union in 2011. The aim in promoting and implementing CSR was the partnership cooperation to overcome social development barriers and to promote responsible leadership, conducive to effective coordination of undertaken activities. Under the Polish Presidency was among others the conference "Shared responsibility—how to increase the presence and effectiveness of CSR in the EU" organized, during which the following conclusions were presented:

- the need to define the clear role of state administration and leadership in shaping the direction of further CSR institutionalization;
- the need to increase cross-sectoral cooperation for CSR;
- ensuring the coherence of the CSR agenda with the direction of other development policies;
- significance of the selection of flexible instruments supporting the development of CSR (ibidem).

During this period, further governmental actions for CSR were planned, including participation in the preparation of the EC Communication, implementation of the recommendations developed by the CSR Team, strengthening of the regional dimension of CSR, integration of the various public administration bodies in favor of particular CSR areas.

Pointing to the above example as a good practice of institutionalizing the CSR concept in Poland, it is worth noting that, despite the significant contribution made by the CSR Team in this regard, it remained detached from the overall state policy as such due to their placement at the lower and middle level of government administration. The initiatives undertaken by the Team were not linked to a wider political vision: the issue of Corporate Social Responsibility was not a priority for political leadership in the period 2004–2015 and was treated as marginal. Such positioning of CSR issues in the institutional system has also made the first decade of Poland's presence in the European Union not efficient as a basis for integrated implementation of CSR concept. It has not become rooted in key public policies.

In addition, the Polish administration, being one of the most important organizations in the public space, has failed to introduce into its governance structures solutions that take into account socially responsible standards. This applies both to the internal and external dimension of CSR implementation: public sector HR management (including employment and dismissal policies, development and improvement of qualifications, employee involvement in decision-making processes, fair pay distribution, staff support in work-life-balance, equal opportunities); codes of conduct (including the adoption and respect of ethical codes, the appointment of ethics spokesman, the development of procedures in case of breach of the code of conduct); environmental management (e.g. reduction of water consumption, energy, waste management, recycling, rational use of resources, procedures, programs to systematically reduce the environmental nuisance of public organizations); public governance principles (transparency of information, ways of appointing and remunerating members of supervisory boards); social actions (e.g. cooperation with non-governmental organizations, volunteering, social investments). It is worth mentioning that while being a buyer of goods and services, only 1.8% of calls for tenders announced by the public administration covered the requirements for employment conditions. Environmental considerations are only included in every fifth order (Piskalski 2015). Despite the introduction of EU standards, such as social clauses providing decent working conditions, the activation of vulnerable groups (e.g. disabled people, dependents and refugees) and environmental clauses aiming at reducing the environmental impact of the institution (e.g. ecological paper, biodegradable detergents, energy-saving electronic equipment, etc.), the Polish administration as such does not use these tools, still considering the lowest price as the criterion (ibidem). Also, the other standards of socially responsible organizations are not widely implemented in the public sector. The current postulate remains the improvement of human resources policies, including the prevention of mobbing (Sojka and Walczuk 2015). The challenge is also the privatization of domains that should be the domain of the state, and as a result of over-marketing they are removed from public control. This applies not only to widely understood public policies, e.g. in the area of health, but also to the privatization of public utility or security maintenance tasks, where the promotion of outsourcing while awarding tenders based on the lowest price criterion awards companies socially irresponsible to provide employees with decent working conditions and taxpayers of services at decent levels.

The unambiguous assessment of the concept of institutionalization of the CSR concept in Poland after its accession to the European Union retains its ambivalent character: on the one hand, national regulations are drawn up, which ultimately serve the purpose of CSR implementing. The recognition of this approach among the representatives of the public administration and the initiatives undertaken during this period should be also appreciated. The initiative of setting up a CSR Team in the Ministry of Economy deserves here a special attention. On the other hand, the issues of the democratization of the state institutions as such, in which important CSR values, such as responsibility, cooperation, partnership, sustainable development, have been internalized, remain unresolved. The current postulate emphasizes the need to strengthen civil society and promote democratic standards as a prerequisite for subsequent CSR initiatives. Despite the slow adoption of the European Commission's guidelines in Polish institutional solutions, the implementation of CSR is still a marginal phenomenon.

4 Nihil Novi Sub Sole? Institutionalization of CSR Concept in Polish Public Policy After 2015

The change of the government in Poland as a result of the parliamentary elections in the autumn of 2015 resulted in a "Good Change" (the main election slogan of the party Law and Justice; Pol. Prawo i Sprawiedliwość) in the exercise of power. The newly introduced "Good Governance" standards, promoted by the European Commission, which are hardly anchored in the Polish institutional solutions developed so far, ceased to function in favor of other, still undefined interpretations of key concepts for public order (European Governance 2001). The period after 2015 is characterized by a sharp introduction to public discourse of key values for democracy such as "freedom", "solidarity" and "responsibility", but these are subject of a violent redefinition, which is matched to the interpretation imposed by the ruling party. This new defined "responsibility" is also at the heart of the new vision of Poland's development strategy, a key document from the perspective of the state, adopted by the government and presented to the European Commission. The "Responsible Development Strategy", focusing on macro-economic dimension, i.e. on the organization of the socio-economic system, as a guiding principle, balances the development of the whole country in economic, social, environmental and territorial dimensions. The particular attention has been paid to the inclusion of all social groups and all territories in development processes, including social and economic partners, academics and local governments. In the context of the CSR concept, it is worth emphasizing that, while the document itself has received a lot of attention from the public, the way it has been implemented has reaffirmed the growing dichotomy in government standards and strengthened the conviction that in the current approach—even radical changes in the ideology of government transition from neo-liberalism to conservative social populism—new public governance standards are not expected. Some of the ideas and initiatives included in this strategy have been warmly evaluated. However, their lack of respect for the culture of dialogue and participation, dichotomy in adopted visions and their implementation, further restrictions on economic freedom, and unethical public governance standards (e.g. the growing unpredictability of legal changes, in particular tax changes, and government policy aimed at pushing the private sector through state-owned companies and institutions) affect the backlash of post-socialist attitudes, consolidate homo sovieticus, and do not encourage a culture of friendly implementation of CSR. In the wide context of analyzing the way of institutionalizing CSR in Poland, there has not been a significant breakthrough in the desired direction of action. On the contrary, phenomena are present, which are the premise of a possible thesis about strengthening the model of authoritarian populism. Professor Jerzy Hausner wrote about this in the 1990s, pointing to its distinctive features that can be observed in the current socio-political reality: "the elimination or marginalization of all the intermediary structures of interests, especially those characteristic of parliamentary democracy" (Hausner 1995a, b, 324). Social activity begins to "manifest itself only through the activities of licensed and politically controlled mass organizations, whose task [is] to mobilize masses and public opinion, in accordance with the needs of a exclusive political leadership" (ibidem). By analyzing the way of institutionalizing the CSR concept in current public policy development and implementation, it should also be emphasized that governmental administrations continue to disregard the standards for managing socially responsible organizations: public governance pathologies identified before 2015 have not yet been eliminated and replaced by new, desirable in democracy solutions.

However, in order to be able to accurately examine their influence and formulate a change theory, it must take some time. Increasing dichotomy further complicates this process: with the "socially irresponsible" actions of the current government, there are also those that are a direct positive response to the requirements of institutionalization of CSR formulated by the European Commission. In this context, it is important to mention the continuation of the work carried out by the CSR Team in the new formula as the Team for Sustainable Development and Corporate Social Responsibility (Zarządzenie 2016), which is an auxiliary body of the Minister of Development and Finance and assists the Minister in actions for responsible conduct of business. Among new tasks to be carried out by the Team are cooperation and dialogue with stakeholders (entrepreneurs, non-governmental organizations, trade unions, public institutions and civil society actors) in the field of sustainable development and responsible business. Additionally, it is worth highlighting other CSR activities undertaken by the current ruling party, such as the launch of the "Vision for Sustainable Development for Polish Business 2050" (Vision 2050), a project implemented since 2015 by the Ministry of Economy, the Responsible Business Forum and the Deloitte. The program document within this initiative is the "Declaration of Polish Business for Sustainable Development" which defines the ten key challenges that business should engage in to support social and economic change. It aims at integrating business around the idea of sustainable development, identifying challenges in this area and strengthening the administrative and business dialogue for the achievement of Poland's development goals. The Polish government has shown a great deal of CSR activity after 2015, resulting in a number of initiatives, e.g. in 2016, the Ministry of Development, in cooperation with the Ministry of Finance, the Responsible Business Forum, Polish Employers and other companies, was the organizer of series of debates on CSR development in Poland. During these meetings, the issue of implementation in Poland of EU Directive 2014/95/EU on non-financial data reporting was discussed. This act of law considers among others labor standards and human rights. In 2017 a conference "Responsible business activities in the light of the new regulations on disclosure of non-financial data" was held for board members and supervisory boards of large companies, a seminar on socially responsible public procurement, and a conference on "Agenda 2030 for sustainable development. Goals for the world, challenges for the country, responsibility for all" (https://www.mr.gov.pl).

In addition, in the wider context of institutionalization of CSR concept, it is worth highlighting that in the period after 2015, social campaigns aimed at the mitigating of the idea of Corporate Social Responsibility and the re-evaluation of key issues related to it were reinforced: the first period of CSR dissemination in Poland after 2000 consisted in adopting the assumption referring to its positive valuation, Corporate Social Responsibility is simply "good" as such: not only for the companies themselves, but also for employees, other stakeholders, including local communities. Meanwhile, the policy pursued by the current government has significantly contributed to changing perceptions of some companies, mainly those with Western capital, which operated as "socially responsible" companies in Poland, acted unethically against employees and suppliers (mainly popular discounters but also Polish branches of international producers of food and household chemicals). Negative image campaigns have proved to be a weakening of the belief in the added value of CSR and reinforce the conviction of unorthodox business practices. They also reinforce the polarized vision of the social order that consists of disadvantaged "us" and privileged "them". Even if one can say that assumptions of the CSR concept are being challenged in the sense that it functions in the Western European states, it would be worth conducting a public debate on the native standards of social business engagement. However, at the same time, governmental actions towards business (such as fiscal control, taxation or fair competition regulations) are not designed to build a culture of cooperation and dialogue between the state and the business, and undermine the credibility of the concept of CSR in general acceptance.

The introduction of new standards for institutional setting for CSR after 2015 is in statu nascendi. Their direction, however, allows to make a cautious thesis that one should not expect the desired quality in this regard. The new model of social responsibility should be characterized by openness to a range of stakeholders, greater cooperation, better coordination, e.g. in the provision of information on sustainable development, and the generation of innovation in public governance. However, the current practice shows that such activities do not take place.

5 Conclusion

The institutionalization of the CSR concept in public policies is extremely important because it involves external stakeholders, internal stakeholders, social and environmental impacts, and the costs that the state pays. Although Polish formal and legal regulations are gradually being adapted to world and European, including EU, CSR requirements, there is still a significant dichotomy in many areas of interest for CSR.

In Poland, there is still no clear vision for the functional framework of social concepts, especially in terms of their funding, which may be explained by the initial fascination with the neo-liberal approach, the lack of an educated democratic-oriented political culture, and the growing authoritarian populism that is not interested in the strengthening of socially responsible regulations. In addition, the existing institutionalization of the CSR concept in Polish public policies is characterized by ambiguity, disorganization and hybridity. CSR, as it expected by EU regulations, has not been sufficiently embedded in the set of values and internalized. Although the existence of façade structures and procedures can be stated, they are not accompanied by the appropriate institutional setting in the form of a consolidated corporatist system. In the meantime, the creation of a base for institutionalizing CSR should be based not only on companies and entrepreneurs but also on strong institutions and state organizations and on partnerships—which also include other stakeholders, including civil society and the media—between them. However, the state and its structures, as demonstrated by the current practice of public governance in Poland, turn out to be entities that do not implement social responsibility standards and do not support other stakeholders in this regard. Also during the last decades after the political transformation—and Europeanisation has not turned out to be the expected breakthrough here—there has been no effective vision of the functioning of the state in which the business has sufficient conditions for the CSR implementation. So far developed arrangements in this area as well as established structures, irrespective of their constituting governments, prove to be ineffective and marginal for the policy pursued.

Looking for the metaphor that reflects the nature of CSR institutionalization in Polish public policy, one can recall the image of the archipelago: groups of islands located close together, of common origin, as they are emerging from post-transformation culture and similar political culture. Due to their arrangement, it is possible in the case of evaluating of the CSR concept to talk about the swarms of islands, that is, archipelagos, which are composed of many neatly arranged small enclaves of solutions and practices. Created more and more intensively after 2007, however, they do not generate a clear and understandable system of links. In addition, they tend to modify, and their structure often turns out to be a façade for image and political action.

References

Börzel, T., & Risse, T. (2007). Europeanization: The domestic impact of European Union politics. In K. Jørgensen, M. Pollack, & B. Rosamond (Eds.), *Handbook of European Union politics*. London: Sage Publication.

COM. (2011). *A renewed EU strategy 2011–14 for corporate social responsibility*, 681.

Corruption Perception Index, Transparency International. (2016). Accessed September 19, 2017, from https://www.transparency.org

Dylus, A. (2002). Erozja standardów etycznych w biznesie. In J. Mariański (Ed.), *Kondycja moralna społeczeństwa polskiego* (pp. 273–287). Kraków: Wydawnictwo WAM Polska Akademia Nauk.

European Governance. (2001). A white paper, COM (2001) 428 final.

Friedman, M. (1999). Społeczna odpowiedzialność biznesu to zwiększanie zysków. In G. Chryssides & J. Kaler (Eds.), *Wprowadzenie do etyki biznesu* (pp. 260–265). Warszawa: PWN.

Graziano, P., & Vink, M. (Eds.). (2007). *Europeanization: New research agendas*. Basingstoke: Palgrave.

Hausner, J. (1992). Postsocjalistyczne państwo w procesie ustrojowej transformacji. In J. Hausner (Ed.), *Studia nad systemem reprezentacji interesów* (Vol. 1, pp. 9–37). Kraków: Akademia Ekonomiczna w Krakowie.

Hausner, J. (1995a). Formowanie się systemów stosunków pracy i reprezentacji interesów w Polsce w warunkach transformacji ustrojowej. In T. Kowalak (Ed.), *Negocjacje. Droga do paktu społecznego* (pp. 289–314). Warszawa: Doświadczenia, treść, partnerzy, formy, Polskie Towarzystwo Polityki Społecznej & Instytut Pracy i Spraw Socjalnych.

Hausner, J. (1995b). Modele systemu reprezentacji interesów w społeczeństwach postsocjalistycznych. In J. Hausner (Ed.), *Studia nad systemem reprezentacji interesów* (Vol. 3, pp. 317–333). Kraków: Akademia Ekonomiczna w Krakowie & Fundacja „Praca Polska".

Hausner, J. (2007). *Pętle rozwoju. O polityce gospodarczej w latach 2001–2005*. Warszawa: Scholar.

Horváth, P., & Pütter, J. M. (2017). Status quo and future development of sustainability reporting in central and Eastern Europe. *Journal of East European Management Studies, 22*, 221–243.

Kudlak, I., & Ogórek, W. (2017). CSR influence on modern corporations – Examples from Poland. Вісник Національного університету "Львівська політехніка". *Логістика, 863*, 279–284.

Piskalski, G. (2015). *Społeczna odpowiedzialność biznesu w polskich realiach. Teoria a praktyka*. Warszawa: Fundacja CentrumCSR.PL.

Radaelli, C. (2006). Europeanization: Solution or problem? In M. Cini & A. Bourne (Eds.), *Advances in European Union studies* (pp. 56–76). Bansingkoke: Palgrave.

Rosamund, B. (2004). New theories of European integration. In M. Cini (Ed.), *European Union politics* (pp. 109–127). Oxford: Oxford University Press.

Śmigielska, G., & Oczkowska, R. (2017). Retailers' responsibility towards consumers and key drivers of their development in Poland. *Administrative Science, 2017*(7), 1–15. https://doi.org/10.3390/admsci7010003.

Sojka, W., & Walczuk, K. (2015). Mobbing – Socjologiczno-prawny zarys zagadnienia. *Rocznik Administracji Publicznej, 2015*(1), 145–161.

Sroka, J. (2003). Dialog społeczny w Polsce. In J. Sroka (Ed.), *Administracja i polityka. Ewolucja stosunków przemysłowych w Europie Środkowej i Wschodniej*. Wrocław: Wydawnictwo Uniwersytetu Wrocławskiego.

Stawicka, E. (2017). Social responsibility model of the sustainable development in enterprises in Poland. [Polonya'daki İşletmelerin Sürdürülebilir Gelişmesinin Sosyal Sorumluluk Modeli]. *The Journal of Human and Work* [İş ve İnsan Dergisi], *4*(2), 143–149.

Szreder, J. (2006). Portret przedsiębiorcy wyobrażenia a rzeczywistość. In Z. A. Walo (Ed.), *Przedsiębiorca i jego odbiór społeczny*. Słupsk: Słupska Izba Przemysłowo-Handlowa.

Wołodźko, K., & Woźniak, J. (2017). The use by large polish organizations of information about CSR activities in e-recruitment. *Economics & Sociology, 10*(2), 47–60.
www.mg.gov.pl. Accessed September 20, 2017.
www.mr.gov.pl. Accessed September 20, 2017.
Zarządzenie Ministra Gospodarki z dnia 9 lipca 2014 r. (Dz. Urz. Ministra Gospodarki z dnia 21 lipca 2014 r., poz. 13).
Zarządzenie Ministra Rozwoju z dnia 21 września 2016 r. (Dz. Urz. Ministra Rozwoju z dnia 22 września 2016 r., poz. 49).
Zarządzenie Nr 38 Prezesa Rady Ministrów z dnia 8 maja 2009 r. (Dz. Urz. Ministra Gospodarki z dnia 11 maja 2009 r., poz. 17).

Corporate Social Responsibility in Visegrad Countries (Poland, Czech Republic, Slovakia, Hungary) Overall Landscape

Sylwia Przytuła, Sylvie Formánková, Iveta Ubrežiová, and Anna Dunay

1 Introduction

In 1990s, Central Europe transformed politically and economically what influenced their policies regarding social, cultural and natural environment. In 1991, the Visegrad Three (V3) countries established formal sub-regional groupings with a common agenda that mainly emphasized political cooperation for the strategic goals of European Union (EU) and NATO membership. In 1993, Czechoslovakia split into the Czech Republic and Slovakia and the V3 Group became V4 (Fig. 1).

Today, Poland, Czech Republic, Slovakia, Hungary is a member of both the European Union and NATO. In fact, cooperation within the EU is the most important subject for them and regardless of whether they are discussing infrastructure or CSR policy, it is all encompassed within the EU political and financial context. However, the founding leaders did not decide to institutionalize the V4, and even today it has only one permanent institution, the *International Visegrad Fund*, which was set in Bratislava in 2000.

S. Przytuła (✉)
Katedra Zarządzania Kadrami, Uniwersytet Ekonomiczny we Wrocławiu, Wrocław, Poland

S. Formánková
Provozně ekonomická fakulta, Ústav managementu, Mendelova univerzita v Brně, Brno, Czech Republic
e-mail: sylvie.formankova@mendelu.cz

I. Ubrežiová
DMan FEM, Slovak University of Agriculture, Nitra, Slovakia
e-mail: iveta.ubreziova@uniag.sk

A. Dunay
Szent István University, Gödöllő, Hungary
e-mail: dunay.anna@gtk.szie.hu

A. Długopolska-Mikonowicz et al. (eds.), *Corporate Social Responsibility in Poland*, CSR, Sustainability, Ethics & Governance,
https://doi.org/10.1007/978-3-030-00440-8_20

Fig. 1 Visegrad region

The total population of the Visegrad countries is around 13% of the whole EU population, the combined GDP of the group already makes it the world's 15th largest economy (2018). According to *Global Competitiveness Ranking* (2017) encompassing 138 countries worldwide, Poland and Czech Republic are placed in the top 30 of that ranking and Slovakia and Hungary are placed in 65 and 69 accordingly.

This chapter presents the current state of CSR practices in Czech Republic, Slovakia and Hungary.

Sylvie Formánková presents an insight to the CSR in **Czech Republic**. She focuses on the general awareness and perception of CSR in this country. Both, the companies and customers´ point of view are reflected. The general overview is supported by results obtained from Author's research conducted in automotive and chemical industry.

Iveta Ubrežiová presents the current state of CSR in **Slovakia.** The literature review of Slovakian scientific achievements reflects the three interrelated pillars—economic, environmental and social CSR. Then this theoretical part is supplemented with empirical research showing the application of CSR concept in the selected business entities.

Anna Dunay in her part of this chapter shows the CSR activities in **Hungary**. Based on a wide document analysis and literature review the author presens the importance of CSR in different sized enterprises and sectors of the Hungarian economy. This part summarizes the legislative background of CSR in Hungary, the main CSR programs and activities of the large Hungarian and multinational companies, NGOs and the representatives of the SME sector.

1.1 Towards Sustainable Development and CSR in Central Europe and Visegrad Countries

In the socialist era, sustainability was often neglected and the governments were more focused on economic development and achieving economic targets. Thus the business and social environment, society welfare were not equally included in the strategies. After the political changes in 1989–1990, all the Visegrad countries addressed sustainable development in the national strategies and legal acts and they implement this concept into various policies (in economy, ecology, social and cultural field) highlighting the role of human and social capital, as well as the institutions. Since 2004 the analyzed countries share some of the challenges, namely institutional support, public-private cooperation, social cohesion (*Joint Statement...* 2018). For instance, in Poland, a major focus was brought to education and stakeholder cooperation. In the Czech Republic, Slovakia and Hungary, common challenges were demographic issues and economic stability (The V4 is interested in an economically stabile EU 2018; *Trends of business ethics* 2015).

The transition to a market economy in Central Eastern Europe during the past 25 years requires some more specific approach to the CSR, since in that area some institutionalized (authoritarian) social solidarity were embedded with the ideological aspects of the planned economies.

Business is changing Central Europe for the better, corporate social responsibility (CSR) and sustainable development are all new concepts. CSR became a “kaleidoscopic” tool permitting to appreciate and evaluate the perception of the company and its role in the society, useful also for implementing and monitoring business sustainability. At glance, it appears clear that in the former socialist countries there is still a certain reluctance to an uncritical implementation of CSR, both in SMEs and LE (Popowska 2015; Piasecki and Gudowski 2017). Moreover, the findings indicate that CSR reporting practices are not widespread among V4 countries and suggest some area of improvements (Hąbek 2017).

Today we can see a growing number of companies in Central Europe region that consistently follow in the footsteps of the most advanced global enterprises, implementing strategic initiatives and thus increasing their contribution to sustainable social and economic growth. CSR activities bring visible internal and external benefits to companies (Fig. 2).

According to the CSR managers in Central Europe (179 managers from 10 countries—Bulgaria, Czech Republic, Hungary, Kosovo, Latvia, Lithuania, Romania, Serbia, Slovakia and Slovenia), internally CSR activities ensure a stronger employee involvement (65%) and a better ethical awareness (46%). Externally, they

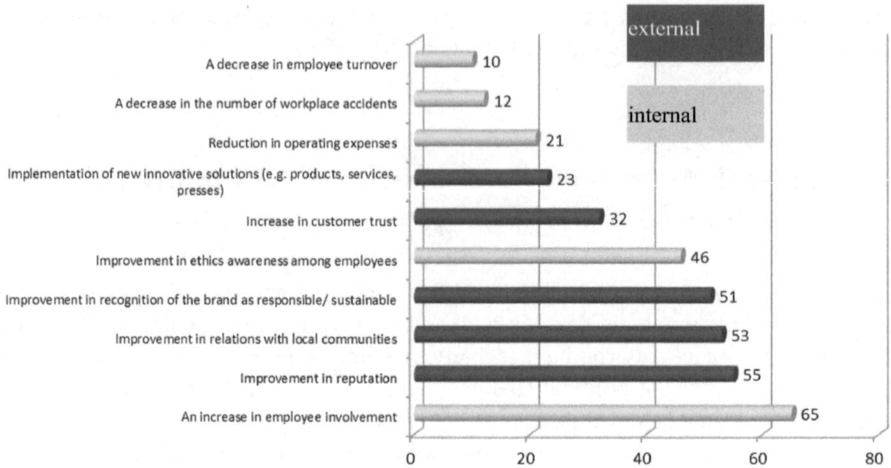

Fig. 2 What benefits has your company derived from CSR action implementation over the past 2 years?. Source: (*CSR Managers Survey* 2015)

improve the companies' reputation (65%) and contacts with local communities (53%), recognition of the brand as socially responsible (51%).

The same population of respondents (84%) believe business has played a role in solving social and economic problems in individual countries, in particular protecting the natural environment, supporting education and counteracting unemployment. CSR managers see the best chances for social and economic development of their countries in the impact of business on the growth in the competitiveness of the economy, the positive impact of enterprises on employment rates and aligning work availability with actual needs as well as their impact on the knowledge-based economy and intellectual capital growth (*CSR Managers Survey* 2015). Also other research proved that the main sources of economic growth for each of these countries includes innovations, human capital, capital accumulation and total factor productivity growth (Konya 2018) and common treasury centre operating in V4 region (the centre of expertise, knowledge and innovation in financial management (Šarkanová and Krištofík 2018).

For 25 years we have been observing in Poland the development of idea of corporate business responsibility. It can be manifested through the interest of that theme out of idle curiosity but also in building the business strategies based on well-balanced development, which we increasingly could observe in Polish companies. Those effects have been caused by the engagement and cooperation between a lot of people, companies and institutions. Polish CSR managers witnessed positive change within business itself (81%). The most significant changes found regard conducting business (no more business than usual—61%), tackling environmental problems (30%) and solving social challenges (29%). An important result we found to be very meaningful was that 53% respondents indicated personal engagement as one of the most important factors for CSR promotion in Poland (*CSR Managers Survey* 2015, p. 8; Molendowski 2017).

With such a strong support from individuals, no wonder that Polish CSR managers strongly believe in the future of CSR. They claim that business models are likely to change in the coming years (80%), and to a great extent, the change will be the result of pressure from customers (79%). However, the CSR experts from Czech Republic, Slovakia and Hungary are more gloomy in this respect.—one third or more of the respondents from these countries foresee stagnation in CSR (Fig. 3).

The managers from Visegrad countries are reliable in that field, which is proven by the fact that 73% of them measure the effectiveness of their CSR initiatives. They are also very realistic when it comes to the main obstacles (Fig. 4).

The "weakest link" for Hungarian managers seems to be lack of incentives from the state administration (70%), 65% indicated on lack of proper understanding of the

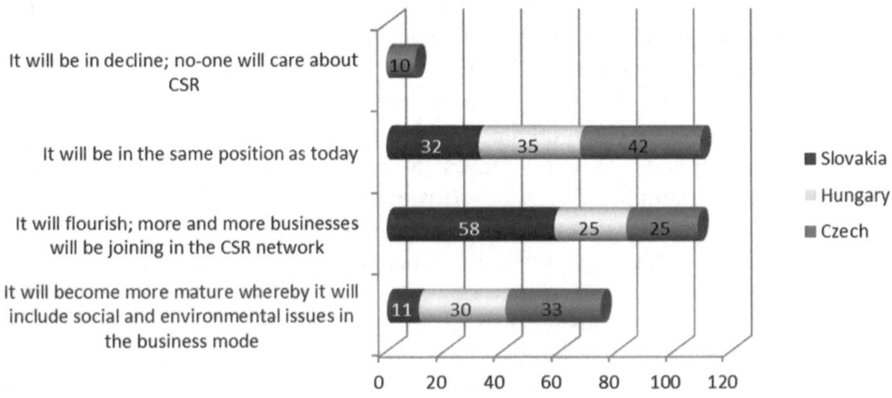

Fig. 3 What do you think the position of corporate social responsibility will be in the coming years?. Source: (*CSR Managers Survey* 2015, p. 17)

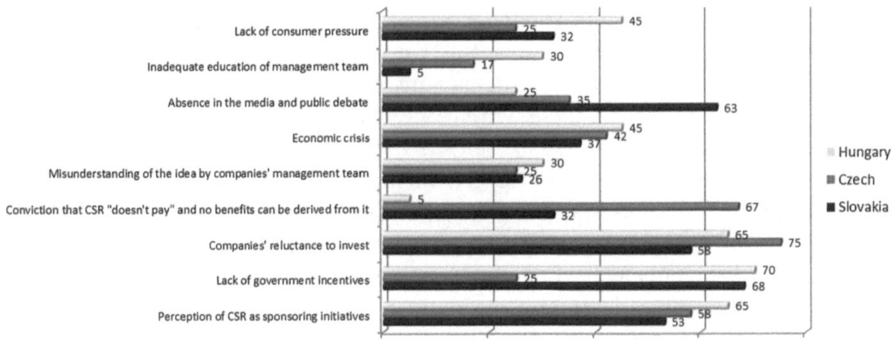

Fig. 4 What do you think will be the main problems or obstacles to CSR implementation in your country in the next 15 years? Source: (*CSR Managers Survey* 2015, p. 16)

idea and companies' reluctance to invest in CSR. According to 75% of Czech respondents companies' reluctance to invest in this idea is the main obstacle, conviction that CSR doesn't pay in business (67%), and 58% of them claim that perception of CSR as sponsoring initiatives might be a problem in spreading the idea in the future. In Slovakia, lack of government incentives was indicated by 68% of CSR managers, absence of topic in media and public debate (63%) and companies reluctance to invest (58%).

2 CSR in Czech Republic

2.1 Development of CSR in Czech Republic

The Commission has defined CSR as the "responsibility of enterprises for their impact on society. Public authorities can play a supporting role through a smart mix of voluntary policy measures and, where necessary, complementary regulation. (*European Commission* 2011) The simplest way for implementing CSR in business is to maximize its positive and minimize the negative impact on the society.

The Czech Republic currently makes up for the CSR deficit caused by the 40 years of Communist dictatorship. However, there is a tradition to follow. According to Skácelík (2010) the current situation in CSR in the Czech Republic may be described as: "the wake-up stage", when the CSR theme is becoming part of public awareness and general discussions and when companies realize more and more not only that there is such a thing as CSR but also that it is useful and absolutely necessary for successful future progress (Skácelík 2010). Kašparová believes that: "corporate social responsibility is no longer the domain of foreign companies in the Czech Republic but that Czech companies also attempt at implementation of CSR in their corporate culture "(Kasparová and Kunz 2013).

In the Czech Republic Internet is the main source of information and communication between corporate representatives and the general public. The information mainly includes annual reports, CSR reports or separate CSR sections on company websites. As reported by Kasparová and Kunz (2013), recently more and more companies come to believe that ethical and responsible conduct of companies towards their employees and the environment in which they operate results in their long-term prosperity, thus representing one of the conditions for long-term growth of the companies. Also in the eyes of the general public corporate CSR activities begin to substantially affect consumer behaviour (Kasparová and Kunz 2013).

This conclusion is confirmed by a survey performed by GFK, where 82% of Czechs stated that company behaviour in compliance with the principles of corporate social responsibility was important to them (GFKCzech 2010). Czech companies also see the area of application of CSR activities as an opportunity to become a demanded employer and to attract most talented employees.

Ipsos Tambor research findings proved that consumers do not have any theoretical knowledge about the concept of CSR and CSR is not known under this acronym.

However, the survey also shows that the population has a great interest in its extension and the shopping behaviour of almost two out of three Czech citizens is affected by a company's corporate social responsibility. Moreover, it follows that 81% of employees also favour employers who are socially responsible and 75% of Czechs assert that they would be disposed to pay more for products that cause no harm to nature (Formánková and Mikusová 2014). Also in the pharmaceutical industry the majority of respondents believe that the implementation of CSR rules would help their work and moreover foster success. They believe that ethical behaviour would be beneficial for the organization (Lorinczy and Formánková 2015).

According to Kašparová the share of corporate philanthropy in total income of the non-profit sector also increases. The numbers of corporate and individual donors grow. However, this proportion is still very low in comparison to the global average (Kasparová and Kunz 2013). Large companies in the Czech Republic usually focus on long-term partnership with a certain NGO. The second most common activity is the establishment of their own corporate foundation or endowment fund. The areas most often supported are children, family and young people. Large companies also support education. The three areas that were not supported (or just minimally) by the analyzed large companies were animals, the homeless and old people (Válová and Formánková 2014; Čanik 2007), however, critically comments on the situation in the Czech Republic: "I do not think that most Czech companies are as responsible and fair as many of them say about themselves. Some companies have just adopted what foreign sources call "pinkwashing" or "greenwashing"; i.e. creation of an impression of a socially responsible company" (Čanik 2007). This view of Čaník is also shared by a number of other experts who perceive corporate behaviour in some cases mainly as self promotion and PR support rather than as inclusion of CSR in the corporate strategy.

According to Pavlik and Bělčik (2010), the unethical business climate is one of the major handicaps of the Czech Republic in comparison with Western Europe. There is no doubt that an improvement would help the Czech Republic, its economy and individuals. The question is under which conditions ethical and socially responsible behaviour would become a natural part of the Czech economy.

2.2 CSR in Chemical Industry

In this sub-chapter, we follow the definition mentioned at the beginning of the chapter. This research project investigated CSR activities and public non-financial corporate information with the following features:

All companies of both groups have an implemented Code of Conduct. Agrofert companies have taken over their parent company Code of Conduct. In the case of **PKN Orlen** the parent company Code of Conduct strongly inspired the individual subsidiaries—who adopted it for the domestic Czech environment. Agrofert operates an anonymous ethical line Tell Us (resolving dozens of cases every year across the concern). All analysed companies are similar as concerns transparency of their general communication, but the willingness to publish and communicate adverse events is rather visible in the case of **Unipetrol** and **Spolana**, i.e. in companies controlled by a foreign entity. On the other hand, **Agrofert** is a holder of the Fraud and Corruption Resistance Profile certificate, manifesting the holder's resistance to corrupt practices. Although the analysed concern members to not hold this prestigious certificate, their parent company systematically requires consistent adherence to the anticorruption activity concept.

All the analysed companies invest large amounts of money in to their emission reduction, which is manifested by their decreasing trend, and are members of the Responsible Care as the main initiative in chemical industry focused on sustainable and transparent corporate behaviour. All analysed companies visibly try to keep the general public informed about impact of their production by way of both unilateral communication in the form of press releases and contributions to the Integrated Pollution Register (IPR) maintained by the Ministry of the Environment, and dialogue in the form of meetings with representatives of non-profit organisations and local self-governing bodies. Communications concerning particular accidents and substance leaks are detailed in all of the companies and include warnings of surrounding villages by SMSs sent to their mayors. The city and journalists are rather informed when the event steps beyond the company boundaries.

While one company of the Agrofert group—**Precheza**—prepares for certification of its system of energy management pursuant to ISO 50001, **Unipetrol** and **Spolana** of **PKN Orlen** group are already allowed to use the certificate. The companies make efforts to obtain the available management system certifications for the purpose of reduction of impact of their activities. All of them are ISO 9001 and ISO 14001 certified. The certification standard OHSAS 18001 defining requirements for occupational health and safety is only used by Synthesia, Precheza and Unipetrol. Agrofert companies, according to the press speaker of one of them, are not interested in the level of CSR of their suppliers.

Companies welcome the recycling efforts of their employees. For that purpose the companies offer recycling of basic materials—paper, plastics, aluminium—in their offices and manufacturing plants. They also offer collection of used batteries. There is the unique and significant project organised by **Unipetrol** since 2000 and concerning support for activities of the Most Ecology Centre. The centre considerably contributes to the dialogue between companies and the general public in the area of environment protection, provides data on pollution measurements and organises cross-boundary communication with the neighbouring Saxony. In 2007 a similar Ecology Centre operation was commenced in Kralupy nad Vltavou, serving the Kralupy region. All companies provide subsidies to non-profit organisations working in the environmental area, to competitions and educational events on environmental themes, in all cases focusing about 90% of the subsidies to the nearest surroundings of the company (within 20 km radius).

The employment issue gains importance with decreasing unemployment due to which the companies experience a lack of candidates for employment. The social pillar in its internal form acts on motivation and satisfaction of the current employees, though, while its external part focuses on regional support, both financial and by way of volunteering and philanthropy. All analysed companies offer their employees the today already standards, or in some cases also super-standard benefits—meal contributions, an extra week of paid holiday, contributions to pension schemes and life insurance, discounts for products made by the holding.

Development training offers language courses. As for training in general, the approach of **Synthesia** is unique. The company offers its employees free requalification in the subject of chemist. Employees can extend their education by taking the Course in Industrial Chemistry, prepared in cooperation with the Secondary Chemical School in Pardubice. They can study in their working hours to extend their apprenticeship certificates or sit for the school-leaving examination. Satisfaction surveys are performed by all companies among their customers and suppliers, with employee satisfaction surveys only organised by **Unipetrol** and **Spolana** that is the companies of **PKN Orlen** group. **Synthesia** concerns care of former employees—old-age pensioners—the Agrofert companies approach is more favourable. These companies invite their former employees to various cultural or tourist events, trying to keep contact between them and the company as such and its current employees. The external social pillar is represented by cooperation with schools and local associations. **PKN Orlen** group, unlike **Agrofert** companies, can boast of corporate volunteering, mainly implemented in the activities of **Unipetrol.** The Unipetrol group has organised common volunteer days since 2012. In 2015 nearly seven dozen volunteers from among the company employees helped the regional non-profit organisations on the volunteer days. Every chemical plant of the Agroferst group holds annual family runs with the route among its chemical plants, in which hundreds of runners participate. The comparison with the survey of 2016 conducted among more than 1000 entities (CMKOS 2017), reveals that employee benefits provided by the companies do compare to the national average. The conclusion is that all analysed largest chemical companies in the Czech Republic act with social responsibility.

2.3 *CSR in Automotive Industry*

Analysis of the position of the automotive industry in the Czech Republic and in respect to small and medium-sized enterprises (SMEs) is the initial part of this sub-chapter. Quantitative research has been conducted among 37 out of 341 contacted Czech automotive SMEs, giving a return ratio of 10.9%. Considering the awareness of CSR, Czech automotive SMEs representatives claimed that the issue of CSR is important for them and consider it a tool towards improving the company's reputation. Slightly better awareness of CSR principles was expressed by enterprises operating internationally. The reason behind this might be the higher demands on responsibility from the side of suppliers—typically car manufacturers. For Czech main exporting partners, the higher importance that is given to CSR in Western European countries such as Germany or Austria. The larger the company, the greater the impact is that it has towards society and therefore the need to invest more into sustainability.

Research showed that enterprises pay the least attention to the social area. On one hand, they considered some issues of this area as non-problematic; this can be demonstrated through no interest in supporting human rights and very positive responses connected to usage of principles respecting human rights. On the other hand, enterprises expressed interest in order to improve working conditions. Nonetheless, most of the selected enterprises provide good working conditions.

Moreover, fulfilment and improvement of working standards is more demanding for manufacturing firms than for firms offering services. Therefore, all manufacturing enterprises included SMEs in the Czech automotive industry need more financial resources, which seems to be the largest obstacle in order to improve the social area of corporate social responsibility (see also: Hąbek 2017). In particular, the lack of time and finances play a crucial role, which was confirmed in both researches within the reasons that limit enterprises to behave more responsibly. Norms and guidelines are perceived by Czech automotive SMEs as necessary. But solely acquiring of certificates will be insufficient for the future. More important will be the ability of companies to strive for innovations and to tackle current social and environmental issues (Visser 2016).

According to the Global Reporting Initiative (2015) the future of CSR is connected with the evolution of technologies. These will enable data to be collected more efficiently, which will lead to more accurate reporting with higher transparency. Moreover, companies will be held more accountable and will have to consider their business activities more profoundly. Considering the level of communication of CSR activities among enterprises, this research showed its insufficiency (Zysk 2017). The highest benefit for Czech SMEs in the automotive industry is represented by an improved reputation, which was claimed by more than 80% respondents. The concept itself is receiving not only more attention but also more importance. However, there are certain restraints which limit responsible activities. This research revealed that it is not distrust from the management of enterprises which slows them down in activities towards CSR, but it is simply a lack of time and expert knowledge. Companies without a CSR strategy often do not know how to apply it. Even if they knew how, the time and finances necessary for implementation of CSR are insufficient.

In other words, these enterprises state that even if they consider CSR principles important in general, the implementation of those principles is difficult and time-consuming. Nevertheless, it does not necessarily have to be true because CSR can be implemented in a simple, cost-effective way. The high importance of employees is a visible fact of SMEs in the automotive industry. CSR can certainly bring a positive influence of CSR on human rights, law and policy. In particular, multinational corporations have adopted CSR as a means of altering ethics together with human standards and thereby transforming the role of business and social relations (Molendowski and Żmuda 2011). Effectively managed CSR projects lead to remarkable benefits in terms of reputation. They do not just strengthen relationships with business partners but also bring higher motivation, loyalty of employees and finally financial return. The results of the research confirmed that every enterprise in the Czech automotive sector is specific, and it is not easy to generalize a strategy for the whole industry.

This part of the chapter focused on Czech Republic might be inspiring for business units of different kinds. They can: get inspiration in CSR implementation, get better awareness of the CSR situation in the Czech Republic—in case of starting a new business, opening a new branch in CR, better understand the CSR topic and see different perspectives and find out that CSR is applicable and suitable for both large companies and SMEs.

3 CSR in Slovakia

3.1 Developing CSR Concept and Practice in Slovakia

There are many definitions of Corporate Social Responsibility. The Green Paper issued by European Commission in 2001 provides two definition of CSR. Firstly, it states that the corporate social responsibility is essentially a concept whereby companies decide voluntarily to contribute to a better society and a cleaner environment. Secondly, CSR is described as a concept which indicates that companies integrate social and environmental concerns in their business operation and in their interaction with their stakeholders on a voluntary basis. According to the European Union, CSR improves mutual trustworthiness between the companies and society (Káposzta and Nagy 2015). On the other hand, the Green Paper states that being socially responsible does not mean only fulfilment of legal expectations, but also going beyond compliance and investing more into human capital, the environment and the relations with stakeholders (European Commission 2001).

Based on the model developed by Carroll (1979), the definition of CSR should emphasize on the whole range of social obligations that a company has in that society. A company can be succeed in CSR only after the accomplishment of all these obligations. These responsibilities include economic, legal, ethical and philanthropic (discretionary) responsibilities (see Fig. 5):

Economic responsibility is the responsibility of a company to provide the goods and services that the company wants and to maximize profits for its owners and stakeholders**. Legal** responsibilities define what the society deems to be important, taking into account appropriate corporate behaviour. **Ethical** responsibilities include behaviours that are not necessarily codified into law and may not serve the corporation's direct economic interests. Organizations' decision makers should act with equity, fairness, respect to the rights of individuals. **Discretionary** responsibilities are purely voluntary and are guided by a company's desire to make social contributions not mandated by economics, law, or ethics. These responsibilities involve generous philanthropic contributions without any payback expected.

Veber (2011) and Merhaut (2013) adds that companies applying the concept of CSR integrate social and environmental considerations into their business activities

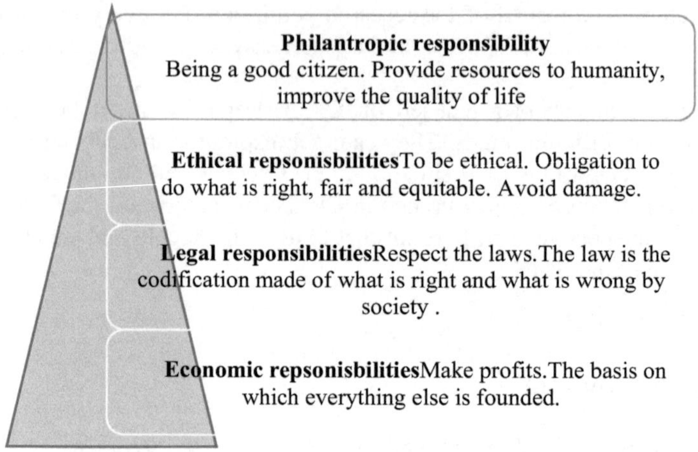

Fig. 5 Carroll's Pyramid of CSR. Source: adapted from Kašparová (2006)

in collaboration with their stakeholders on a voluntary basis. Firms complement their traditional economic objectives by the objectives that take into consideration the benefits of society (expansion of the company's strategy by environmental, social, ethical and community aspects) (CSR Managers... 2015).

Although the history of CSR in Slovakia as an individual business model is not very long, there have always been activities of the organizations that have had impact on society and environment. Such NGOs institutions as **Pontis Foundation**, **PANET** and **Integra Foundation** have begun to specialize in CSR and developed several initiatives in this field focusing on educational, informational and consulting activities. There have not been so many actions undertaken by government or universities. However, one minor initiative was developed by the Ministry of Labor, Social Affairs and Family when is established the Family-Friendly Employer Award.

Taking into consideration the university environment, there only one specialized course exists dedicated to CSR which is held by the Comenius University in cooperation with Pontis Foundation. PANET, one of the most active organizations dealing with CSR, prepared a strategic action plan for the period 2002–2004 entitled Support and Development of Social Responsibility, Corporate Philanthropy and Inter-Sector Cooperation in Slovakia.

Since there was a missing platform for businesses operating in Slovakia that could serve as a model and guarantor of CSR in our country, **BLF Slovakia** was established. 18th November 2004, founding members signed the Memorandum about CSR and declared their interest to integrate voluntarily the set of principles reflecting social, ethical and environmental aspects into their corporate strategy. They also declared to promote and inspire other businesses operating in Slovakia to apply principles of CSR. Comparing the situation of BLF after 10 years of functioning there is the evident the increase of 11 founding members to 32 current members. In 2004 the first survey was carried out by the World Bank, Pontis Foundation, Integra Foundation and BLF to determine the attitude of the Slovak companies towards CSR concept (Moravčíková 2016).

On the other hand, Corporate social responsibility can be evaluated on the basis of selected quantitative indicators (ex. Social Progress Index 2017). Since 2002 such evaluation has been carried out by the international non-governmental, non-profit initiative called Global Reporting Initative (GRI).

Kita (2005) focused the evaluation of the CSR on three pillars:

1. *The economy* and its indicators: management and organizational security of CSR, economic performance, market presence, indirect economic impacts, responsibility for the product.
2. *Social responsibility* and its indicators: employment and decent working conditions, human rights, community, public policy and corruption, competitive behavior, compliance of legislation.
3. *Ecology* and its indicators: materials, energy, water, biodiversity, emissions, waste, waste water, environmental impact of products, transportation, other influences.

Based on the research of Jaďuďová (2013), the level of implementation of CSR concept in the sectors of national economy. The survey was realized in 2012 and altogether, 50 companies completed the questionnaire. Firstly, the author focused on determination whether companies are familiar with CSR. Out of 50 respondents, almost half (24 companies) knew CSR, 22 knew it partly and only 4 have never heard about it. Then, they concentrated attention on ***economic pillar***. The results indicate that the most important factor for the companies is transparency, since 98% of respondents provide transparent information about their activities. When dealing with anti-corruption measures, the answers were as following: 48% of companies measure such activities, and 46% do not do it. The rest does it partly (2.4%) or did not answer to the question (2%). Within the ***social pillar***, authors focused on philanthropy, involving the employees into decision-making and employees´ benefits. (See Table 1) The answers proved that philanthropy is the most important activity, performed in 46% of companies. However, the survey did not study philanthropic activities more closely.

Table 1 Forms of employee benefits

Forms of employee benefits	Number of organizations
Contribution to the diet	39
Education allowance	22
Company car	30
Company mobile phones	42
Gift vouchers	22
Contribution to superannuation	24
Travel allowance	16
Contribution to culture and sport	18
Others	6

Source: Jaďuďová (2013)

In terms of ***environmental pillar***, she studied implementation of cleaner technologies and on the use of recyclable materials. The vast majority of companies (80%) take into consideration the potential impacts of their operation on the environment and they are also willing to invest in new (environmentally more friendly) technologies. In case of the use of recyclable materials, half of the companies answered positively (52%), 13% use them at least partly and 14% marked negative answer to the question.

Ubrežiová and Horská (2011) considered the offer of the high-quality products and services as the primary CSR principle and most important activity, 41% of SMEs consider acting according to law as the second main CSR principle and 40% answered that under the third most important CSR activity could be providing employment and salary. For instance, only 3% consider volunteering as one of the most important CSR activities and only 24% would deal with training of employees. Moreover, 63% want to keep the current level of CSR activities, 33% would like to extend their CSR activities in the future, whereas in the Czech Republic the number represent only 23%, 1% would like to start with CSR activities, and 3% of the companies answered they want to reduce the range of current CSR activities provided or they do not want to start at all. They concluded that currently more than 96% of the Slovak SMEs have more than positive approach towards the concept of CSR in connection to their business. Many of Slovak SMEs decided to include CSR principles also due to the fact that it is becoming an important competitive advantage on the market and they want to keep pace with their competition (52%). More than 42% wants to attract attention of new customer and almost 42% of Slovak SMEs use CSR as a public relation tool to improve their image as a responsible company, which as we stated at the beginning is one of the misunderstanding approaches that CSR is facing today. Building up an image of a company is also the most important benefit, Slovak SMEs see in the implementation of CSR (49%). Other benefits SMEs expect from CSR approach is improvement of economic performance (44%), loyalty of employees (41%), increase of company culture (35%) and better cooperation with business partners (34%).

However, the implementation of CSR concept in SMEs is not resulting only from all international movements, but the pressure and requirements to act more sustainable towards triple bottom line comes from governments, NGOs, local communities, stakeholders and shareholders as well. As 99% of European companies are small and medium sized enterprises, implementation of CSR in their business activities is more than crucial. However, the development is different among European countries. On the other hand, Hupková et al. (2009) focused on consumers´ purchasing decisions confirm the fact that recently, non-economic factors are becoming part of this process and are more and more important to consumers.

Ubrežiová et al. (2013) realized the project: *Perception of CSR in Companies of Eastern Slovakia Region*. The research involved 164 companies in Eastern Slovakia which were divided as following: 8 micro companies (0–9 employees), 142 SMEs (10–249 employees), and 14 large companies (250 and more employees). Out of 164 companies, only 34% of them are familiar with the CSR concept, 42% have never heard about it and 24% are not sure whether they know it or not. The research questions formulated in the questionnaire dealt with the opinions of companies about the importance of implementation of CSR and with the activities the companies realize. In the next part of the survey, companies were asked to sort the list of offered activities according to their importance in CSR. 73% of companies stated that the most important activity is *business honesty*, followed by 13% related to *training of employees* and the rest—8% chose *environmental protection.*

As the second most important activity, respondents selected *training of employees* (38%), *environmental protection* (27%) or *work-life balance* (20%). The third place belonged to *environmental protection* (35%) or training of employees. According to the results, was concluded that companies in Eastern Region of Slovakia have just started to be aware of their social responsibility. The inner interest of companies—their own profit, is becoming a strong incentive to fulfil the activities of CSR. However, the respondents stated what within CSR, it is necessary to focus on the activities directly connected with the core of their business or on those which could be implemented in the company. Business honesty, employee training, work-life balance are activities of the biggest interest for surveyed companies. When dealing with activities outside the company, the interest is smaller—environmental protection (4%) and volunteering (1%).

Partial research shows us that companies in eastern region perceive the concept of Corporate Social Responsibility. Based on the importance of engaging in the activities of CSR, the companies clearly agreed that it is important to engage in activities that are beneficial to them.

3.2 CSR in Selected Agri-Food Companies

The next and the main aim was to compare the situation in the selected agro-food companies and to evaluate the level of knowledge about CSR, describe means the companies use to apply CSR and activities undertaken within the policy.

The research was a part of the dissertation thesis *"Corporate Social Responsibility of the Slovak agri—food enterprises in the competitive environment."* (Moravčíková 2016). The results of the survey were compared with the other surveys, which were conducted within Slovakia Research realized by Prokeinová and Paluchová (2012) are primarily source for comparing the results since they deal with the CSR in food industry. Author also compared the state of CSR with research by Ubrežiová et al. (2013) and other general surveys considering CSR in SMEs.

The main aim of the questionnaire was to gain insights into CSR situation of agri-food companies in Slovakia. Altogether, 81 companies took part in the survey. The activities undertaken by the surveyed companies in three main areas of CSR—environmental, social (policy towards community/at the workplace) and economic.

3.2.1 Environmental Pillar

Moravčíková (2016) confirmed that companies were offered nine activities (see Table 2) in this pillar and also the extent to which they can realize them.

The most common activities are saving the energy (almost 72% of companies indicate it to great or quite a great extent), optimizing the transportation (around of 68% of companies undertake them to great or quite a great extent) and saving water (almost 68% of companies realize it in great or quite a great extent). A very high percentage was achieved also by environmentally friendly processes, as 70.4% of companies are involved in this activity to quite a great or great extent. On the other hand, only 47% of companies indicated involvement in the mutual cooperation with other subjects, which represents the lowest portion. The purchase of environmentally friendly machinery and increasing knowledge of employees about environmental issues appear to be not so common activities of agro-food companies as well, since only 58% of companies indicated involvement to great or quite a great extent.

The results correspond with the findings of similar survey realized by Prokeinová and Paluchová (2012) in which the most common activities undertaken by companies in the food industry, were recycling, waste minimization, energy savings, prevention against pollution. The similar data can be found in the study by FOCUS (2010), where most of the companies indicated involvement in economical use of raw materials, energy and water, waste reduction, its separation and recycling. In the survey of Ubrežiová et al. (2013), the environmental protection was prioritized on the third place within the selected companies, however, the survey was not focused specifically on environmental pillar but activities in CSR as a whole. Since this section consists of scaling questions, it was necessary to measure the internal consistency and find out how closely are individual activities related as a group.

Table 2 The activities of companies in relation to the CSR

	Activity	Companies in %
1	Mutual cooperation with other subjects	47
2	Recycling	43
3	Increase employees´ knowledge about environmental protection	58
4	Purchase of machinery and equipment friendly to environment	58
5	Waste reduction	69
6	Saving the water	68
7	Optimizing the transportation	68
8	Environmentally friendly work processes	70
9	Saving the energy	72

Source: Own work

3.2.2 Economic Pillar

There were analyzed five areas: (1) protection of intellectual property rights, (2) providing information to stakeholders, (3) transparency of company"s activities, (4) fair trade and (5) relations with shareholders, suppliers and business partners. The companies assigned the extent to which they realize them and also free space, where they can describe additional activities or comments related to economic pillar. More than 80% of the companies consider **relations with other stakeholders** as very important (to great or quite a great extent). Based on the answers, companies put great emphasis on **fair trade** and more than two thirds (75.3%) of respondents realize it to great or quite a great extent. On other hand, surveyed agro-food companies realize actions to protect intellectual property rights on very low extent.

This results were compared with survey of FOCUS (2010), where almost all respondents carry out principles of honor market behavior and principles of transparency. Ubrežiová et al. (2013) also found out that for most of the companies the most important activity within all the pillars of CSR is business honesty.

3.2.3 Social Pillar (Policy Towards Community/at the Workplace)

The most of companies (64.2%) indicated the safety and protection of *health of the employees*. This could be caused by obligatory conditions for employers to instruct employees about this issue. Other activity indicated as important and realized to great or quite a great extent was *communication of management with employees* (66.7%). The highest percentage (76%) belongs to *the development of long-term careers of the employees*. On the other side, agro-food companies do not devote the main attention to further *carriers of dismissed employees or their re-qualification*—only 16.1% of companies indicate that they realize this activity to great or quite a great extent. It can be assumed that activities such as work benefits would achieve high percentage due to the core of the business. However, only 52.5% companies indicate they undertake actions *to prevent corruption and bribery*.

According to Humieres and Chauveau (2001), employees are seen as the most important stakeholder. The similar results as in our survey were achieved in the one realized by FOCUS (2010). In that case food companies stated the most common activity was support and development of employees. Question of safety and health standards was formulated with addition beyond the law, which could cause its position as the third mostly realized activity. Second part of social pillar includes *the policy at the workplace*. This section consists of four main activities (1. charity donations, 2. cooperation with local community, 3. supporting the community in organization of events and 4. purchasing goods from local suppliers) followed up by the space where company can express their opinion on the thesis or describe additional activities related to community.

Primary analysis indicates that most common activity in agro-food companies is purchasing goods and resources from local suppliers. This may be connected also to nature of the companies, since most of them are SMEs. Very popular is also the support of community in organizing events of various nature (sports, cultural, etc.) This is in comparison with results of the survey, which was realized by Prokeinová and Paluchová (2012). The survey confirmed that more two thirds of the companies have bought resources from local suppliers. The rest of the companies is trying to do so, but limited offer makes it sometimes not possible. The next most common activity was providing the regular financial support for projects in local community.

Based on the presented research findings some recommendations and ideas supporting CSR in the Slovak Republic can be given. It is necessary to ensure successful and continuous activities with the aim to increase awareness and familiarity of the CSR concept. The perception of this concept is usually limited to philanthropy or charity, but the involvement of stakeholders in the CSR is still very low. However, there is not much attention dedicated to charity donations in the case of surveyed agri-food companies. Overall familiarity and involvement in CSR in business sector is relatively low (around 50% and in agro-food sector around 40%). Some of the companies do not know the term and do not undertake CSR activities. On the other side, there is group of companies that realize activities of CSR nature even though they are not aware of it. There are also companies familiar with the term but not practicing it in reality.

Therefore, there arises a need to enable dialogue between companies and other subjects about adoption and implementation of CSR. It would help the stakeholders to enhance the knowledge and gain better understanding of CSR. On the other hand, stakeholders are not clearly distinguished—usually it is business partners or employees. If the company is able to identify its stakeholders and to lead proper dialogue with them, it would ensure more smooth implementation of CSR. As it is already mentioned, it is NGOs who are the key bearers of CSR promotion in Slovakia.

4 CSR in Hungary

There are various definitions of CSR, which differ greatly for each company and country. CSR covers topics from human rights, healthy and safe working conditions, corporate governance, environmental effects, and contribution to economic and social development. There are differences between the various definitions, but the purpose of CSR is common: to guide activities and business intentions towards economic, social and environmental sustainability. CSR means more than compliance with legislation, CSR is a special way of thinking which puts ethical values and respecting the society, individuals and the natural environment into focus.

4.1 From Economic to Responsible Thinking: Evolution of CSR in Hungary

CSR is a relatively new concept in Hungary, as its history was highly influenced by the socialist system between 1948 and 1989 and the changing political, social and economic conditions after the political transition. In the era of socialism, the state and state-owned companies provided the social care using the financial support of the state. Huge state-owned companies did not have experiences in environmentally and socially sustainable corporate governance methods and were not forced to work in an economically sustainable way, as the national economy was the part of the political system. Their main function was not to produce profit, but to provide full employment (Győri 2010).

A main objective of the economic transformation process was to separate the economic and social role of the companies and to make a clear difference between public welfare responsibilities and the HR policies of the companies. As in the socialist system the different supports were financed by the state, the transformation to market economy offered several possibilities for the management of the new companies to get rid of the former obligatory social spending. In these so called "paternalistic" companies employers enjoyed the benefits of well-built social and cultural infrastructure (for example nurseries, holiday-resort places, libraries and sport sites, health services, etc.) and provisions (canteens, special credits to housing, transport support, etc.) that were granted by state funds, independently from the company profits (Segal et al. 2003).

After the economic changes, the attention of the new market players has turned towards the importance of social and human investments. Foreign investors, who made long-term investments, imported their corporate social responsibility models and adopted their human resource patterns in their Hungarian firms. Companies with Hungarian ownership—after stabilizing their market positions—have also changed their former welfare models to comply with the new market conditions and introduced certain elements of CSR (Angyal 2009).

A comprehensive research was conducted by Győri (2010) to analyse the development of CSR and CSR attitudes in Hungary focusing on the different levels or stages of responsibility. Goodpaster (2006) described the different levels of responsibility of individuals and organizations, and summarized three different mind-sets from (1) focusing on own goals through (2) rational thinking, where companies consider other players in an instrumental way, to (3) thinking rationally while respecting others. The first mind-set represents a clearly economic rationality (focusing exclusively on economic factors), the second represents organizations which consider the rules, regulations and other market players, while the third means a responsible company, Győri (2010) added. Zadek (2001) described three generations of CSR based on their main central questions. The question of the 1st generation is "*Can corporations be responsible in ways that do not detract from and may add commercial value to their business*?" (Zadek 2001) These organizations use CSR as a tool for defence in such cases when it cannot endanger profit realization process (e.g. philanthropy). Question of the 2nd generation is "*Are responsible companies more likely to prosper in the future*?" (Zadek 2001) They focus beyond on the short-term frame of cost-benefit analysis and considered ethical questions as well (focusing on win-win situations). 3rd generation's question is (Zadek 2001): "*Is corporate citizenship likely to be significant in addressing growing levels of poverty, exclusion, and environmental degradation*?" Third Generation Corporate Citizenship should involve such processes which move beyond individual initiatives and build alliances between business, NGOs and government in raising global performance.

Győri (2010), using the abovementioned methodology, pointed out that in Hungary 3 main groups may be distinguished: (1) companies without CSR activities, (2) companies with "instrumental" CSR activities and (3) companies committed to CSR. Győri (2010) highlighted that CSR attitudes are different depending on the size of the companies. Large companies without CSR activities are focusing only on profit as the main company goal. The size of this group is decreasing by spreading of CSR concept and companies shall keep up with the others. Instrumental CSR is represented by large companies (mainly companies from the financial or chemical sector) who know and use CSR tools (focusing primarily on standards and qualifications) in order to increase their reputation. In these companies, PR and HR departments deal with CSR activities focusing on the communication of CSR. There is a special group of companies called "Paternalists", which are rooted from the former socialist era, they are mostly large, state-owned companies, they have a well-built social system for their employees, and now they call it CSR. These activities are restricted for their employees' wellbeing, for example training and implementing company health funds and programs (see Simonyi 1999).

SMEs have to face more challenges in the market competition, as they are smaller in size. Most of them thinks that they have no opportunity for responsible thinking, they, of course, consider ethical questions, but it is not built into their strategy. They start responsible thinking when it is forced by external parties, for example when they start to act as a supplier of a large company, or when they must achieve different qualifications required by their partners. These companies apply "instrumental" CSR, and they think that responsible behaviour is only a cost-increasing factor, as they cannot feel the respect of consumers.

The group of companies committed in CSR consists of SMEs and large companies that realized value-based management principles, i.e. they use a management philosophy which states that management should foremost consider the interests of shareholders in its business actions. These companies integrate their ethical behaviour into their business processes and consider that they shall correspond to the needs of the social and environmental system, not only to business issues (Győri 2010). Nevertheless, in many cases, CSR activities of SMSs are not connected to company strategies and sometimes they even do not know that they perform CSR activities (Szlávik 2009).

In Hungary, at the present, there are not official statistics on CSR activities, the main data are available from research results of individual researchers, research institutions (e.g. GKI Economic Research Company), NGOs committed to CSR such as Hungarian Business Leaders Forum (est. in 1992), KÖVET Egyesület (Association for Sustainable Economies est. in 1995), EMVFE (First Hungarian CSR Association, est. in 2005), the sustainability reports of different companies and the reports of well-known international consulting companies.

4.2 Legal Background of CSR in Hungary

The legislative background of CSR in Hungary is mostly affected by the fact that CSR principles were brought by foreign investors and multinational companies into the country. The CSR policies (European Commission 2007, 2011) of the European Union were automatically adopted and companies—based on ethical and environmental considerations—built in these principles and objectives voluntarily into their business processes and the communication with their stakeholders.

There are different acts, rules and regulations dealing with CSR activities, but it is mentioned directly by only a few of them. The rights of the employees are mentioned in the Act XCIII of 1993 on Labour Safety describing working conditions and the right for labour safety education. Act XXVI of 1998 on the Rights and Equal Opportunities of Persons with Disabilities describes the rights of people with disability and it is also underlined in the Fundamental Law of Hungary.

The first government decision regarding CSR was adopted in 2006, (1025/2006. decision on the social responsibility of employers), which fully complies with the relevant issues of EU legislation. Many laws deal with environmental protection, the most important are the Act LXXXIX of 2003 on the environmental load fee, Act LVI. of 1995 on Environmental Protection Product Charges which enables companies to reduce their environmental load fee when they increase re-use level or decrease emissions (Szuchy 2013). An important step in Hungarian CSR legislation is the Act CLXXXV of 2010 on Media Services and Mass Communication, which declares that information concerning the corporate social responsibility of an undertaking shall not qualify as commercial communication, but such reports may only contain the name, logo and trademark of the product or service or the company.

Szegedi and Mélypataki (2016) underlined that CSR is not described in the Hungarian law as an independent legal term, but for example, Government Decree 1201/2015 describes the priorities of CSR and provides an Action Plan. Previously, the Act IV of 2006 on Business Associations laid down the rules for annual reports (Article 312) of limited companies admitted for trading on the Budapest Stock Exchange, where the management board shall present to the annual general meeting the company governance and management report together with the annual financial report. This report shall contain the management board's conclusions on the company's adopted policy with a view to sound governance and management. The new Civil Code of Hungary (Act V of 2013)—like former laws—consider CSR activities as a part of responsible corporate governance, and does not deal with this topic directly. It means that the original EU principles shall be applied, i.e. CSR should be a company led activity (Szegedi and Mélypataki 2016).

From 2018, a change will occur regarding the CSR or Sustainability Reports of companies. According to the Hungarian Act LXVI of 2016, where the principles of the EU Directive No. 2014/95/EU were implemented, certain large undertakings and groups shall prepare a mandatory report on their non-financial and diversity information containing environmental, social and employment matters. This report is mandatory for those public interest entities, which employee number exceeds 500 heads and balance sheet total exceeds 6 billion HUF (approx. 20 million EUR) or turnover exceeds 12 billion HUF (40 million EUR). These rules shall be applied from the 2017 financial period, which means the reports shall be attached to the financial reports in 2018.

4.3 CSR Business Practices in Hungary

Development of CSR in Hungary shows a slow, but continuous tendency, which is caused by the attitudes and activities of large multinational companies, the EU tendencies and by the movements of civil organizations. The main fields, where CSR is manifested are the following: companies' own social welfare policies, health and safety conditions and regulations at work places, labour relations and social dialogue at the work places, engagement of the management in training activities and investments in life-long learning, investments in career planning within the firms and investments in community development and on cultural fields. Table 3 summarizes the most recent researches conducted in Hungary in the past 10 years.

Mazurkiewicz et al. (2007) explored the Hungarian situation based on a sample of 150 companies, where respondents of the survey assessed the risks of adopting CSR practices. The highest share of respondents (35%) indicated the increased operating costs as the highest risk. The second most frequently mentioned risk (12%) was the adverse impact on profitability, while 10% of firms mentioned competitive disadvantages as third risk. According to nearly 20% of the sample answered that there are no risks in the implementation of CSR into their business processes. The respondents were asked about the main actions that improve CSR practices. More than 80% of them indicated tax incentives, more than 60%–60% indicated the recognition (both by business partners and consumers). The role of local government empowerment, subsidized interest rates, reporting regulations and recognition by Civil Society Organizations (CSOs) and the government were also assessed as important actions which might improve CSR activities. The most frequently mentioned benefits of CSR were business stability, increased productivity and quality, HR retention, reputation, sustainable development, clients' loyalty and visibility (Mazurkiewicz et al. 2007).

The survey of Csillag (2008) showed that the managers of the examined sample considered sponsorship and the different measures for own employees as the main aspects of CSR. The main benefits of CSR activities can be measured by the increasing loyalty and commitment of employees and therefore, it affects company's competitiveness directly.

Table 3 CSR researches conducted in Hungary

Researcher	Sample	Preferred CSR activities/aspects
Mazurkiewicz et al. (2007)	150 companies	Stakeholder concerns, ethical issues, correction of social inequalities, environmental issues, health programs, training, education, community development
Csillag (2008)	27 managers of 10 companies	Sponsorship, internal CSR (measures for employees)
GKI (2008, 2009)	1500 companies	Workplace health programs, training, social issues, considering work-life-balance problems, supporting communities, environmental programs, sponsorship (culture, sports, health)
Győri (2010)	55 companies	Internal stakeholders in focus (trainings, health programs), voluntary work, ethical operation
Tardos (2014)	185 companies	Health screening programs, family programs, sport programs, transportation to workplace, trainings, flexible working hours, programs for employees on maternity leave
Málovics (2011)	9 in depth interview +77 companies	Profitability, environmental aspects: waste management, recycling and reusing of materials, environment protection, keeping local values, social sustainability aspects of internal stakeholders and local external stakeholders
Bank (2017)	21 in depth interview +169 companies	Internal stakeholders in main focus (workplace health, trainings, work-life balance, social support), community support, sport and cultural programs, sponsorship, company image, profitability

Source: Author's summary

Until present, GKI Economic Research Company (GKI 2011) has conducted the most comprehensive analysis of CSR in Hungary, in 2008, which proved that CSR is more important and better accepted in large companies, but 85% of the Hungarian companies (despite of their size) assessed CSR as an important activity and agreed with CSR principles and goals. Companies preferred internal CSR aspects, and respondents indicated improving company image, health promotion and programs for employees and using environment-friendly technologies for improving sustainable operation as the first three main CSR goals and benefits (GKI 2011).

The research results of Győri (2010) highlighted that the most active companies in CSR were multinationals or companies with foreign ownership, and they focus mostly on internal CSR while in companies with Hungarian ownership external CSR (primarily sponsorship and voluntary work) were preferred.

According to the research results of Tardos (2014) measures improving equal opportunities, health programs, and family programs were supported by nearly 50% of the examined companies. Sport programs were supported by 30% and transportation of employees were mentioned by 20% of the sample. 50% of the companies are conscious about improving working conditions (flexible working hours, training and educational programs) but for example summer camp or own nursery and kindergarten opportunities were much less frequently mentioned by companies with location in Hungary (Tardos 2014).

Málovics (2011) conducted a quantitative survey among Hungarian companies based on the database of KÖVET Association. Research results underlined that among Hungarian SMEs the from the three main sustainability aspects of business operation (the 3 "P"s, People, Planet, Profit) profit was the most important, as based on the results of this survey company managers ranked company profit and long term operation and providing continuous work as the most important company goal. "Planet" i.e. environmental aspects were the next most important goals with activities connected to waste management, recycling and reusing of materials, environment protection, keeping local values, and "People" were considered the less important, and the social sustainability aspects were focused only on internal stakeholders and local external stakeholders (i.e. citizens and organizations of the local community). The research results revealed that in case of SMEs the corporate responsibility is influenced significantly by the company size and profitability (smaller size represents less CSR activities) and the local relations and binds of the owners/managers.

According to the results of Benedek (2017) company managers preferences related to CSR practices are the following (the averages are calculated by a 6-grade Likert scale): environmental goals: 4.42, social goals: 4.33, employees: 4.26, business environment: 4.02, stakeholders: 3.94: The high priority of environmental considerations may be related to the environmental or partners' or industrial requirements. Research results revealed that there is a strong correlation between the personal attitudes of the managers and their motivations towards CSR activities.

A recently finished research of Bank (2017) showed a similar picture than previous researches (see Table 3), but this research underlined that the most active sector in CSR is bank sector, and showed the differences between the preferences of multinational, state owned and private companies with Hungarian ownership (see Table 4).

Bank sector is one of the most active sector of the Hungarian national economy in CSR aspects. A recent survey of Lentner et al. (2017) proved that the economic

Table 4 External and internal CSR activities of companies located in Hungary with different ownership

Multinational	State owned	Private
External CSR: Environmental friendly operation, sponsorship (sport and culture), supporting education, voluntary work *Internal CSR:* Correctness with employees	*External CSR:* Sponsorship (sport and culture) *Internal CSR:* Insurance (health and life), catering, resort place, health programs, programs for pensioners,	*External CSR:* Environmental friendly operation, sponsorship, supporting education *Internal CSR:* High quality processes, Support for employees

Source: Bank (2017)

responsibility of commercial banks and the central bank has been increased until the end in the examined period. In ethical aspects the authors underlined that the culture of ethics and integrity should be rebuilt in the financial sector. Focusing on legal responsibility in the financial sector the research results highlighted that a more stringent regulation has been built to minimize risk and ensure safety and trust in the bank sector. Many commercial banks are very active in voluntary, charitable and philanthropic responsibility, which activities can contribute to solving some social problems, and may increase the reputation of these institutions.

This situation may be justified by the applications and awards of the CSR Hungary organization: in the past years, different commercial banks applied for the CSR Hungary Award by different special programs and projects. CSR Hungary Award has a 10 year history in 2017, it was established by EMVFE (First Hungarian CSR Association in 2008. The award is given in different categories (for large and small companies) such as sustainable large company, responsible SME, green environment, social and health affairs. The applicants represent multinational companies and companies with Hungarian ownership as well. The most important programs cover environmental and social projects and solutions of the applicants, either for external and their internal stakeholders. These projects may be used as good practices in the future. The award winner can hold the quality seal for 2 years, which may bring a higher reputation and prestige of the company and a competitive advantage as well.

As the introduced research results proved, there is a huge difference between the CSR attitudes and activities of large companies and SMEs in Hungary. As the financial situation of SMEs is less stable, the CSR activities are connected mostly to the present economic status of the enterprises and show a significant relation with the personal attitudes and motivations of their leaders (Benedek 2017; Málovics 2011). CSR attitudes are mostly generated by local social and environmental needs and responsibility is restricted to the internal stakeholders and local external stakeholders. CSR strategy is informal, or less formal, and instead of ethical responsibility CSR activities are represented by industrial norms, professional and legal obligations, as Málovics (2011) underlined. As Skowron-Grabowska et al. (2016) pointed out, CSR should be built into the strategy of—even small and medium-sized—companies.

In most companies CSR is still considered as a PR tool, which is not compulsory, which added value cannot be clearly defined in money terms. For changing this situation, the consultative process between the different parties (firms, NGOs, state representatives, CSR associations and all stakeholders) should be accelerated in the future, in order to clarify the measures to promote acceptance of CSR and to change attitudes and perceptions in the future.

As research results revealed that the personal attitudes, motivations and commitment of company leaders highly determine the CSR attitudes of the firms, CSR issues and CSR strategy should be built into the curricula of business trainings.

5 Conclusions

In the last decade, the Visegrad countries have been diversifying and developing the global trends in their economies, including corporate social responsibility. The presented in this chapter examples of the CSR's maturity in business field shows that these practices are not widespread among companies, yet.

Firstly, it seems that foreign, multinational companies and investors have brought their own CSR models, which national corporations from Poland, Czech Republic, Slovakia and Hungary have begun to adopt.

Secondly, the level of knowledge concerning the importance of CSR practices is increasing due to many initiatives taken by national institutions, business associations and NGO that have positive impact on environment, society and economic development. There are some measures and rankings prepared by consulting companies which help to place the importance of CSR in V4 countries, show the differences and benchmark them. For example, amongst the V4 countries, the highest indicator of CSR reporting was achieved by Hungary, followed by the Czech Republic and Poland. The lowest indicator was achieved by Slovakia. The CSR reports from V4 countries are developed most often by large, private and subsidiary companies.

Thirdly, there is a lack of holistic scope of implementing CSR policies in enterprises. In all V4 countries there are many examples of introducing small puzzles of CSR and the perception of this concept is usually limited to business honesty, philanthropy or charity, improving "green- reputation" of the company, strengthen public relations policy, especially in small-medium enterprises.

Fourth, it is difficult to get official statistics on CSR activities as the main data are available from research results of individual researchers, research institutions, reports of different companies and the reports of well-known international consulting companies, which do not have the overall picture of business in this region. There is a room for such institutions to fill in this gap.

Fifth, CSR should be considered as a PR tool, which is not compulsory, which added value cannot be clearly defined in money terms. It requires more research, promotion activities and support from government of each country which change the people attitudes and perceptions in the future.

References

Angyal, Á. (2009). *Vállalatok társadalmi felelőssége, felelős társaságirányítás* [Corporate social responsibility]. Budapest: Kossuth Kiadó.

Bank, D. (2017). *Implicit és explicit, valamint belső és külső CSR egy kettős függésben lévő piacgazdaságban* (PhD Thesis). Budapest: Corvinus University.

Benedek, A. (2017). A stratégiai CSR személyiségi gyökerei. In Á. Csiszárik-Kocsir (Ed.), *Vállalkozásfejlesztés a XXI. században: VII. tanulmánykötet* (pp. 55–69). Budapest: Óbudai Egyetem Keleti Károly Gazdasági Kar.

Čanik, P. (2007). *Kde začíná a končí společenská odpovědnost firem?* Accessed March 15, 2016, from http://www.canik.cz/2007/09/18/kde-zacina-a-konci-spolecenska-odpovednost-firem

Carroll, A. (1979). A three dimensional model of corporate social performance. *Academy of Management Review, 4*, 497–505.
Csillag, S. (2008). *Vállalatok társadalmi felelőssége és a versenyképesség*. Versenyképesség kutatások műhelytanulmány-sorozat. Budapest: BCE Világgazdaságtan Intézet Versenyképesség Kutató Központ.
CSR Managers Survey in Central Europe. (2015). *How CSR has influenced Central European societies and economies? Lessons learnt and future trends*. Accessed June 28, 2017, from https://www2.deloitte.com/csr-managers-survey-in-central-europe-2015.html
CMKOS. (2017). *Czech-Moravian confederation of trade unions*. https://www.cmkos.cz/obsah/330/documents/14982?language=en
European Commission. (2001). *Green Paper promoting a European framework for corporate social responsibility*. Accessed April 12, 2017, from http://www.europarl.europa.eu/meetdocs/committees/deve/20020122/com(2001)366_en.pdf
European Commission. (2007). *The European alliance for CSR progress review 2007 making Europe a pole of excellence on CSR*. https://www.businesseurope.eu/sites/buseur/files/media/imported/2008-00311-E.pdf
European Commission. (2011). *A renewed EU strategy 2011-14 for corporate social responsibility Brussels*. COM/2011/681.
FOCUS. (2010, October). *A guide to CSR in Europe: Country insights by CSR Europe's national partner organisations*. CSR Europe
Formánková, S., & Mikusová, J. (2014). Companies" decision making about CSR. In *New trends in management in the 21st century*. Czestochowa: Sekcja Wydawnictw Wydzialu Zarzadzania Politechniki Czestochowskiej.
GFKCzech. (2010). http://www.gfk.cz/public_relations/press/press_articles/006700/index.cz.html
GKI. (2011). A társadalmi felelősségvállalás szerepe és szintje a hazai vállalkozások körében.
GKI Economic Research Co. (2008, March). *Monthly business survey*. Accessed April 20, 2018.
GKI Economic Research Co. (2009, March) *Monthly business survey*. Accessed, April 20, 2018.
Global Competitiveness Report. (2017). Accessed June 28, 2017, from http://www.weforum.org/reports/global-competitiveness-report
Global Reporting Initiative. (2015). Sustainability and reporting trends 2025. In: *Global reporting initiative*. Accessed May 19, 2016, from https://www.globalreporting.org/resourcelibrary/Sustainability-and-Reporting-Trends-in-2025-1.pdf
Goodpaster, K. E. (2006). *Conscience and corporate culture*. Malden, MA: Wiley-Blackwell.
Győri, Z. (2010). CSR-on innen és túl. Doktori (PhD) értekezés. Budapesti Corvinus Egyetem, Gazdálkodástani Doktori Iskola.
Hąbek, P. (2017). CSR reporting practices in Visegrad Group countries and the quality of disclosure. *Sustainability, 9*(12), 2322.
Humieres, P., & Chauveau, A. (2001). *Les Pionniers de l'entreprise responsable, Editions d'Organisation*.
Hupková, D., Bielik, P., & Turčeková, N. (2009). Structural changes in the beef meat demand in Slovakia and demand elasticity estimation. *Agricultural Economics – Czech, 55*, 361–367.
Jaďuďová, J. (2013). Perception of corporate social responsibility in selected sector of national economy of the Slovak Republic. *Electronic International Interdisciplinary Conference, 2*, 339–342.
Joint Statement of the Visegrad Group and Croatia. (2018). http://www.kormany.hu
Káposzta, J., & Nagy, H. (2015). *Status report about the progress of the Visegrad countries in relation to Europe 2020 targets*. Łódź: Wydawnictwo Uniwersytetu Łódzkiego.
Kašparová, K. (2006). *Stakeholderská teórie a její propojení s CSR*. Accessed August 26, 2017, from http://www.csronline.cz/Page.aspx?publikace
Kasparová, K., & Kunz, V. (2013). *Moderní přístupy ke společenské odpovědnosti firem a CSR reportování*. Praha: GRADA.
Kita, J. (2005). *Marketing 3. prepracované a doplnené vydanie* (p. 432). Bratislava: Ekonóm.

Konya, I. (2018). *Economic growth in small open economies lessons from the Visegrad countries*. Cham: Palgrave.
Lentner, C., Tatay, T., Szegedi, K., & Chauduri, S. (2017). Corporate social responsibility in the banking sector. *Firma i Rynek, 51*(1), 53–64.
Lorinczy, M., & Formánková, S. (2015). Business ethics and CSR in pharmaceutical industry in the Czech Republic and Hungary? *Acta Universitatis Agriculturae et Silviculturae Mendelianae Brunensis, 63* https://doi.org/10.11118/actaun201563062011.
Lőrinczy, M., Sroka, W., Jankal, R., Hittmár, S., & Szántó, R. (2015). *Trends of business ethics and corporate social responsibility in Central Europe*. Aachen: Shaker Verlag.
Málovics, G. (2011). *A vállalati fenntarthatóság értelmezéséről Szeged*. Szeged: JATEPress Kiadó.
Mazurkiewicz, P., Crown, R., & Bartelli, V. (2007). *What does business think about corporate social responsibility? A comparison of attitudes and practices in Hungary, Poland and Slovakia*. Washington, DC: World Bank.
Merhaut, M. (2013). *Etika jako součást společenské odpovědnosti firem hotelového průmyslu*. Praha: Wolters Kluwer ČR.
Molendowski, E. (2017). An internationally competitive economy: A comparison of Poland and the Visegrad Group countries in the post-accession period. *Comparative Economic Research, 20* (4), 5–21.
Molendowski, E., & Żmuda, M. (2011). Changes in competitiveness among the Visegrad countries after accession to the European Union: A comparative analysis based on a generalized double diamond model. *Comparative Economic Research, 16*(4), 121–153.
Moravčíková, K. (2016). *Corporate social responsibility of the Slovak agro-food enterprises in the competitive environment*. Nitra: Dissertation, Slovak University of Agriculture.
Pavlik, M., & Bělčik, M. (2010). *Společenská odpovědnost organizace: CSR v praxi a jak s ním dál*. Praha: Grada.
Piasecki, R., & Gudowski, J. (2017). Corporate social responsibility: The challenges and constraints. *Comparative Economic Research, 20*(4), 143–157.
Popowska, M. (2015). Europe, between mandatory and voluntary approach to CSR: Some new evidence from the Polish listed companies. *Journal of Yaşar University, 10*, 38–45.
Porter, M., Scott, T., & Green, M. (2017). *Social progress index*. http://www.socialprogressindex.com
Prokeinová, R., & Paluchová, J. (2012). Social responsibility of food companies in Slovakia – A selected tool of sustainability. *Annals of Agricultural Economics and Rural Development, 99*(3), 41–50.
Šarkanová, B., & Krištofík, P. (2018). Innovation through treasury centralization: The potential of the Visegrad countries for establishment of corporate treasury centres. In A. Dias, B. Salmelin, D. Pereira, & M. S. Dias (Eds.), *Modeling innovation sustainability and technologies. Economic and policy perspectives*. Cham: Springer.
Segal, J. P., Sobczak, A., & Triomphe, C. A. (2003). *Corporate social responsibility and working conditions*. Dublin: European Foundation for the Improvement of Living and Working Conditions.
Simonyi, A. (1999). Labour and social welfare in competitive firms. In C. Makó & C. Warhurst (Eds.), *The management and organization of firm in the global context*. Budapest: University of Gödöllő.
Skácelík, P. (2010). *Společenská odpovědnost organizací*. Praha: Grada.
Skowron-Grabowska, B., Tomski, P., Dunay, A., & Illés, B. C. (2016). Multidimensionality of decision-making of corporate social responsibility in the strategy of enterprises. *Zeszyty Naukowe Politechniki Częstochowskiej Zarządzanie, 24*(2), 7–16.
Szegedi, K., & Mélypataki, G. (2016). A vállalati társadalmi felelősségvállalás (CSR) és a jog kapcsolata. *Miskolci Jogi Szemle, 11*(1), 51–70.
Szlávik, J. (2009). *A vállalatok társadalmi felelősségvállalása*. Budapest: Complex Kiadó Kft.

Szuchy, R. (2013). A gazdasági társaságok társadalmi felelősségvállalása. In *Magyar Tudomány Napja a Délvidéken*. Vajdasági: Vajdasági Magyarok Tudományos Társasága. http://www.vmtt.org.rs/mtn2013/287_304_Szuchy_A.pdf.

Tardos, K. (2014). *Esélyegyenlőség és családbarát vállalati gyakorlatok*. Budapest: mtd Tanácsadói Közösség MTA Társadalomtudományi Kutatóközpont Szociológiai Intézet.

The V4 is interested in an economically stabile EU. (2018, February 8). *The V4 connect*. http://v4.gov.hu/the-v4-isinterested-in-an-economically-stabile-eu

Ubrežiová, A., & Horská, E. (2011). *Perception and approach towards corporate social responsibility in SMEs: Case study of Slovak and Czech Republic*. Brno: European Scientific Conference of Ph.D. Students.

Ubrežiová, I., Stankovič, L., Mihalčová, B., & Ubrežiová, A. (2013). Perception of corporate social responsibility in companies of Eastern Slovakia region in 2009 and 2010. *Acta Universitatis Agriculturae et Silviculturae Mendelianae Brunensis, 61*(7), 2903–2910.

Válová, A., & Formánková, S. (2014). Corporate philanthropy in the Czech Republic. In *17th international conference enterprise and competitive environment*. Amsterdam: Elsevier Science Bv.

Veber, J. (2011). *Management Základy – moderní manažerské přístupy – výkonnost a prosperita*. Praha: Management Press.

Visser, W. (2016). *Future trends in CSR*. Waynevisser. Accessed May 21, 2016, from http://www.waynevisser.com/articles/fu-ture-trends-in-csr

Zadek, S. (2001). *Third generation corporate citizenship – Public policy and business in society*. London: The Foreign Policy Center.

Zysk, W. (2017). Trends in the area of responsibility and justice in the activities of enterprises. In *Proceedings of the 5th international conference on innovation management, entrepreneurship and sustainability*. Prague: IMES.